Current Topics in
Insect Endocrinology
and Nutrition

Current Topics in
Insect Endocrinology and Nutrition

A TRIBUTE TO GOTTFRIED S. FRAENKEL

Edited by

Govindan Bhaskaran

Texas A & M University
College Station, Texas

Stanley Friedman

University of Illinois
Urbana, Illinois

and

J. G. Rodriguez

University of Kentucky
Lexington, Kentucky

Plenum Press · New York and London

Library of Congress Cataloging in Publication Data

Endocrinology Symposium, Denver, 1979.
  Current topics in insect endocrinology and nutrition.

  Proceedings of a symposium held during the annual meeting of the Entomological
Society of America in Denver, Colo., Nov. 1979.
  Includes index.
  1. Insects—Physiology—Congresses.   2. Insects—Food—Congresses.   3. Endocrin-
ology—Congresses.   4. Fraenkel, Gottfried Samuel, 1901-   —Congresses.
I. Fraenkel, Gottfried Samuel, 1901-   II. Bhaskaran, Govindan, 1935-
III. Friedman, Stanley. IV. Rodriguez, J. G., 1920-   V. Entomological Society
of America. VI. Title.
QL495.E48 1979   595.7'01   80-24274
  ISBN-13: 978-1-4613-3212-1   e-ISBN-13: 978-1-4613-3210-7
  DOI: 10.1007/978-1-4613-3210-7

Proceedings of the symposium honoring Gottfried S. Fraenkel held at the
Annual Meeting of the Entomological Society of America, November, 1979,
in Denver, Colorado

© 1981 Plenum Press, New York
Softcover reprint of the hardcover 1st edition 1981
A Division of Plenum Publishing Corporation
227 West 17th Street, New York, N.Y. 10011

CONTENTS

GOTTFRIED S. FRAENKEL

GOTTFRIED S. FRAENKEL: AN APPRECIATION

In examining the corpus of G. S. Fraenkel's achievement, one is struck by the breadth of his subject matter as well as his mastery of the experimental method. It is to honor this diversity of interest that we have bound together in this volume two symposia of very different thematic content. However, even these titles, insect nutrition and endocrinology, hardly indicate the range of his effort, which encompasses such other areas as invertebrate behavior, coordination of function, temperature adaptation and host selection. Mindful of this large number of very different subjects, we might well derive some interesting insights from an examination of his scientific life and the contexts in which his various contributions were made.

Gottfried Samuel Fraenkel was born on April 23, 1901 in Munich, Germany, and as a young man became a member of one of the Zionist youth groups which in the period after the first World War provided the ideology and manpower that ultimately led to the birth of the state of Israel. It was at this time that he made a commitment to go to Palestine as a teacher, and he pursued his course vigorously, completing the primary step in late 1925 with a Ph.D. in Zoology at the University of Munich. His major interest as a student was animal behavior and orientation, and his thesis provided the first real proof that the statocysts in medusae (jellyfish) really functioned as such.

Before taking the plunge into the unsettled conditions of the middle East, he decided on a postdoctoral stint, and was lucky enough to obtain an International Education Board fellowship for a year, permitting him to work through 1927 at the Marine Zoological Stations in Naples (Italy), Roscoff (France), and Plymouth (England),

1

with a short time in Alfred Kühn's laboratory in Göttingen. All
of these studies were directly concerned with orientation, or, as
in the case of his paper with Kühn (on the spectral photosensitivity
of bees) based upon orientation reactions.

His European journeys finished, Fraenkel went to Palestine to
become a teacher, but was hardly settled in when he was introduced
to F. S. Bodenheimer, an outstanding entomologist and the newly
appointed head of the Department of Zoology at the Hebrew University. This meeting changed Fraenkel's life, since with the offer
of an assistantship by Bodenheimer, he was taken from the world
of a school teacher into the world of research. Within a few months,
a locust outbreak occurred (the first in Palestine in 15 years), and
Fraenkel plunged into the investigations which were to culminate in
his classical papers on the physiology and behavior of the migratory
locust and his more general studies on migratory behavior and
coordination of function leading to and during flight. It is well
to point out that it was while working on locust behavior that he
discovered the tarsal reflex (removal of the substrate from tarsal
contact will cause certain insects to assume a flight posture and
begin wing flapping), which opened the way for all further laboratory studies on insect flight. . (It was also the Palestine work
coupled with that done earlier in Naples which led to the classical
text, "Orientation of Animals", written with Donald Gunn in 1940.)

As a result of this independent success, his interactions with
Bodenheimer were not easy. His scientific life became so con-
stricting and uncomfortable, that ideals and interest notwithstand-
ing, he felt impelled to leave, moving back to Germany in 1931, where
in spite of being a Jew he attained the position of Privatdozent in
the University of Frankfurt. He had hardly begun his academic
career, when in 1933, under the pressure of the Nazis, he and all
of the other Jewish faculty were fired. He emigrated to England,
and, through the efforts of English scientists who were financially
helping to support their displaced colleagues, was able to secure
a post as a Research Associate in University College of the University of London. He was at University College for only two years,
but it was in this position, under the pressures of a new language,
a new culture, a new family, and almost no salary, that the classical fly larva ligation experiments were performed. This work,
establishing the presence of a moulting hormone in flies, and that
of Wigglesworth on Rhodnius (which, as Fraenkel later found, was
taking place at the same time just around the corner in the London
School of Tropical Hygiene and Medicine), form two of the corner-
stones of modern insect endocrinology.

In 1935, he was hired into a post in insect physiology (perhaps
the first full time teaching position in this discipline in the
world), in the Department of Zoology and Applied Entomology at the
Imperial College of London University. It was expected by the

administration that he would begin to think in practical terms, but this was not to be, since his interest in flies, stimulated by his work on the hormonal control of development, became even stronger, and his next four years were spent examining the morphological and physiological adaptations of their various life stages. In this job he was meeting many other biologists at scientific gatherings, and it was at one of these that he met young Pringle, whose work on insect campaniform organs and unique insights led to their joint publications on the halteres of Diptera. It was here also that he met and described to the physical chemist, Rudall, the strange changes occurring in the larval fly integument at pupariation, a discussion which was to provoke a fruitful collaboration on the cuticle, and Rudall's lifelong interest in cuticle structure.

When, in 1939, England went to war, the Zoology-Entomology Department of the Imperial College was evacuated to Slough and the Pest Infestation Lab was inaugurated. Munro, the Professor and head of department, a toxicologist by interest, intended that Fraenkel would work on insecticides, but, still independent, Fraenkel chose to take the view that understanding the nutritional requirements of stored grain pests would develop the intelligence with which to deal with them successfully. It is doubtful whether Fraenkel's work contributed directly to the war effort, but his experiments and results then and later have shaped the views of a generation of insect nutritionists and applied entomologists. Needless to say, his approach to his task did not endear him to Professor Munro.

In 1947 he was invited by Glenn Richards to give a series of lectures at the University of Minnesota, and, as a result of his visit, he came to the attention of entomologists at the University of Illinois. In 1948 he received an offer from that department, and with encouragement frum Munro, decided to emigrate once again and take up a position as Professor of Entomology at Illinois. From this time on, he was "free" to choose the directions of his work (although his previous histbry leaves us with no doubt that he had always been free in his own mind to do just that).

Just prior to leaving England, he discovered a soluble factor from yeast with vitamin activity for tenebrionid beetles, which he carried with him to Illinois. On the campus he found an interested biochemist in Herbert Carter who, with some of his students, collaborated with Fraenkel in 1952 to isolate, crystallize and identify vitamin B_T (Tenebrio) as carnitine. Further work with various colleagues (Bhattacharyya, Friedman) established the universality of occurrence and first indications of the importance of carnitine in Coenzyme A transfer reactions. These ideas and reaction mechanisms have since been incorporated into the common body of biochemical information to the extent that their origins have all but disappeared from the literature.

At the same time, given room to stretch his mind, Fraenkel began to ask questions about the meaning of his nutritional observations. Coming at the problem from two directions, he and Dethier had both concluded as early as 1951 that the similarities in nutritional requirements across the class Insecta, and the fact that plant leaves generally contained all of the required compounds, meant that host specificity must be based upon the presence in plants of other compounds which served to attract or repel individuals of different species. By 1958, Fraenkel had examined enough of the literature to recognize that the so-called "secondary" plant compounds, of many different structures and apparently not moving through the major pathways, might provide a clue to the evolution of host selection. For years, botanists and chemists had been isolating different classes of these compounds and associating them with different plant families, but had been unable to establish functions for most of them. With what some consider a flash of insight, but is documented in his 1959 Science paper as a long, thoughtful process tempered by his extensive experiences in nutrition and behavior, Fraenkel opened the field of insect-plant coevolution based upon chemical and sensory interactions between the groups, by describing the raison d'être of secondary plant compounds "as solely for the PURPOSE to repel and attract insects" (caps. mine).

It was in 1961, when this idea was beginning to cause ferment in ecological circles that Fraenkel received one of the few Research Career Awards given in his field by the U. S. Public Health Service. Then, with many of his students (Soo Hoo, Hsaio, Nayar, Waldbauer, Yamamoto) engaged in an examination of the chemical basis of host selection, and the pressures of his career relaxed, he was able to return to the cyclorraphous Diptera, whose adaptations had so intrigued him in the 1930's, and which by force of circumstance had been excluded from his laboratory. There were also good scientific reasons for his return to the field of endocrinology, since he had never been able to continue his pioneering experiments on ecdysone, and others were now showing the way to interesting and important aspects of the hormonal control of development.

His long memory took him back to 1936 and experiments he had done on newly emerged adult blowflies, and it was from information originally developed in that paper that, in 1962, he reported a new hormone which was responsible for adult tanning. By 1965 the hormone had been purified, generally characterized, and named "bursicon", and for the next few years, Fraenkel and those of his students not doing host selection experiments (Seligman, Fogal) spent their time delineating its activity. As others became aware of Fraenkel's initial observation that bursicon was present in insects in a number of orders, interest grew, so that by 1968 many persons were engaged in corroborating and extending his work. It was time to move on to other questions concerning fly development, and it was also at this juncture that Fraenkel's excellence was formally recognized by the scientific community in the U.S. with election to the National Academy of Sciences.

The anecdote introducing Dr. Denlinger's paper in this symposium provides us with an explanation of how Fraenkel's serendipity led to his 1968 incursion into the physiology of diapause, a bridgehead which Denlinger has brilliantly expanded, but it was not chance which made him pick up the fly puparium once again.

If bursicon was the tanning hormone in the adult fly, what replaced it in the tanning of the puparium? This obvious question had to be answered in Fraenkel's lab, and it prompted an overall review of the nature of puparium formation under the influence of ecdysone. In so doing, Fraenkel in the early 70's established the presence in late larvae of several protein mediators of puparial morphogenesis and tanning, themselves released as neurosecretions through the action of ecdysone. For these investigations, he recruited Dr. J. Zdarek of the Department of Entomology of the Czech Academy of Sciences, who he had met sometime before at a symposium in Czechoslovakia. (This was, indeed, a lucky meeting, since their mutual affection and cooperation on this problem continues to this day.) Dr. P. Sivasubramanian, who came to Fraenkel's lab as a student, having worked in India as a junior colleague to Dr. Bhaskaran, another of Fraenkel's associates, was also involved in this project. And the international character of the laboratory was capped off with the addition of Dr. Nalini Ratnasiri, a Sri Lankan student. She, together with Dr. Fraenkel, cleared up a number of questions concerned with the uncertainties surrounding the larval ligation work and its use as a hormone bioassay by demonstrating, in 1974, the importance of the tracheal system in the tanning reaction and its fragility when subjected to constriction.

Dr. Fraenkel became Emeritus Professor of Entomology at the University of Illinois in 1972, but, with the exception of a reference in the annual departmental newsletter, there were no indications of retirement. Work continued on the puparial model for endocrinologically linked events, and new experiments were underway on fly oogenesis. In an examination of the interactions between nutritional states and developmental hormones in promoting egg production, Fraenkel re-evaluated some unpublished work on larval nutrition, and as a result, embarked on a new study concerned with the relationship between nutrition, metamorphosis, and aging processes. These experiments are evolving now, and it makes us wonder whether at 79 Fraenkel has decided that his flies may have something to add to what he has already taught us about aging properly.

Having begun this biographical sketch with the statement that we might learn something from a perusal of the scientific journey of this highly productive man, I come to its end with the feeling

that there are real lessons to be derived from his wanderings. The
first is that good questions and answers may be developed in the
least scientifically crowded places. The second is that virtue
sometimes triumphs over adversity. The third, more serious than
the first two, is that Fraenkel provides a role model for the young
scientist thrust into a position where problem choices are limited.
Intelligence and imagination can turn the worst of situations into
sources of important information from which broad and useful
generalizations may be made.

Fraenkel has been physically "home" for the past twenty-five
years, but his intellectual voyage continues unabated.

GOTTFRIED S. FRAENKEL - PUBLICATIONS

Fraenkel, G., 1925, Der statische Sinn der Medusen, Zeitschr. vergl. Physiol. 2:658-690.

Fraenkel, G., 1927, Phototropotaxis bei Meerestieren, Die Naturwissenschaften 14:117-122.

Fraenkel, G., 1927, Beiträge zur Biologie eines Arkturiden, Zool. Anzeiger 69:219-222.

Fraenkel, G., 1927, Beiträge zur Phototaxis und Geotaxis von Littorina, Zeitschr. vergl. Physiol. 5:585-597.

Fraenkel, G., 1927, Die Grabbewegungen der Soleniden, Zeitschr. vergl. Physiol. 6:167-220.

Fraenkel, G., 1927, Über Photomenotaxis bei Elysia viridis, Zeitschr. vergl. Physiol. 6:385-401.

Fraenkel, G., 1927, Biologische Beobachtungen an Janthina, Zeitschr. Morph. Oekol. d. Tiere. 7:597-608.

Kühn, A., and Fraenkel, G., 1927, Über das Unterscheidungsvermögen der Bienen für Wellenlängen im Spektrum, Nachr. Ges. d. Wiss. in Göttingen. Math. Phy. Kl:1-6.

Fraenkel, G., 1928, Über den Auslösungsreiz des Umdrehreflexes bei Seesternen und Schlangensternen, Zeitschr. vergl. Physiol. 7:365-378.

Fraenkel, G., 1929, Über die Geotaxis von Convoluta roscoffensis, Zeitschr. vergl. Physiol. 10:237-247.

Fraenkel, G., 1929, Untersuchungen über Lebensgewohnheiten, Sinnesphysiologie und Sozialpsychologie der wandernden Larven der afrikanischen Wanderheuschrecke Schistocera gregaria, Biol. Zentralblatt. 49:657-680.

Fraenkel, G., 1930, Der Atemmechanismus des Skorpions, Zeitschr. vergl. Physiol. 11:656-661.

Fraenkel, G., 1930, Die Orientierung von Schistocerca gregaria zu
    strahlender Wärme, Zeitschr. vergl. Physiol. 13:300-313.
Fraenkel, G., 1930, Beiträge zur Physiologie der Atmung der Insecten,
    Atti dell XI Congresso Internationale di Zoologie, Padova:905-
    921.
Bodenheimer, F. S., Fraenkel, G., Reich, K., and Segal, N., 1930,
    Studien zur Epidemiologie, Ökologie und Physiologie der
    afrikanischen Wanderheuschrecke, Zeitschr. angew. Entomol.
    15:435-557.
Fraenkel, G., 1931, Die Mechanik der Orientierung der Tiere im Raum,
    Biol. Rev. and Biol. Abstr. Cambridge Phil. Soc. 6:36-87.
Fraenkel, G., 1932, Untersuchungen über die Koordination von
    Reflexen und automatisch-nervösen Rhythmen bei Inseknen. I.
    Die Flugreflexe der Insekten und ihre Koordination, Zeitschr.
    vergl. Physiol. 16:371-393.
Fraenkel, G., 1932, II. Die nervöse Regulierung der Atmung während
    des Fluges, Zeitschr. vergl. Physiol. 16:394-417.
Fraenkel, G., 1932, III. Das problem des gerichteten Atemstromes in
    den Tracheen der Insekten, Zeitschr. vergl. Physiol. 16:418-442.
Fraenkel, G., 1932, IV. Über die nervösen Zentren der Atmung und
    die Koordination ihrer Tätigkeit, Zeitschr. vergl. Physiol.
    16:443-462.
Fraenkel, G., 1932, Die Wanderungen der Insekten, Ergebnisse der
    Biologie 9:1-238.
Fraenkel, G., 1934, Der Atemmechanismus der Vögel während der
    Fluges, Biol. Zentralblatt. 54:96-101.
Fraenkel, G., 1934, Pupation of flies initiated by a hormone, Nature
    133:834.
Fraenkel, G., 1935, A hormone causing pupation in the blowfly
    Calliphora erythrocephala, Proc. Roy. Soc. Ser. B. No. 807,
    118:1-12.
Fraenkel, G., 1936, Observations and experiments on the blowfly
    (Calliphora erythrocephala) during the first day after
    emergence, Proc. Zool. Soc. London, pp. 894-904.
Fraenkel, G., 1936, Utilization of sugars and polyhydric alcohols
    by the adult blowfly, Nature 137:273.
Fraenkel, G., and Herford, G. V. B., 1938, The respiration of in-
    sects through the skin, J. Exp. Biol. 15:266-280.
Fraenkel, G., and Pringle, J. W. S., 1938, Halteres of flies as
    gyroscopic organs of equilibrium, Nature 141:919.
Fraenkel, G., 1938, Temperature adaptation and the physiological
    action of high temperatures, Kongressbericht II, 16th Internat.
    Physiol. Congr. Zurich.
Fraenkel, G., and Harris, J. L., 1938, Irregular abdomina in
    Calliphora erythrocephala, Proc. R. Ent. Soc. London (A) 13:7-9.
Fraenkel, G., 1938, The evagination of the head in the pupae of
    cyclorrhaphous flies (Diptera), Proc. R. Ent. Soc. London (A)
    13:137-138.
Fraenkel, G., 1938, The number of moults in the cyclorrhaphous flies
    (Diptera), Proc. R. Ent. Soc. London (A) 13:158-160.

Fraenkel, G., 1939, The function of the halteres of flies (Diptera),
    Proc. Zool. Soc. London (A) 109:69-78.
Fraenkel, G., 1940, Utilization and digestion of carbohydrates by
    the adult blowfly, J. Exp. Biol. 17:18-29.
Fraenkel, G., and Rudall, K. M., 1940, A study of the physical and
    chemical properties of the insect cuticle, Proc. Roy. Soc.
    London (B) 129:1-35.
Fraenkel, G., and Hopf, H. S., 1940, The physiological action of
    abnormally high temperatures on poikilothermic animals. I.
    Temperature adaptation and the degree of saturation of the
    phosphatides, Biochem. J. 34:1085-1092.
Fraenkel, G., and Herford, G. V. B., 1940, The physiological action
    of abnormally high temperatures on poikilothermic animals. II.
    The respiration at high sublethal and lethal temperatures, J.
    Exp. Biol. 17:396-398.
Fraenkel, G., and Davis, R. A., 1940, The oxygen consumption of
    flies during flight, J. Exp. Biol. 17:402-407.
Fraenkel, G., and Gunn, D. L., 1940, "The Orientation of Animals",
    Clarendon Press, Oxford, 352 pp.
Fraenkel, G., Reid, J. A., and Blewett, M., 1940, The sterol require-
    ments of the larva of the beetle, Dermestes vulpinus, Biochem.
    J. 35:712-720.
Fraenkel, G., and Blewett, M., 1941, Deficiency of white flour in
    riboflavin (tested with the flour beetle Tribolium confusum),
    Nature 147:716.
Fraenkel, G., and Blewett, M., 1942, Biotin as a possible growth
    factor for insects, Nature 149:301.
Fraenkel, G., and Blewett, M., 1942, Biotin, B$_1$, riboflavin, nico-
    tinic acid, B$_6$, and pantothenic acid as growth factors for
    insects, Nature 150:177.
Fraenkel, G., and Blewett, M., 1943, Vitamins of the B-group required
    by insects, Nature 151:703.
Fraenkel, G., and Blewett, M., 1943, Intracellular symbionts of
    insects as a source of vitamins, Nature 152:1943.
Fraenkel, G., 1943, Insect Nutrition, J. Roy. College of Science
    13:59-69.
Fraenkel, G., and Blewett, M., 1943, The basic food requirements
    of several insects, J. Exp. Biol. 20:28-34.
Fraenkel, G., and Blewett, M., 1943, The vitamin B-complex require-
    ments of several insects, Biochem. J. 37:686-692.
Fraenkel, G., and Blewett, M., 1943, The sterol requirements of
    several insects, Biochem. J. 37:692-695.
Fraenkel, G., and Blewett, M., 1943, The natural foods and the
    food requirements of several species of stored products in-
    sects, Trans. R. Ent. Soc. 93:457-490.
Blewett, M., and Fraenkel, G., 1944, Intracellular symbiosis and
    vitamin requirements of two insects, Lasioderma serricorne and
    Sitodrepa panicea, Proc. Roy. Soc. (B) 132:212-221.
Fraenkel, G., and Blewett, M., 1944, Stages in the recognition of
    biotin as a growth factor for insects, Proc. R. Ent. Soc.
    London (A) 19:30-35.

Fraenkel, G., and Blewett, M., 1944, The utilization of metabolic
    water in insects, Bull. Ent. Res. 35:127-139.
Fraenkel, G., and Blewett, M., 1945, Linoleic acid, α-tocopherol
    and other fat-soluble substances as nutritional factors for
    insects, Nature 155:392.
Fraenkel, G., and Blewett, M., 1946, The dietetics of the clothes
    moth, Tineola biselliella Hum., J. Exp. Biol. 22:156-161.
Fraenkel, G., and Blewett, M., 1946, The dietetics of the cater-
    pillars of three Ephestia species, E. kuehniella, E. elutella
    and E. cautella, and of a closely related species, Plodia
    interpunctella, J. Exp. Biol. 22:162-171.
Fraenkel, G., and Blewett, M., 1946, Linoleic acid, vitamin E and
    other fat-foluble substances in the nutrition of certain
    insects (Ephestia kuehniella, E. elutella, E. cautella and
    Plodia interpunctella, J. Exp. Biol. 22:172-190.
Fraenkel, G., and Blewett, M., 1946, Folic acid in the nutrition
    of certain insects, Nature 157:697.
Fraenkel, G., 1946, Britain's nutritional requirements, in: "Towards
    a Socialist Agriculture.   Studies by a Group of Fabians",
    F. W. Bateson, ed., Victor Gollancz, London, pp. 42-76.
Fraenkel, G., and Rudall, K. M., 1947, The structure of insect
    cuticles, Proc. Roy. Soc. (B) 134:111-143.
Fraenkel, G., and Blewett, M., 1947, The importance of folic acid
    and unidentified members of the vitamin B complex in the
    nutrition of certain insects, Biochem. J. 41:469-475.
Fraenkel, G., and Blewett, M., 1947, Linoleic acid and arachidonic
    acid in the metabolism of two insects, Ephestia kuehniella
    and Tenebrio molitor, Biochem. J. 41:475-478.
Ellinger, P., Fraenkel, G., and Abdel Kader, M. M., 1947, The
    utilization of nicotinamide derivatives and related compounds
    by mammals, insects and bacteria, Biochem. J. 41:559-568.
Fraenkel, G., 1948, B_T, a new vitamin of the B-group and its rela-
    tion to the folic acid group and other anti-anaemia factors,
    Nature 161:981-983.
Fraenkel, G., 1948, The effects of a relative deficiency of lysine
    and tryptophane in the diet of an insect, Tribolium confusum,
    Biochem. J. 43:Proceedings XIV.
Fraenkel, G., 1948, Evidence for the need, by certain insects, for
    three chemically unidentified factors of the vitamin B-complex,
    Brit. J. Nutrition 2, Abstracts of Communications 1:ii.
Fraenkel, G., 1949, Unidentified vitamins of the B-complex,
    Feder. Proc. 8:382.
Fraenkel, G., 1950, The nutrition of the meal worm, Tenebrio molitor
    Physiol. Zool. 23:92-108.
Pant, N. C., and Fraenkel, G., 1950, The function of the symbiotic
    yeasts of two insect species, Lasioderma serricorne F. and
    Stegobium (Sitodrepa) paniceum L., Science 112:498-500.
Fraenkel, G., and Stern, H. R., 1951, The nicotinic acid require-
    ments of two insect species in relation to the protein con-
    tents of their diets, Arch. Biochem. 34:468-477.

Fraenkel, G., 1951, Vitamin B_T. I. Deficiency symptoms, method of testing, distribution, isolation procedures and some properties, Feder. Proc. 10:183-184.

Carter, H. E., Bhattacharyya, P. K., and Fraenkel, G., 1951, Vitamin B_T. II. Purification and chemical studies, Feder. Proc. 10: 170.

Fraenkel, G., 1951, Effect and distribution of vitamin B_T, Arch. Biochem. Biophys. 34:457-468.

Fraenkel, G., 1951, Isolation procedures and certain properties of vitamin B_T, Arch. Biochem. Biophys. 34:468-477.

Carter, H.E., Bhattacharyya, P.K., Weidman, K., and Fraenkel, G., Fraenkel, G., 1952, The identity of vitamin B_T with carnitine, Arch. Biochem. Biophys. 35:241-242.

Cooper, M. I., and Fraenkel, G., 1952, Nutritive requirements of the smalleyed flour beetle, Palorus ratzeburgi Wissman (Tenebrionidae, Coleoptera), Physiol. Zool. 25:20-28.

Fraenkel, G., 1952, Function of vitamin B_T (carnitine), Feder. Proc. 11:443.

Fraenkel, G., 1952, A function of the salivary glands of the larvae of Drosophila and other flies, Biol. Bull. 103:285-286.

Fraenkel, G., 1952, The role of symbionts as sources of vitamins and growth factors for their insect hosts, Tijdsch. Entomologie 95:183-195.

Fraenkel, G., 1952, The nutritional requirements of insects for known and unknown vitamins, Trans. Ninth Int. Congr. Entomol. Amsterdam, 1951, Vol. 1:277-280.

Fraenkel, G., 1953, The nutritional value of green plants for insects, Trans. IXth Intern. Congr. Entomol. Amsterdam, 1951, Vol. 2:90-100.

Gray, H. E., and Fraenkel, G., 1953, Fructomaltose, a recently discovered trisaccharide isolated from honeydew, Science 118:304-305.

Fraenkel, G., 1953, Studies on the distribution of vitamin B_T (carnitine), Biol. Bull. 104:359-371.

Friedman, S., Bhattacharyya, P. K., and Fraenkel, G., 1953, Function of carnitine (B_T), Feder. Proc. 12:414-415.

Fraenkel, G., and Brookes, V. J., 1953, The process by which the puparia of many species of flies become fixed to a substrate, Biol. Bull. 105:442-449.

French, E. W., and Fraenkel, G., 1954, Carnitine (vitamine B_T) as a nutritional requirement for the confused flour beetle, Tribolium confusum Duval, Nature 173:173.

Gray, H. E., and Fraenkel, G., 1954, The carbohydrate components of honeydew, Physiol. Zool. 27:56-65.

Fraenkel, G., and Chang, P. I., 1954, Manifestations of a vitamin B_T (carnitine) deficiency in the larvae of the mealworm, Tenebrio molitor L., Physiol. Zool. 27:40-56.

Fraenkel, G., and Printy, G. E., 1954, The amino acid requirements of the confused flour beetle, Tribolium confusum, Biol. Bull. 106:149-157.

Moorefield, H. H., and Fraenkel, G., 1954, The character and ulti-
    mate fate of the salivary secretion of Phormia regina Meig.
    (Diptera, Calliphoridae), Biol. Bull. 106:178-184.
Friedman, S., Bhattacharyya, P. K., and Fraenkel, G., 1954, Acetyl
    carnitine as an acetyl donor, Feder. Proc. 13:214.
Lipke, H., and Fraenkel, G., 1954, Toxicity of corn germ for
    Tenebrio molitor, Feder. Proc. 13:464.
Fraenkel, G., 1954, Inhibitory effects of sugars, Feder. Proc. 13:
    457-458.
Lipke, H., Fraenkel, G., and Liener, I. E., 1954, The effect of
    soybean inhibitors on the growth of Tribolium confusum,
    J. of Agric. and Food Chemistry 2:410-414
Chang, P. I., and Fraenkel, G., 1953, Histopathology of vitamin
    B_T (carnitine) deficiency in larvae of the mealworm,
    Tenebrio molitor, Physiol. Zool. 27:259-267.
Fraenkel, G., 1954, The distribution of vitamin B_T (carnitine)
    throughout the animal kingdom, Arch. Biochem. Biophys. 50:486-495.
Rasso, S. C., and Fraenkel, G., 1954, The food requirements of the
    adult female blowfly, Phormia regina (Meigen) in relation
    to ovarian development, Ann. Ent. Soc. Amer. 47:636-645.
Pant, N. C., and Fraenkel, G., 1954, Studies on the symbiotic yeasts
    of two insect species, Lasioderma serricorne F. and Stegobium
    paniceum L., Biol. Bull. 107:420-432.
Pant, N. C., and Fraenkel, B., 1954, On the function of the intra-
    cellular symbionts of Oryzaephilus surinamensis L. (Cucujidae,
    Coleoptera), J. of the Zool. Society of India 6:173-177.
Fraenkel, G., 1955, Inhibitory effects of sugars on the growth of
    the mealworm, Tenebrio molitor L., J. Cell Comp. Physiol. 45:
    393-408.
Bhattacharyya, P. K., Friedman, S., and Fraenkel, G., 1955, The
    effect of some derivatives and structural analogues of
    carnitine on the nutrition of Tenebrio molitor, Arch. Biochem.
    Biophys. 54:424-436.
Fraenkel, G., Friedman, S., Hinton, Taylor, Laszlo, Sylvia and
    Noland, Jerre L., 1955, The effect of substituting carnitine
    for choline in the nutrition of several organisms, Arch.
    Biochem. Biophys. 54:432-439.
Lipke, H., and Fraenkel, G., 1955, The toxicity of corn germ to the
    mealworm, Tenebrio molitor, J. Nutrition 55:165-178.
Brust, M., and Fraenkel, G., 1955, The nutritional requirements of
    the larvae of a blowfly, Phormia regina Meigen, Physiol. Zool.
    28:186-204.
Friedman, S., and Fraenkel, G., 1955, Reversible enzymatic
    acetylation of carnitine, Arch. Biochem. Biophys. 59:491-501.
Friedman, S., McFarlane, J. E., Bhattacharyya, P. K., and Fraenkel,
    G., 1955, Quantitative separation and identification of
    quaternary ammonium bases, Arch. Biochem. Biophys. 59:484-490.

Ito, T., and Fraenkel, G., 1956, Anti-carnitine effect of γ-butyrobetaine on the development of the chick embryo, Feder. Proc. 15:558.

Friedman, S., Galun, A. B., and Fraenkel, G., 1956, Isolation and physiological action of (+) carnitine, Feder. Proc. 15:256.

Lipke, H., and Fraenkel, G., 1956, Insect nutrition, in: "Annual Review of Entomology", Vol. 1:17-44.

Fraenkel, G., 1956, The Tenebrio assay for carnitine, in: "Methods of Enzymology", S. P. Colowick and N. O. Kaplan, ed., Vol. 3, pp. 662-667.

Fraenkel, G., 1956, Insects and plant biochemistry. The specificity of food plants for insects, Proc. XIV Intern. Congr. Zoology Copenhagen, pp. 383-387.

Fraenkel, G., and Leclercq, J., 1956, Nouvelles recherches sur les besoins nutritifs de la larve du Tenebrio molitor, Arch. Internat. de Physiologie et de Biochimie 64:601-622.

Bloch, K., Langdon, R. G., Clark, A. J., and Fraenkel, G., 1956, Impaired steroid biogenesis in insect larvae, Biochim. Biophys. Acta 21:176.

Friedman, S., Galun, A. B., and Fraenkel, G., 1957, Isolation and physiological action of (+) carnitine, Arch. Biochem. Biophys. 66:10-15.

Fraenkel, G., 1957, Zinc and another inorganic growth factor required by Tenebrio molitor, Feder. Proc. 16:386.

Ito, T., and Fraenkel, G., 1957, γ-butyrobetaine as a specific antagonist for carnitine in the development of the early chick embryo, J. Gen. Physiol. 41:279-288.

Fraenkel, G., and Friedman, S., 1957, Carnitine. Vitamins and Hormones 15:73-118.

Galun, R., and Fraenkel, G., 1957, Physiological effects of carbohydrates in the nutrition of a mosquito, Aedes aegypti and two flies, Sarcophaga bullata and Musca domestica, J. Cell. and Comp. Physiol. 50:1-23.

Brookes, V. J., and Fraenkel, G., 1958, The nutrition of the larva of the housefly Musca domestica, Physiol. Zool. 31:208-223.

Fraenkel, G., 1958, The effect of zinc and potassium in the nutrition of Tenebrio molitor, with observations on the expression of a carnitine deficiency, J. Nutrition 65:361-396.

Fraenkel, G., 1958, The basis of food selection in insects which feed on leaves, 18th Ann. Meeting of the Ent. Soc. of Japan 5 pp.

Fraenkel, G., 1959, A historical and comparative survey of the dietary requirements of insects, Ann. New York Acad. Sci. 77:267-274.

Fraenkel, G., 1959, The raison d'être of secondary plant substances, Science 129:1466-1470.

Fraenkel, G., 1959, The chemistry of host specificity of phyto-
      phagous insects, Fourth Internat. Congr. of Biochem. Vol.
      XII, Biochemistry of Insects, Pergamon Press, London, pp. 1-14.
Ito, T., Horie, Y., and Fraenkel, G., 1959, Feeding on cabbage and
      cherry leaves by maxillectomized silkworm larvae, J. Sericult.
      Science of Japan 28:107-113.
Yamamoto, R. T., and Fraenkel, G., 1959, Common attractant for the
      tobacco hornworm, Protoparce sexta (Johann.) and the Colorado
      potato beetle, Leptinotarsa decemlineata (Say), Nature 184:
      206-207.
Yamamoto, R. T., and Fraenkel, G., 1960, The specificity of the
      tobacco hornworm Protoparce sexta (Johann.) to solanaceous
      plants, Ann. Ent. Soc. Amer. 53:503-507.
Yamamoto, R. T., and Fraenkel, G., 1960, The isolation and assay
      of the principal gustatory attractant for the tobacco horn-
      worm, Protoparce sexta Johann. from solanaceous plants, Ann.
      Ent. Soc. Amer. 53:499-503.
Yamamoto, R. T., and Fraenkel, G., 1960, The suitability of tobaccos
      for the growth of the cigarette beetle, Lasioderma serricorne
      Fab., J. Econ. Ent. 53:381-384.
Fraenkel, G., 1960, Lethal high temperatures for three marine in-
      vertebrates, Limulus polyphemus, Littorina littorea, and
      Pagurus longicarpus, Oikos 11:171-182.
Friedman, S., McFarlane, J. E., Bhattacharyya, P. K., and
      Fraenkel, G., 1960, (-) - Carnitine chloride, in:"Biochemical
      Preparations", John Wiley, New York, pp. 26-30.
Fraenkel, G., 1961, A new type of negative phototropotaxis observed
      in a marine isopod, Eurydice, Physiol. Zool. 34:228-232.
Fraenkel, G., 1961, Resistance to high temperatures in a
      Mediterranean snail, Littorina neritoides, Ecology 42:604-606.
Fraenkel, G., 1961, Quelques observations sur le comportement de
      Convoluta roscoffensis, Cahiers de Biologie Marine 2:115-160.
Waldbauer, G. P., and Fraenkel, G., 1961, Feeding on normally re-
      jected plants by maxillectomized larvae of the tobacco horn-
      worm, Protoparce sexta (Lepidoptera, Sphingidae), Ann. Ent.
      Soc. Amer. 54:477-485.
Fraenkel, G., and Gunn, D. L., 1961, The Orientation of Animals.
      Kineses, Taxes and Compass Reactions, Dover Publications,
      Inc., New York, 376 pp.
Galun, R., and Fraenkel, G., 1961, The effect of low atmospheric
      pressure on adult Aedes aegypti and on housefly pupae, J.
      Insect Physiol. 7:161-176.
Fraenkel, G., 1961, The physiology of insect nutrition. Symposia
      Genetica et Biologica Italica, Vol. IX., Atti del Simposio
      Internazionale di Biologi Sperimentale, Celebrazione
      Spallanzaniana, Reggio Emilia-Pavia 2-7 Maggio, 9 pp.
Fraenkel, G., 1961, Die biologische Funktion der sekundären
      Pflanzenstoffe im Allgemeinen und solcher Stoffe in Solanazeen
      im Besonderen, in: "Chemie und Biochemie der Solanum-

Alkaloide", Tagungsberichte Nr. 27., Intern. Symp. d. Deutsch. Akad, d. Landw. Berlin, pp. 297-307.

Yamamoto, R. T., and Fraenkel, G., 1962, The physiological basis for the selection of plants for egg-laying in the tobacco hornworm, Protoparce sexta (Johann.). XI. Internat. Congr. Entomology Vienna, Verhandlungen, Bd. III, pp. 127-133.

Fraenkel, G., Nayar, J., Nalbandov, O., and Yamamoto, R. T., 1962, Further investigations into the chemical basis of the insect-host plant relationship, XI, Intern. Congr. Entomology, Vienna, Verhandlungen, Bd. III, pp. 122-126.

Fraenkel, G., 1962, Tanning of the adult fly: a new hormone action, Amer. Zool. 2:524.

Nayar, J. K., and Fraenkel, G., 1962, The chemical basis of host plant selection in the silkworm, Bombyx mori (L.), J. Insect Physiol. 8:505-525.

Fraenkel, G., and Hsiao, C., 1962, Hormonal and nervous control of tanning in the fly, Science 138:27-29.

Nayar, J. K., and Fraenkel, G., 1963, The chemical basis of host selection in the Catalpa Sphinx, Ceratomia catalpae (Lepidoptera, Sphingidae), Ann. Ent. Soc. Amer. 56:119-122.

Nayar, J. K., and Fraenkel, G., 1963, The chemical basis of host selection in the Mexican bean beetle, Epilachna varivestis (Coleoptera, Coccinellidae), Ann. Ent. Soc. Amer. 56:174-178.

Nayar, J. K., and Fraenkel, G., 1963, Practical methods of year-round laboratory rearing of the silkworm, Bombyx mori (L.) (Lepidoptera, Bombycidae), Ann. Ent. Soc. Amer. 56:122-123.

Fraenkel, G., and Hsiao, C., 1963, Tanning in the adult fly. A new function of neurosecretion in the brain, Science 141:1057-1058.

Nalbandov, O., Yamamoto, R. T., and Fraenkel, G., 1964, Insecticides from plants. Nicandrenone, a new compound with insecticidal properties, isolated from Nicandra physalodes, Agric. Food Chem. 12:55-59.

Soo Hoo, C. F., and Fraenkel, G., 1964, The resistance of ferns to the feeding of Prodenia eridania larvae, Ann. Ent. Soc. Amer. 57:788-790.

Soo Hoo, C. F., and Fraenkel, G., 1964, A simplified laboratory method for rearing the southern armyworm, Prodenia eridania, for feeding experiments, Ann. Ent. Soc. Amer. 57:798-799.

Fraenkel, G., and Hsiao, C., 1965, Bursicon, a hormone which mediates tanning of the cuticle in the adult fly and other insects, J. Insect Physiol. 11:513-556.

Hamby, R. J., and Fraenkel, G., 1965, Effects of high temperatures on the prosobranch snail, Littorina littorea, Biol. Bull. 129:406-407.

Fraenkel, G., Hsiao, C., and Seligman, I. M., 1965, The proteinaceous nature of an insect hormone, bursicon, Amer. Zool. 5:674.

Fraenkel, G., 1965, A brief survey of the recognition of carnitine
    as a substance of physiological importance, in: "Recent Re-
    search on Carnitine. Its Relation to Lipid Metabolism",
    G. Wolf, ed., The M.I.T. Press, Cambridge, pp. 1-3.
Fraenkel, G., Hsiao, C., and Seligman, I. M., 1966, Properties of
    bursicon, an insect hormone that controls cuticular tanning,
    Science 151:91-93.
Pausch, R. D., and Fraenkel, G., 1966, The nutrition of the larva
    of the oriental rat flea, Xenopsylla cheopis (Rothschild),
    Physiol. Zool. 39:202-222.
Hsiao, C., and Fraenkel, G., 1966, Neurosecretory cells in the
    central nervous system of the adult blowfly, Phormia regina
    Meigen (Diptera, Calliphoridae), J. Morph. 119:21-38.
Soo Hoo, C. F., and Fraenkel, G., 1966, The selection of food
    plants in a polyphagous insect, Prodenia eridania (Cramer),
    J. Insect Physiol. 12:693-709.
Soo Hoo, C. F., and Fraenkel, G., 1966, The consumption, digestion
    and utilization of food plants by a polyphagous insect, Prodenia
    eridania (Cramer), J. Insect Physiol. 12:711-730.
Ito, T., and Fraenkel, G., 1966, The effect of nitrogen starvation
    on Tenebrio molitor L., J. Insect Physiol. 12:803-817.
Fraenkel, G., 1966, The heat resistance of intertidal snails at
    Shirahama, Wakayama-ken, Japn. Publ. Seto Mar. Biol. Lab.
    14:185-195.
Fraenkel, G., and Hsiao, C., 1966, Pupal diapause in Sarcophaga
    falculata (Diptera), Amer. Zool. 6:576-577.
Fraenkel, G., and Hsiao, C., 1967, Calcification, tanning, and the
    role of ecdyson in the formation of the puparium of the facefly,
    Musca autumnalis, J. Insect Physiol. 13:1387-1394.
Fraenkel, G., and Hsiao, C., 1968, Manifestations of a pupal dia-
    pause in two species of flies, Sarcophaga argyrostoma and S.
    bullata, J. Insect Physiol. 14:689-705.
Fraenkel, G., and Hsiao, C., 1968, Morphological and endocrinologi-
    cal aspects of pupal diapause in a fleshfly, Sarcophaga
    argyrostoma, J. Insect Physiol. 14:707-718.
Fraenkel, G., 1968, The heat resistance of intertidal snails at
    Bimini, Bahamas; Ocean Springs, Mississippi; and Woods Hole,
    Massachusetts, Physiol. Zool. 41:1-13.
Hsiao, T. H., and Fraenkel, G., 1968, The influence of nutrient
    chemicals on the feeding behavior of the Colorado Potato
    Beetle, Leptinotarsa decemlineata (Coleoptera: Chrysomelidae),
    Ann. Entom. Soc. Amer. 61:44-54.
Hsiao, T. H., and Fraenkel, G., 1968, Isolation of phagostimulatory
    substances from the host plant of the Colorado potato beetle,
    Ann. Ent. Soc. Amer. 61:476-484.
Hsiao, T. H., and Fraenkel, G., 1968, The role of secondary plant sub-
    stances in the food specificity of the Colorado potato beetle,
    Ann. Ent. Soc. Amer. 61:485-493.

Hsiao, T. H., and Fraenkel, G., 1968, Selection and specificity of the Colorado potato beetle for solanaceous and nonsolanaceous plants, Ann. Ent. Soc. Amer. 61:493-503.

Fogal, W., and Fraenkel, G., 1969, Melanin in the puparium and adult integument of the flesh-fly, Sarcophaga bullata, J. Insect Physiol. 15:1437-1447.

Fogal, W. H., and Fraenkel, G., 1969, The role of bursicon in melanization and endocuticle formation of the cuticle of the adult fleshfly, Sarcophaga bullata, J. Insect Physiol. 15: 1235-1240.

Seligman, M., Friedman, S., and Fraenkel, G., 1969, Hormonal control of turnover of tyrosine and tyrosine phosphate during tanning of the adult cuticle in the fly, Sarcophaga bullata, J. Insect Physiol. 15:1085-1102.

Seligman, M., Friedman, S., and Fraenkel, G., 1969, Bursicon mediation of tyrosine hydroxylation during tanning of the adult cuticle of the fly, Sarcophaga bullata, J. Insect Physiol. 15:553-562.

Hsiao, T. H., and Fraenkel, G., 1969, Properties of leptinotarsin: A toxic hemolymph protein from the Colorado potato beetle, Toxicon 7:114-130.

Berreur, P., and Fraenkel, G., 1969, Puparium formation in flies: Contraction to puparium induced by ecdyson, Science 164:1182-1183.

Zdarek, J., and Fraenkel, G., 1969, Correlated effects of ecdyson and neurosecretion in puparium formation (pupariation) of flies, Proc. Nat. Acad. Sci. 64:565-572.

Fraenkel, G., 1970, Evaluation of our thoughts on secondary plant substances, Ent. Exp. and Appl. 12:473-486.

Fraenkel, G., and Zlotkin, E., 1970, Acceleration of puparium formation in Sarcophaga argyrostoma by electrical stimulation or scorpion venom, J. Insect Physiol. 16:1549-1554.

Fraenkel, G., and Zdarek, J., 1970, The evaluation of the "Calliphora test" as an assay for ecdysone, Biol. Bull. 139:138-150.

Fogal, W. H., and Fraenkel, G., 1970, Histogenesis of the cuticle of the adult flies, Sarcophaga bullata and S. argyrostoma, J. Morph. 130:137-158.

Zdarek, J., and Fraenkel, G., 1970, Overt and covert effects of endogenous and exogenous ecdysone in puparium formation of flies, Proc. Nat. Acad. Sci. 67:331-337.

Zdarek, J., and Fraenkel, G., 1971, Neurosecretory control of ecdysone release during puparium formation of flies, Gen. Comp. Endocrin. 17:483-489.

Zlotkin, E., Fraenkel, G., Miranda, F., and Lissitzky, S., 1971, The effect of scorpion venom on blowfly larvae - A new method for the evaluation of scorpion venom potency, Toxicon 9:1-8.

Denlinger, D. L., Willis, J. H., and Fraenkel, G., 1972, Rates
    and cycles of oxygen consumption during pupal diapause in
    Sarcophaga flesh flies, J. Insect Physiol. 18:871-882.
Zdarek, J., and Fraenkel, G., 1972, The mechanism of puparium
    formation in flies, J. Exp. Zool. 179:315-324.
Fraenkel, G., Zdarek, J., and Sivasubramanian, P., 1972, Hormonal
    factors in the CNS and hemolymph of pupariating fly larvae which
    accelerate puparium formation and tanning, Biol. Bull. 143:
    127-139.
Friedman, S., and Fraenkel, G., 1972, Carnitine, in: "The Vitamins",
    W. H. Sebrell and R. S. Harris, ed., Vol. V., Academic Press,
    New York and London, pp. 329-355.
DeGuire, D. M., and Fraenkel, G., 1973, The meconium of Aedes aegypti
    (Diptera: Culicidae), Ann. Ent. Soc. Amer. 66:475-476.
Fraenkel, G., and Bhaskaran, G., 1973, Pupariation and pupation in
    flies: terminology and interpretation, Ann. Ent. Soc. Amer.
    66:418-422.
Ratnasiri, N. P., and Fraenkel, G., 1973, Inhibition of pupariation
    in Sarcophaga bullata, Nature 243:91-93.
Ratnasiri, N., and Fraenkel, G., 1974, Anterior inhibition of
    pupariation in ligated larvae of Sarcophaga bullata and other
    fly species: Incidence and expression, Ann. Ent. Soc. Amer.
    67:195-203.
Ratnasiri, N., and Fraenkel, G., 1974, The physiological basis of
    anterior inhibition of puparium formation in ligated fly
    larvae, J. Insect Physiol. 20:105-119.
Sivasubramanian, P., Ducoff, H. S., and Fraenkel, G., 1974, Effect
    of X-irradiation on the formation of the puparium in the
    fleshfly, Sarcophaga bullata, J. Insect Physiol. 20:1303-1317.
Sivasubramanian, P., Friedman, S., and Fraenkel, G., 1974, Nature
    and role of proteinaceous hormonal factors acting during
    puparium formation in flies, Biol. Bull. 147:163-185.
Fraenkel, G., 1975, Interactions between ecdysone, bursicon, and
    other endocrines during puparium formation and adult emergence
    in flies, Amer. Zool. 15:29-38.
Fraenkel, G., 1976, Molting and development in undersized fly
    larvae, in: "The Insect Integument", H. R. Hepburn, ed.,
    Elsevier Scientific Publishing Company, Amsterdam, pp. 323-338.
Fraenkel, G., Blechl, A., Blechl, J., Herman, P., and Seligman, M.,
    1977, Cyclic AMP and the hormonal control of puparium formation
    in the fleshfly Sarcophaga bullata, Proc. Natl. Acad. Sci.
    74:2182-2186.
Pappas, C., and Fraenkel, G., 1977, Nutritional aspects of oogenesis
    in the flies Phormia regina and Sarcophaga bullata, Physiol.
    Zool. 50:237-246.
Pappas, C., and Fraenkel, G., 1977, Endocrinological aspects of
    oogenesis in the flies Phormia regina and Sarcophaga bullata,
    J. Insect Physiol. 24:75-80.

Seligman, M., Blechl, A., Blechl, J., Herman, P., and Fraenkel, G., 1977, The endocrinological control of the formation and tanning of the puparium of the fleshfly, Sarcophaga bullata, Proc. Natl. Acad. Sci. 74:4697-7401.

Fraenkel, G., and Hollowell, M., 1979, Actions of the juvenile hormone, ecdysone and the oostatic hormone in the oogenesis of the flies Phormia regina and Sarcophaga bullata, J. Insect Physiol. 25:305-310.

Zdarek, J., Slama, K., and Fraenkel, G., 1979, Changes in internal pressure during puparium formation in flies, J. Exp. Zool. 207:187-195.

Fraenkel, G., 1980, The proposed vitamin role of carnitine, in: "Carnitine Biosynthesis, Metabolism and Functions", R. E. Frenkel and J. D. McGarry, ed., Academic Press, New York, pp. 1-6.

SYMPOSIUM ON INSECT ENDOCRINOLOGY

INTRODUCTION

Stanley Friedman

University of Illinois
Urbana, Illinois

A perusal of the table of contents of this short symposium on Insect Endocrinology might lead those who are generally uninstructed in Fraenkel's activities to wonder as to our choice of papers. Our justification is based upon our effort to concentrate the symposium upon a single area in which he has had an extended involvement: namely, the metamorphic molt in flies. In so doing, we immediately recognized the enormous humoral and physiological ramifications, both direct and indirect, of such an interest. Our selection of these papers is an attempt to present the broad picture with a few quick strokes.

The first offering, by J. H. Willis, et al., is a direct outgrowth of Fraenkel's work with Rudall on the chemistry of the puparium. Their investigations on the structure of fly cuticle and its changes at pupariation (Fraenkel and Rudall, 1940, Proc. Roy. Soc. London (B) 129:1; 1947, ibid, 134:111), provided quantitative data which have, since that time, influenced our ideas concerning cuticular changes at metamorphosis and are presently at the center of the controversy over the mechanism of cuticular hardening and darkening (see Hillerton and Vincent, 1979, J. Ins. Physiol. 25:957). The studies described by Dr. Willis lay to rest certain of the interpretations of the data obtained in those early papers, and at the same time revise our thinking regarding cuticular chemistry at the metamorphic molts. (Dr. Fraenkel's continued involvement in this manifestation of metamorphosis may be seen in the presence of a student from his laboratory as a co-author on the paper.)

Fraenkel's preoccupation with the formation of the puparium certainly transcends the cuticle, as we well know from his early

23

studies describing the hormonal activation of the process (Fraenkel, 1935, Proc. Roy. Soc. London (B) 118:1). Therefore, it is quite appropriate that a major part of the symposium be devoted to the regulation of the metamorphic molt. In view of the fact that there are a number of difficulties in demonstrating juvenile hormone intercession in metamorphosis in flies, it is proper from a conceptual viewpoint to describe recent advances in our knowledge of JH activity and regulation in other insects.

In this vein, the paper by Scharrer describes hypertrophic morphological changes in the corpus allatum of gonadectomized females of Leucophaea maderae, and, more importantly, interactions within the corpus allatum between neurosecretory elements deriving from the corpus cardiacum. In her paper, she suggests that these interactions lead to "mutual exchange of information presumably related to the control of the corpus allatum." To give this thought further expression, there are two papers, one by Bhaskaran, describing surgical intervention in Manduca sexta larvae leading to the turning-on or off of the corpus allatum by the brain, as shown by the presence or absence of supernumerary larvae; and a second, by Granger, et al., describing organ culture techniques for corpus allatum and "gland complexes" (brain-CC-CA) of Manduca sexta larvae which permit a study of the control of JH synthesis in "complexes" of various developmental stages, and in CA alone under the influence of brains taken from larvae of different ages. Finally, there is a paper by Kunkel, in which a case is made for establishing a fundamental role for JH in metamorphosis by focusing on a tissue at the metamorphic molt and examining the manner in which the hormone effects stage specific protein synthesis.

As an entre-acte we have Denlinger's paper on pupal diapause in flies, describing the enormous complexity underlying the diapause state and deriving from Fraenkel's interest in the physiology of pupariation and pupation. The paper also provides us with a description of Fraenkel's happening upon a situation and taking advantage of it to instruct us all in the meaning of thoughtful observation (Fraenkel and Hsaio, 1968, J. Ins. Physiol. 14:689,707). The subject matter of Denlinger's contribution is of great concern to us and his paper is an excellent review of an important field, but its presence in this symposium equally reflects our desire to call attention to one of the indirect spinoffs from Fraenkel's focus on the metamorphic molt.

The final pair of papers are involved with endocrine material directed to various metamorphic activities: the second echelon of metamorphic hormones.

Fraenkel's discovery of bursicon in 1962 (the adult tanning hormone in flies, liberated after eclosion from the puparium) (Fraenkel and Hsaio, 1963, Science 141:1057), paved the way for a

number of similar findings in other insects.   Then, in 1969, he
and Zdarek came upon (once again a "chance" observation) the
substances which Zdarek discusses further in this symposium: the
pupariation factors (Zdarek and Fraenkel, 1969, Proc. Natl. Acad.
Sci. 64:565).   These are proteins released in response to ecdysone
action, each responsible for an individual part of the program
which in its totality results in the puparium.   Thus, one of these
compounds activates a series of muscle contractions, while a second
(different from bursicon) is involved with tanning.

Truman captures the same spirit of hierarchies of hormonal
activity in his description of eclosion hormone in Manduca sexta.
This protein, produced after the liberation of ecdysone, itself
responsible for the last larval apolysis, has a number of
behavioral effects which finally result in ecdysis to the pupa,
among them the activation of the secretory cells responsible for
the liberation of bursicon.

THE METAMORPHOSIS OF ARTHROPODIN

Judith H. Willis[1], Jerome C. Regier[2], and Bettina A. Debrunner[3]

[1]Department of Genetics and Development
University of Illinois
Urbana, Illinois 61801

[2]Cellular and Developmental Biology
The Biological Laboratories
Harvard University
Cambridge, Massachusetts 02138

[3]Department of Entomology
University of Illinois
Urbana, Illinois 61801

INTRODUCTION

Some modern elementary biology texts inform their readers that arthropod cuticles are composed exclusively of the polysaccharide, chitin. But it has been known since 1823 that there is another major nitrogenous component which we now know to be protein. The first to measure the amount of protein in the cuticle was Gottfried Fraenkel. His publication with Rudall in 1940 alerted the scientific community to the fact that about 40% of the dry weight of fly larval cuticle is protein, much of which is soluble in boiling water. They also discovered that less than 10% of the dry weight remained soluble in boiling water after the puparium was tanned, but that protein could still be extracted with 5% KOH, leaving the chitin behind.

In 1947, Fraenkel and Rudall introduced the term "arthropodin" to refer to readily extractable cuticular protein. Their work on arthropodin's solubility characteristics and studies by Fraenkel's student Trim (1941) of its amino acid composition led them to

27

suggest that arthropodin was a distinct protein, characteristic of soft cuticle throughout the phylum Arthropoda. They even speculated that " . . . its presence is responsible for the free movability of the limbs . . ."

Our task in this paper will be to describe what has happened to the concept of arthropodin in the subsequent three decades.

By 1953, it was apparent that arthropodin was not a single species of protein. Hackman (1953), in Australia, found several discrete zones when he subjected cuticular proteins from scarab beetle larvae to electrophoresis in a Tiselius apparatus. At about the same time, Herbert Moorefield, one of Fraenkel's graduate students at Illinois, performed similar analyses using Sarcophaga. He demonstrated even greater complexity, but, intimidated by Hackman's publication priority, never published his results. These remain in his doctoral thesis (Moorefield, 1953).

ISOLATION OF ARTHROPODIN

Since 1953, with the use of improved electrophoretic techniques, we have seen a dramatic increase in the apparent biochemical complexity of arthropodin. Results from one such technique, isoelectric focusing in polyacrylamide slab gels, are displayed in Fig. 1. In collaboration with Dr. Fraenkel, we have analyzed cuticular proteins from mature Sarcophaga bullata larvae prepared eight different ways. In lane 1 are displayed cuticular proteins extracted by brief homogenization in water. Isoelectric points of the major proteins vary from approximately pH 4 to 7, and at least 35 discrete polypeptides are apparent. This in only one dimension! In lanes 1-6 are water extracts of cuticle, differing as to exact method, duration, and temperature of extraction. The amount of protein extracted is indicated at the bottom of each lane and is expressed as a percentage of the total dry weight of unextracted cuticle. The four most gentle extraction procedures (lanes 1 to 4) give sharp bands with similar but non-identical patterns. Prolonged boiling (lanes 5 and 6) results in considerable smearing of bands, presumably due to peptide bond scission, deamination reactions, and other modifications. Boiling increases the amount of protein extracted, with a maximum of 28% after 48 hours.

Cuticular proteins can be extracted very efficiently at moderate temperatures using urea or guanidine hydrochloride (lanes 7 and 8). These reagents are particularly effective at breaking hydrogen bonds without affecting covalent bonds. About as much protein is extracted with guanidine hydrochloride as with boiling water, and, in addition, the proteins extracted give sharp electrophoretic patterns. At least 50 discrete cuticular polypeptides are resolved.

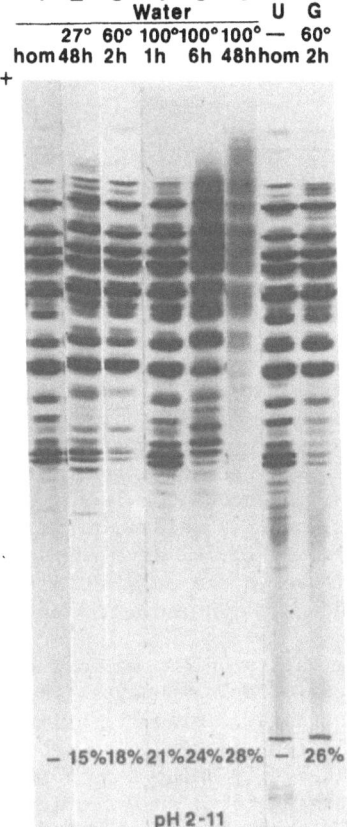

Fig.1 Electrophoretic comparison of extraction procedures on
       cleaned Sarcophaga larval cuticles. All lanes come from a
       single gel. The numbers on the top line indicate the lane.
       The next line designates the extraction medium: water, 8M
       urea (U) or 8M guanidine hydrochloride (G), the latter two
       buffered with Tris-HCl (pH 8.4) and containing 1% mercapto-
       ethanol and 2 mM phenylmethyl-sulfonylfluoride. The third
       line gives the temperature at which extraction was carried
       out. The fourth line indicates whether cuticles were homo-
       genized (hom) or soaked for the time indicated. Following
       extraction, dialysis, lyophilization, and resuspension in 6M
       urea buffered with Tris-HCl, proteins were fractionated on an
       isoelectric focusing slab gel. The gel included 6M urea and
       Brinkmann pHisolytes (pH range 2-11). The percentages indi-
       cate the relative amount of cuticular dry (lyophilized)
       weight removed by extraction.

Thus, we now have an extraction technique which is as effective as boiling water in freeing proteins from cuticle, but which better preserves their integrity. We have used this extraction technique in answering some questions about the nature of arthropodin.

CHARACTERIZATION OF ARTHROPODIN

First, we must seek an explanation for the multiplicity of bands found whenever cuticular proteins have been subjected to electrophoresis.

Is Arthropodin Encoded by a Multigene Family?

The multiplicity of components revealed by electrophoresis raises interesting evolutionary as well as functional questions concerning cuticle. How many distinct proteins are required to construct a complex supramolecular structure? Have genes encoding these distinct proteins evolved independently of each other, or are they a multigene family encoding a family of functionally, structurally, and evolutionarily related proteins (Hood et al., 1975)? Hackman and Goldberg (1976) have suggested that genetic polymorphism may contribute to the multiplicity of cuticular proteins, but have provided no direct evidence for this hypothesis.

In order to attack this problem, we have begun structural studies on individual cuticular components from the Cecropia silkmoth (Hyalophora cecropia). Clearly, primary sequence data would provide the most direct answer to the above questions, but so far, we have determined only amino acid compositions of purified components. Nevertheless, this is an essential preliminary step for future sequencing, and it has already produced some interesting information.

Cecropia has many cuticular proteins which can easily be separated by electrophoresis. In work carried out in F. C. Kafatos' laboratory at Harvard University, we fractionated and isolated some of these proteins by a very simple technique. A single extract was applied across the width of an isoelectric focusing slab gel, and the extract was subjected to electrophoresis. The gel was then fixed in ammonium sulfate solution and individual bands were cut out, extracted, dialyzed, hydrolyzed, and their amino acid compositions determined.

Amino acid compositions of three larval, five pupal and four adult cuticular components are shown in Table 1. The bands share several features. Cysteine and methionine are low or absent in all (except P2), a feature of arthropodin first reported by Trim in 1941. They are enriched in glycine, a not uncommon feature

Table 1. Amino Acid Compositions of Individual Bands Isolated From IF Gels (Residues/100 Residues)

	L1	L2	L3	P1	P2	P3	P4	P5	A1	A2	A3	A4
CM-CYS	0.0	0.1	0.1	0.7	3.6	1.7	1.2	0.1	0.1	0.1	0.0	1.0
ASP*	12.7	11.6	11.3	9.6	10.6	10.3	9.2	10.3	12.5	12.6	12.0	7.0
THR**	5.9	6.9	7.5	7.2	5.0	5.9	5.7	5.4	8.0	8.0	7.1	5.7
SER**	5.3	4.8	8.0	13.5	5.6	12.1	15.7	12.4	7.0	6.8	5.2	6.7
GLU*	16.6	17.2	14.0	8.6	7.8	11.0	13.1	7.9	13.2	12.9	16.9	21.8
PRO	7.4	5.1	6.3	7.1	17.9	5.6	6.0	7.1	11.4	11.2	6.4	9.2
GLY	15.9	16.3	10.7	11.6	5.5	12.0	14.6	10.3	8.9	8.7	14.3	11.3
ALA	5.3	4.9	6.7	12.7	14.0	14.0	12.3	14.8	10.1	9.9	6.1	9.4
VAL	8.2	9.6	10.6	5.3	6.1	5.9	4.9	6.5	3.4	3.4	8.0	4.2
MET	0.0	0.0	0.0	0.1	0.1	0.3	0.4	0.0	0.1	0.1	0.0	0.0
ILE	4.1	5.7	3.9	4.3	4.9	4.2	3.4	4.6	6.8	6.5	5.5	6.2
LEU	5.4	3.7	5.2	5.8	4.7	5.0	4.4	5.9	3.9	4.0	3.7	4.2
TYR	3.9	4.9	4.2	2.2	2.2	0.8	0.5	1.0	2.4	3.5	4.4	2.6
PHE	3.2	3.4	2.7	4.5	4.3	3.5	2.8	5.7	3.2	3.4	3.9	3.4
HIS	1.5	2.3	1.7	1.1	1.1	1.3	1.3	0.6	1.8	1.8	2.1	1.2
LYS	3.3	3.5	4.6	3.0	3.0	3.9	3.1	3.2	6.3	6.2	3.1	3.0
ARG	1.1	1.1	2.5	2.5	3.8	2.4	1.4	4.3	0.9	0.8	1.2	3.1

*Includes the amidic forms.

**Corrected for losses during hydrolysis (Regier et al., 1978b).

for structural proteins. One of the main conclusions from Table 1
is that at least 11 of the 12 proteins are probably encoded by
distinct genes. Only A1 and A2 have amino acid compositions which
are virtually indistinguishable, except possibly for tyrosine. With
respect to the others, one can see many shared features, especially
within a stage, and other features which make each protein distinct.

The degree of evolutionary relatedness of distinct polypeptides is best quantified through direct sequence comparison, although even this method is not perfect, due to the possibility of
convergent evolution. Having no sequence data, we can still estimate the degree of similarity in compositions of cuticular components. A statistical test, developed by Marchalonis and Weltman
(1971), makes pairwise comparison of amino acid compositions and
estimates the probability that two proteins are evolutionarily related. Proteins of known sequence were used to develop this test.
Differences in concentration of each amino acid from two proteins
are squared separately and then summed; the number obtained, called
the SΔQ value, is an indicator of the degree of similarity between
two compositions. The smaller the number, the more similar the
compositions. Marchalonis and Weltman (1971) found that in 98%
of their 5000 comparison pairs " . . . unrelated proteins differed
by more than 100 SΔQ units. In no case did proteins thought to
be unrelated by comparison of sequence, differ by less than 50
SΔQ units."

Concerned that their limits might not apply to our structural
proteins, in which a small number of amino acids make large contributions to the total composition, we have used compositional and
sequence data from silkmoth chorion (eggshell) proteins (Regier
et al., 1978a; 1978b; Jones et al., 1979) as a more appropriate
reference for silkmoth cuticular proteins. Like cuticular proteins,
chorion proteins are biochemically complex and have a high percentage of glycine and alanine. Published sequences show the existence
of two major chorion protein-encoding multigene families, and at
least three more are strongly suggested (Regier, unpublished observations). Compositional comparisons from selected members of
the chorion "A" family are shown in Table 2 (top); their SΔQ
values range from 9 to 160. Thus in this system, members of a
single multigene family may have values over 100.

At least two chorion subfamilies are recognized from sequence
data, and compositional comparisons (Table 2, top) are consistent
with this. SΔQ values within subfamily II ranged from 9-22.

SΔQ values of polypeptides from different chorion multigene
families are considerably larger. For example, comparison of a
specific "B" and a specific "C" family member yields a value of
481 (Regier, unpublished observations). Despite this large

Table 2. Pairwise comparisons of amino acid compositions by the
Marchalonis and Weltman method (SΔQ values).

CHORION MULTIGENE FAMILY A[*]

		subfamily I	subfamily II		
		A1,1 -- a2	A4 -- c1	A4 -- d1	A2,3 -- d5,6
subfamily II	A4--c1	92			
	A4 -- d1	98	9		
	A2,3 -- d5,6	160	18	12	
	A3,5 -- d9	134	22	9	9

CECROPIA CUTICULAR PROTEINS

		LARVAL			PUPAL					ADULT		
		L1	L2	L3	P1	P2	P3	P4	P5	A1	A2	A3
LARVAL	L2	18										
	L3	59	71									
	P1	232	278	139								
	P2	407	500	311	237							
PUPAL	P3	198	238	123	20	257						
	P4	211	252	166	48	380	33					
	P5	283	336	172	20	205	29	90				
	A1	151	205	115	135	158	126	188	161			
ADULT	A2	147	200	109	135	157	131	197	163	2		
	A3	12	11	47	216	392	182	206	265	126	122	
	A4	132	151	155	254	355	204	207	292	139	145	105

[*] Data from Regier et al., 1978b.

difference, it is known that these two proteins contain regions of similar sequence, suggesting regional homology (see Jones et al., 1979).

Comparisons of cuticular proteins are shown in Table 2 (bottom). Numbers enclosed in the three boxes represent intra-metamorphic stage comparisons; the numbers outside the boxes are inter-stage comparisons.

SΔQ values of intra-stage comparisons are within the range of a single multigene family, except for P2. Excluding P2, values range from 2 to 145.

Inter-stage comparisons show that larval and adult proteins are generally more similar to each other than larval and pupal or pupal and adult proteins. Mean values and standard deviations for the three comparisons are $\bar{x} = 120$, $s = 65$; $\bar{x} = 256$, $s = 102$; $\bar{x} = 206$, $s = 73$, respectively. Using chorion protein SΔQ's as our standard, we can say that it is possible that larval and adult cuticular protein-encoding genes belong to the same or to a closely related multigene family. The SΔQ values for specific larval and pupal, and pupal and adult constituents suggests that certain of these proteins also may be closely related.

Fristrom et al. (1978) have shown larval cuticular proteins from Drosophila to be encoded on at least two chromosomes. In addition, amino acid compositions of three electrophoretically homogeneous cuticular proteins yield SΔQ values of 36, 80, and 131.

The evidence from these limited studies suggests that many genes encoding cuticular proteins are members of one or more multigene families. However, certain cuticular proteins may be unrelated to these groups.

What Happens to Arthropodin During Tanning?

The importance of arthropodin tanning to cuticle stabilization has long been recognized. With the more recent appreciation of cuticular protein complexity, it is important to ask whether all or only selected components are modified during this process. In collaboration with Professor Fraenkel, we have extracted larval cuticular proteins from Sarcophaga at different times during larval life and following the onset of puparium formation. Extracts were fractionated on a narrow range (pH 4-6) isoelectric focusing slab gel, which facilitates resolution of these acidic proteins. Since there was a five-fold decline in the amount of protein which could be solubilized as tanning progressed (Fig. 2), accurate comparison

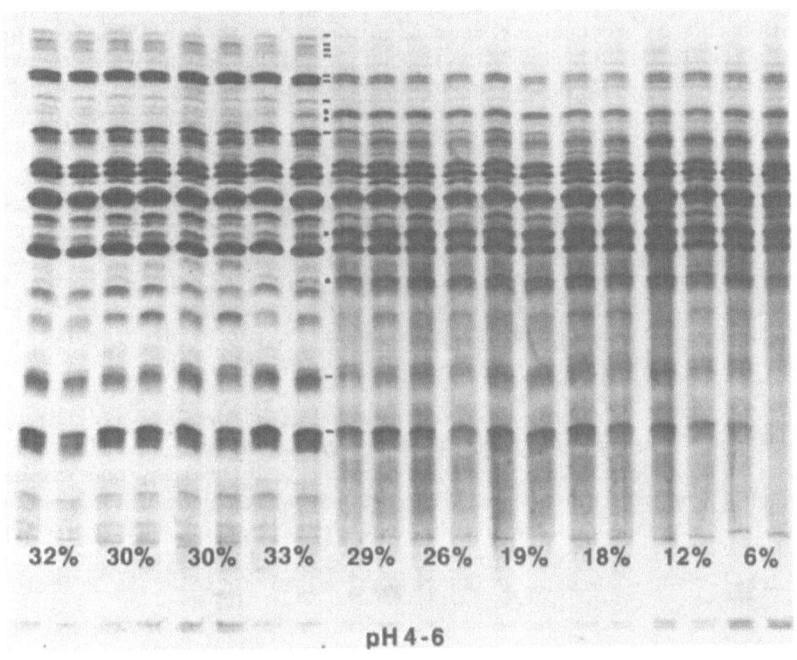

Fig. 2 Electrophoretic comparison of proteins extracted with buf-
        fered 8M guanidine hydrochloride from cleaned Sarcophaga
        cuticles of different ages relative to puparium formation and
        tanning. Abbreviations: L, larva, RS, red spiracle stage;
        WP, white puparium, T, tanning puparium. Larval age (days)
        and puparium age (hours post tanning) are shown. Other pro-
        cedures as in Fig. 1 except for concentration of pHisolytes
        (9 parts pH 4-6, 1 part pH 2-11). During the first hour
        of tanning 10 bands decrease (-) while 4 increase (·) in
        relative intensity.

of relative changes for specific bands required that equal amounts
of protein be loaded in each lane.

In contrast to the gradual decline in overall protein solubility during tanning, the change in banding pattern was abrupt,
occurring within one hour after color was first visible in the
puparium, and thereafter little change occurred (Fig. 2). The most
obvious change within that first hour was the marked decrease in
relative intensity of 10 of approximately 50 bands (see -, Fig. 2).
It is unlikely that these decreases were due to modifications which
alter the isoelectric points of the proteins, moving them to new
positions on the gel, since no prominent new bands appeared and
only four pre-existing bands increased in intensity (see ·, Fig.
2). These declining bands may represent a particular subset of
proteins which are selectively rendered insoluble at the onset of
puparium tanning. Interestingly, after the first hour of tanning
these bands showed no further differential changes.

Another feature which accompanies tanning is an increase in
background staining in each lane. We suspect that this represents
modification of many proteins, producing material of such heterogeneity that it is not resolved into individual bands.

Puparium formation is unusual in that a completed larval cuticle
is tanned. In more typical situations, exocuticle is laid down
first. The animal then ecdyses, tans its exocuticle, and simultaneously begins laying down endocuticle which does not become tanned. We have examined cuticular proteins in which this more typical
tanning regime is followed. These studies were part of Dr. Elaine
Roberts' thesis, and utilized the abdominal cuticle of Tenebrio
molitor (Roberts and Willis, 1980).

Soluble cuticular proteins from progressively more mature
Tenebrio were fractionated on an SDS slab gel, which separates
proteins on the basis of molecular weight (Fig. 3). Two samples of
each metamorphic stage were analysed, at 1 and 24 hours post-
ecdysis. By 24 hours, exocuticular tanning has occurred and
endocuticle deposition is well under way. The results show that
some of the early exocuticular proteins become relatively less
prominent or disappear altogether by 24 hours, suggesting that
they are rendered insoluble. There is also an increase in high
molecular weight bands at 24 hours, and some new bands appear which
are unique to the more mature cuticles. The qualitative changes
are more dramatic in this case than they are in Sarcophaga, which
tans a previously completed cuticle.

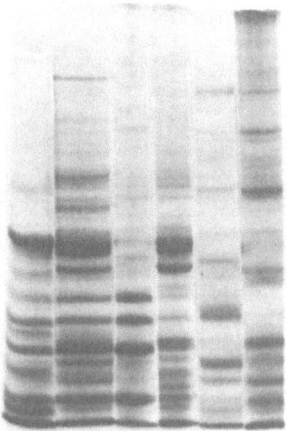

L₁   L₂₄   P₁ P₂₄ A₁ A₂₄

Fig.3   Electrophoretic comparison of banding patterns of buffer-
        soluble extracts from abdominal cuticles of larval (L),
        pupal (P) and adult (A) <u>Tenebrio</u>. The insects were main-
        tained for 1 hour or 24 hours after ecdysis prior to cuticle
        removal. These samples were run on a single SDS (sodium
        dodecyl sulfate) slab gel which was cut into separate channels
        prior to photography. (Modified from Roberts and Willis,
        1980).

Do Different Species Use the Same Arthropodin?

We have discussed intraspecific evolutionary relationships
among cuticular proteins, but a more complete understanding of
cuticle gene evolution will eventually require interspecific comparisons at the sequence level. Fraenkel originally thought that all
arthropods would have the same cuticular protein, arthropodin,
and it certainly would be tempting to postulate homology between
the very acidic proteins in <u>Sarcophaga</u> and Cecropia cuticles.
However, valid comparisons are going to have to wait for complete
sequencing or at least for some precise comparative work with
specific antibodies. In the meantime, we think that electrophoresis
can be used to compare the soluble cuticular proteins of related
species. Such studies have been initiated by Jack Rabin in JHW's
laboratory with two species of <u>Hyalophora</u>: <u>H. cecropia</u> and <u>H.
gloveri</u>, at two metamorphic stages. In Fig. 4, which shows an
isoelectric focusing gel, each lane contains a cuticular extract
from one animal, so we can examine population polymorphism as well
as inter-species differences. In Cecropia, the abundant and
moderately abundant larval and adult proteins do not appear polymorphic. The few minor differences noted between lanes are due to

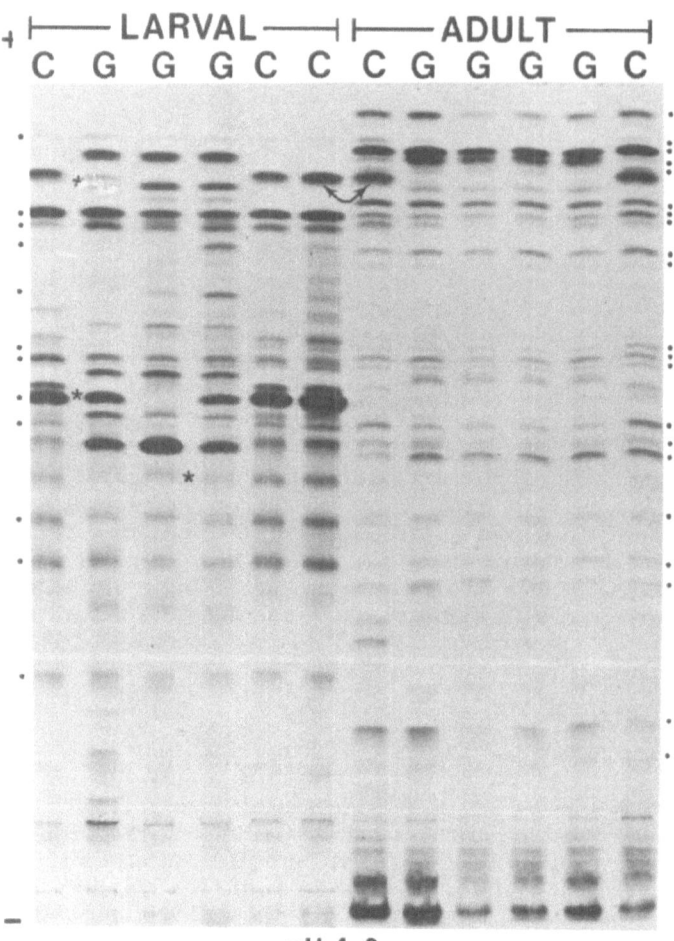

pH 4-6

Fig.4   Electrophoretic comparison of cuticular proteins from larvae
and adults of Hyalophora cecropia (C) and H. gloveri (G).
The arrows indicate bands which have identical electrophoretic mobilities in extracts from Cecropia larvae and
adults. (*) indicates bands indicative of population
polymorphism, (·) indicates bands common to both species;
symbol for larval extracts on left, for adult on right.
Extraction and gel conditions as in Fig. 2.

age differences among the larvae used. In Gloveri, the adult
proteins show no polymorphism, but two prominent larval proteins
are polymorphic (see * in Fig. 4).

Inter-specific comparisons also show considerable apparent
conservatism. In larval extracts of the two species, 12 bands have
identical isoelectric points, three bands are unique to Cecropia,
and 7 are unique to Gloveri. In the adult, 21 bands have identical
mobilities. Cecropia has two unique bands, Gloveri three. (Shared
bands are indicated with a · in Fig. 4.)

Since viable hybrids and backcrosses can be obtained from all
4 species of Hyalophora, it should be possible to map cuticular
protein variants, and determine whether cuticular protein genes are
on the same chromosome. Such simple genetics can be combined with
information gathered by other means to test for membership in the
same gene family.

Does Arthropodin Composition Change During Metamorphosis?

In our work with Tenebrio, illustrated in Fig. 3, we showed
that different bands are present in different metamorphic stages.
The same may be said for Cecropia (Fig. 4, center lanes). While
there are many bands which are unique to a particular developmental
stage, several larval and adult cuticular proteins share identical
or nearly identical isoelectric points. This is consistent with
the general similarities in amino acid composition of the components we have purified from the two stages (Tables 1 and 2).
Interestingly, the major band shared by larvae and adults (see
arrows, Fig. 4) is very similar in amino acid composition in the two
stages (see bands L2 and A3, Tables 1 and 2). A question of special
interest is whether a single protein encoded by a single gene can
be expressed during two metamorphic stages, i.e., in cuticles laid
down under quite different hormonal instructions. Protein and
nucleic acid sequencing will be required to answer this question.

Does Arthropodin's Composition Differ in Different Regions?

Fraenkel and Rudall originally thought arthropodin was responsible for cuticle flexibility. As one might have suspected,
knowing Dr. Fraenkel's track record, they were right, although
perhaps not in the way they anticipated.

To answer this question, we compared the electrophoretic banding patterns of several anatomical regions of Cecropia which have
different mechanical properties. In the larva we compared dorsal
sclerite and intersegmental membrane, both flexible areas. In the
pupa we examined dorsal sclerites, the outer wing case, and the
polygonal field zone of the intersegmental membrane. All three

regions are conspicuously tanned, but the first two are rigid, while the intersegmental membrane is quite flexible. In the pharate adult we separated two regions of scale-bearing sclerite, the firm mid-dorsal region and the more flexible lateral regions. These were compared to the flexible, scaleless, intersegmental membrane, as well as to wing cuticle which is destined to become rigid.

The results are complex, but a consistent pattern emerges (Fig. 5). Extracts from all regions of flexible cuticle (lanes 1, 2, 5, 8, 9) contain an assortment (varying with stage) of very acidic proteins, as well as some more basic proteins, which are missing in all the rigid areas examined. By contrast, many proteins found in extracts from rigid cuticles are also present in extracts from flexible cuticles of the same metamorphic stage.

Comparison of pupal wing and sclerite (lanes 3 and 4) shows that an apparently identical array of cuticular proteins can be secreted by cells with quite different developmental histories. Sclerites are derived from epidermis which had previously made larval cuticle, whereas the wing comes from an imaginal disc with no previous experience in cuticle formation.

Which Cells Make Arthropodin?

From the earliest days of electrophoresis, workers have noticed that proteins extracted from cuticle have the same electrophoretic mobilities as some blood proteins. Furthermore, antibodies made against cuticular proteins cross-react with blood proteins (Fox et al., 1972; Koeppe and Gilbert, 1973; Reeder and Willis, unpublished observations). The conclusion from such studies is that cuticular proteins may be made by tissue other than epidermis, e.g., the fat body, and transported through the blood to the cuticle. An obvious analogy would be the fat body's manufacture of vitellogenin and its delivery via the blood to the oocyte. There are no conclusive experiments to establish such a transfer (see Ruh and Willis, 1974).

We approached this question of epidermal synthesis of cuticular proteins by taking advantage of our recent success in insect tissue culture (Willis and Hollowell, 1976). Small fragments of larval integument -- that is, cells plus cuticle -- were placed in Grace's tissue culture medium with ^{3}H-leucine for four hours and were checked for protein synthesis and incorporation. One fragment of integument was processed for histology and autoradiography. Fig. 6 shows that both epidermis and cuticle were well labelled. Minute pieces of muscle which were present did not become labelled under these conditions. For an examination of cuticular proteins, the epidermis was removed from the remaining

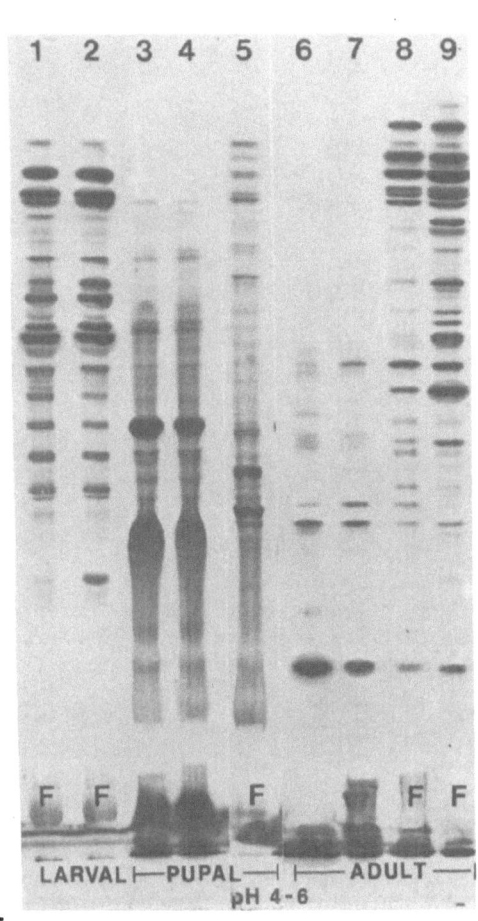

Fig.5   Electrophoretic comparison of cuticular protein extracts from
        different regions of single Cecropia larvae, pupae, and adults.
        Flexible cuticles are indicated (F).   Extraction and gel
        conditions as in Fig. 2, except the gel was made with LKB
        Ampholytes (pH range 4-6).
        Lane 1:   larval sclerite (mid-fifth instar)
        Lane 2:   larval intersegmental membrane
        Lane 3:   pupal sclerite
        Lane 4:   pupal wing (dorsal cuticle of fore-wing)
        Lane 5:   pupal intersegmental membrane (polygonal field zone)
        Lane 6:   adult wing (day 18 pharate adult)
        Lane 7:   adult sclerite, central region
        Lane 8:   adult sclerite, lateral region
        Lane 9:   adult intersegmental membrane

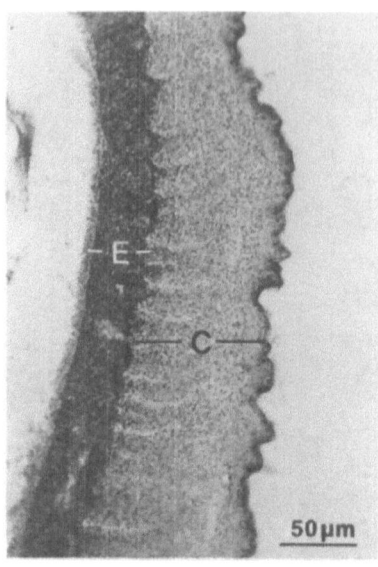

50 µm

Fig.6   Autoradiograph of young fifth-instar Cecropia larval integu-
        ment exposed to tritiated leucine for 4 hours in tissue cul-
        ture.   Note heavy labelling over epidermis (E) and grains
        present throughout cuticle (C).

fragments, leaving clean, labelled cuticles (Fig. 7). Such cuticles were then homogenized and their soluble proteins extracted
and processed for electrophoresis. After electrophoresis, the
gels were dried and covered with sensitive film so that we could
obtain an autoradiograph showing the location of bands which had
incorporated the isotope (Fig. 8).

     When one compares the photograph of the stained gel (lane 1)
with the autoradiograph (lane 2), one sees that in tissue culture
all major larval cuticular proteins are synthesized by the epidermis.
Indeed, in this particular sample, only four bands were unlabelled
(see ·, Fig. 8). These might be synthesized at times other than
the four hour period during which label was present. Alternatively,
they could be non-epidermal products.

     Thus, we have established that the integument is the site of
synthesis of most, if not all, larval cuticular proteins. This
conclusion further suggests that cuticular proteins found in
blood might also be epidermal cell products, either from living
or dead cells.

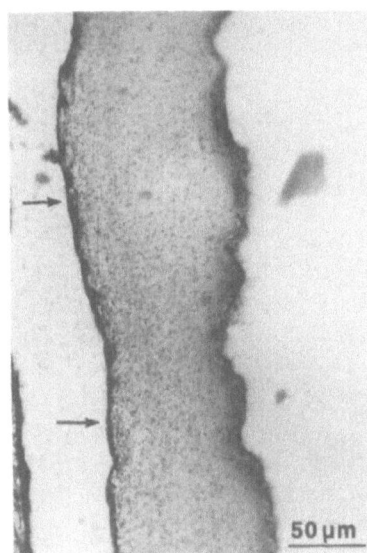

Fig.7   Autoradiograph of cleaned cuticle from the same integument
        as in Fig. 6. Note that incorporation is heaviest on the
        inner surface of cuticle (arrows).

CONCLUSIONS

    We hope this paper has brought the reader up-to-date on the
metamorphosis of the concept of arthropodin, for we have shown
that:

    Cuticular extracts have many discrete proteins, some of which
may belong to one or more multigene families.

    As tanning proceeds the amount of extractable protein diminishes;
some bands disappear or decline in intensity; some new bands appear.

    Different anatomical regions, different species in the same
genus, and different metamorphic stages all have both shared and
unique proteins.

    Cuticular proteins are synthesized by the epidermis.

    Our final conclusion must be that the term arthropodin is
outdated. The soluble cuticular proteins are too numerous and
diverse within and between species, too transient even within a
single developmental stage, and too distinct between anatomical

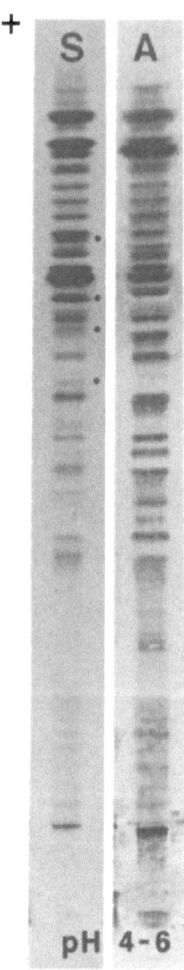

Fig.8   Incorporation of tritiated leucine into individual cuticular
        proteins. Clean cuticles from the same sample as in Fig. 7
        were extracted with buffered 8M urea. The extract was frac-
        tionated on an isoelectric focusing slab gel prepared as for
        Fig. 2. The stained gel was photographed, dried and an auto-
        radiograph was obtained on LKB Ultrofilm exposed for 41 days.
        The left column shows the stained gel (S). The right column
        shows its pattern of incorporation (A). Four stained bands
        which were not labelled are indicated (·).

regions, to be encompassed by any single term. This paper, then, has really been an obituary for the term arthropodin.

But we end on a more positive note: It is our belief that our findings have documented once again Professor Fraenkel's special propensity for doing pioneering work with important systems, and for recognizing and re-initiating activity at the proper time for new technology to provide further advances.

ACKNOWLEDGEMENTS

This research was supported by grant AG-00248 to J.H.W. from the National Institutes of Health. Salary support for J.C.R. came from NIH-GM-24225 to F.C. Kafatos and for B.A.D. from NSF-PCM 77-24247 to G.S. Fraenkel. We acknowledge the excellent assistance of K. Goodhope and D. P. Eisenman.

REFERENCES

Fox, F. R., Seed, J. R., and Mills, R. R., 1972, Cuticle sclero-
    tization by the American cockroach: Immunological evidence
    for the incorporation of blood proteins into the cuticle,
    J. Insect Physiol. 18:2065-2070.
Fraenkel, G. and Rudall, K. M., 1940, A study of the physical and
    chemical properties of the insect cuticle, Proc. Roy. Soc.
    B, 129:1-35.
Fraenkel, G. and Rudall, K. M., 1947, The structure of insect
    cuticles, Proc. Roy. Soc. B, 134:111-143.

Fristrom, J. W., Hill, R. J., and Watt, F., 1978, The procuticle of Drosophila: Heterogeneity of urea-soluble proteins. Biochemistry 17:3917-3924.

Hackman, R. H., 1953, Chemistry of insect cuticle. 1. The water-soluble proteins, Biochem. J. 54:362-367.

Hackman, R. H. and Goldberg, M., 1976, Comparative chemistry of arthropod cuticular proteins, Comp. Biochem. Physiol. 55B:201-206.

Hood, L., Campbell, J. H., and Elgin, S. C. R., 1975, The organization, expression, and evolution of antibody genes and other multigene families, Ann. Rev. Genetics 9:305-353.

Jones, C. W., Rosenthal, N., Rodakis, G. C., and Kafatos, F. C., 1979, Evolution of two major chorion multigene families as inferred from cloned cDNA and protein, Cell 18:1317-1332.

Koeppe, J. K. and Gilbert, L. I., 1973, Immunochemical evidence for the transport of haemolymph protein into the cuticle of Manduca sexta, J. Insect Physiol. 19:615-624.

Marchalonis, J. J. and Weltman, J. E., 1971, Relatedness among proteins: a new method of estimation and its application to immunoglobulins, Comp. Biochem. Physiol. 38B:609-625.

Moorefield, H. H., 1953, Studies on some integumental proteins of insects, Ph.D. Thesis, University of Illinois.

Regier, J. C., Kafatos, F. C., Goodfliesh, R., and Hood, L., 1978, Silkmoth chorion proteins: Sequence analysis of the products of a multigene family, Proc. Natl. Acad. Sci. USA 75:390-394.

Regier, J. C., Kafatos, F. C., Kramer, K. J., Heinrikson, R. L., and Keim, P. S., 1978, Silkmoth chorion proteins -- their diversity, amino acid composition and the NH_2-terminal sequence of one component, J. Biol. Chem. 253:1305-1314.

Roberts, P. E. and Willis, J. H., 1980, The cuticular proteins of Tenebrio molitor. I. Electrophoretic banding patterns during postembryonic development, Develop. Biol. 75:59-69.

Ruh, M. F. and Willis, J. H., 1974, Synthesis of blood and cuticular proteins in late pharate adults of the Cecropia silkmoth, J. Insect Physiol. 20:1277-1285.

Trim, A. R., 1941, Studies in the chemistry of the insect cuticle. I. Some general observations on certain arthropod cuticles with special reference to the characterization of the proteins, Biochem. J. 35:1088-1098.

Willis, J. H. and Hollowell, M. P., 1976, The interaction of juvenile hormone and ecdysone: antagonistic, synergistic, or permissive? In: "The Juvenile Hormones", L. I. Gilbert, ed., Plenum Press, New York, pp. 270-287.

RECENT RESULTS ON THE NEUROENDOCRINE SYSTEM OF <u>LEUCOPHAEA</u>

Berta Scharrer

Department of Anatomy
Albert Einstein College of Medicine
Bronx, New York   10461

INTRODUCTION

The elucidation of the neuroendocrine control mechanisms operating in insects started with the discovery by Kopeć (1917) of a pupation hormone in the brain of a lepidopteran (<u>Pieris</u>). The contributions made by other pioneers in this active area of research, one of whom we honor by this symposium, have been centered around their laboratory insects of choice, <u>Calliphora</u>, <u>Rhodnius</u>, and <u>Cecropia</u> among them. Each of these efforts has yielded specific insights, and for a time it seemed difficult to recognize common denominators in the multitude of control mechanisms geared to the special needs of different insect species.

Yet, in spite of this diversity, certain basic principles of neuroendocrine interaction have now become apparent. They apply not only to the regulation of all important physiological activities of insects but also to analogous control processes throughout the animal kingdom. Much has been learned about these phenomena by the examination of their evolutionary history (see, for example, Scharrer, 1978b), and within this broad framework neuroendocrine systems of insects have acquired a new dimension.

Neuroendocrine Phenomena in Leucophaea

The main objective of my own contributions over the past 40 years, concerned primarily with the neuroendocrine system of <u>Leucophaea</u> <u>maderae</u> and some related cockroach species, has been and continues to be the elucidation of the roles played by the neural and the non-neural parts of this integrative apparatus, and

47

of the manner of their mutual interaction. The essential approach
towards this end has been the structural characterization of this
organ system in relation to known functional states, and the search
for the effects of experimental interventions, e.g., extirpation and
implantation of endocrine structures, gonadectomy, nerve severance,
and stimulation of the release of neurosecretory material.

One of the basic insights gained from these and other studies
is that the endocrine glands proper receive directives from neural
elements some of which are themselves capable of dispatching hormonal
messengers (neurohormones). The more recent and still ongoing
experiments in Leucophaea to be discussed here focus on one of these
non-neural glands of internal secretion, the corpus allatum, and
on the neural structure with which it is intimately associated, the
corpus cardiacum.

Corpus Allatum

L. maderae belongs to the group of insects in which corpus
allatum structure changes qualitatively and quantitatively in
conjunction with periodic fluctuations in the gland's functional
performance. In "active" glands, the cellular and nuclear volumes
(diameters) are significantly larger than in the "inactive" phase,
and the nuclear-cytoplasmic ratio is lower. Pleomorphic mitochon-
dria are more prominent, and so are whorls of ergastoplasmic membranes (RER).

In a recent ultrastructural study of the corpora allata of
gonadectomized adult females (Scharrer, 1978a), these and additional parameters of dynamic cytology were found to surpass those
characteristic of the most active level in normal animals. One of
the striking correlates of the putative hyperactivity of these glands
is an abundance of smooth surfaced endoplasmic reticulum (SER), the
organelle implicated in the biosynthesis of juvenile hormone. Why
does this essential structure, which in the normal corpora allata of
Leucophaea is not nearly as prominent as in a number of other insects (see Cassier, 1979), respond so impressively to the surgical
intervention? A reasonable explanation for this and other ultra-
cytological signs of hyperactivity seems to be the constancy of the
demand on the corpora allata of these experimental animals, since
ovariectomy abolishes the afferent signals in response to which, in
the intact animal, the brain turns off juvenile hormone action during the long periods of gestation. The pleomorphism of the mitochondrial population is even more impressive, and glycogen deposits
seem to be more numerous than in unoperated "active" controls.

The surface of the corpora allta of gonadectomized females
characteristically shows an overaly of hemolymph protein, presumed

to be an excess of vitellogenin, normally destined for yolk formation. These deposits also seem to permeate the acellular sheath
and trabeculae of the gland thus serving as a biological marker
demonstrating this avenue of communication.

Another biological marker, found exclusively in the corpus
allatum and described earlier as C-body material (Scharrer, 1971),
is especially prominent in the stromal compartment of very old
gonadectomized females. The exit of this strikingly structured,
Golgi-derived product from the gland can be visualized particularly
clearly in members of this group. Accumulations of the material
have been caught within and partially outside of the acellular
sheath of the gland, presumably at the moment of their release into
the hemolymph. Also more prominent in the experimental group, are
small, electron dense "nuggets", squeezed between contiguous corpus
allatum cells and apparently on their way to the trabeculae and
sheath. Since these small inclusions lack the characteristic
cribriform structure of C-bodies, their precise relationship with
the latter remains to be determined.

Even though a precise functional interpretation of C-body
material is not yet possible, there is increasing evidence suggesting
a direct or indirect role in the endocrine function of the corpus
allatum. This view is supported by the fact that the formation of
comparable inclusions has now been detected in the corpora allata
of several other insect species.

Corpus Cardiacum

By contrast with the elusive juvenile hormone of the corpus
allatum, the proteinaceous products of the neurosecretory system,
including the extrinsic and intrinsic stores in the corpus cardiacum,
are easily identified at the light and electron microscopic levels.
It was in Leucophaea that the axonal transport of extrinsic neurosecretory material, produced in the pars intercerebralis of the
brain, to its sites of release was first demonstrated by the
severance of the appropriate fiber tract (Scharrer, 1952). Subsequent electron microscopic evidence revealed that some of these
peptidergic fibers, as well as those of the intrinsic corpus
cardiacum cells, discharge their products into the circulation to
act as neurohormones. Others were shown to make direct contact
with effector cells, including the parenchymal elements of the
corpora cardiaca and allata, and thus apparently to exert localized
control in the manner of neurotransmitters.

The synaptoid release sites of these neurosecretory axons and
their modes of operation were given special attention in several
of our recent studies (Scharrer and Wurzelmann, 1974; 1977; 1978).
An issue debated for a long time, i.e., whether or not neurosecretory

granules are exteriorized by exocytosis was clarified by several
experimental approaches. Proof that this process occurs requires
the demonstration of omega-type configurations generally considered
to be indicative of membrane fusion prior to the release of the
secretory product in toto. The reason why, in intact specimens of
Leucophaea, such ultrastructural features are virtually non-existent,
turned out to be the considerable speed with which this process of
extrusion seems to occur.

However, an appropriate stimulus, i.e., the administration of
serotonin prior to fixation of the tissues (Scharrer and Wurzelmann,
1978), yielded images of granules captured at the moment of leaving
the axon. These are especially distinctive where the synaptoid
release sites face the stromal interstitium (acellular matrix forming
the trabeculae and the sheath of these organs). Most of the test
animals responded to the stimulus as early as three minutes after
the injection of the drug. Fully exteriorized secretory structures
appeared in the stroma soon thereafter, and their breakdown was
evident after 25 minutes. In addition to the characteristic time
course of these phenomena, in contrast to that observed in certain
other animals, the neurosecretory system of Leucophaea has an additional distinctive feature. Many of the granules caught in transit
are considerably smaller than the type characteristic of the
respective peptidergic neurons, quite obviously the result of intracellular fragmentation.

Another set of experimental interventions (Scharrer and
Wurzelmann, 1977) revealed membrane phenomena of yet a different
type. Under certain conditions, not necessarily expected to stimulate neurohormone release, e.g., exposure of the tissue to a zinc
iodide mixture without prior fixation, protrusions of bounding
membranes became apparent that are comparable to those demonstrated
by Castel (1977) in the mammalian neurohypophysis. These configurations, presumably too transient to be much in evidence under
physiological conditions, appear to become more prominent not merely
after acceleration of the rate of release, but also after inter-
ference with the regular milieu. Under these conditions, temporary
membrane arrangements become "frozen" and therefore detectable. In
other words, useful artefacts of this kind call attention to an
additional and as yet insufficiently explored facet of the subcellular dynamics leading to the exteriorization of the secretory
product. As Castel suggested, such configurations are consistent
with the concept that special active sites on the bounding membrane of a neurosecretory granule may have the capacity for making
temporary contact with that of an adjacent granule or with the
axolemma.

General Considerations

The recent results in Leucophaea briefly reviewed here have strengthened the long-held view (see E. Scharrer and B. Scharrer, 1963) of the remarkable degree of analogy exhibited by the neuro-endocrine systems of invertebrates and vertebrates. A relatively new concept is that peptidergic mediators are no longer considered to act exclusively as neurohormones. Their capacity to transmit strictly localized stimuli is clearly illustrated by the existence of neurosecretory fibers making direct contact with the cells of the corpora cardiaca, the corpora allata, the salivary glands, muscle, and other somatic elements.

Even more challenging is the demonstration of neurosecretory junctions with other neurons, be they conventional or themselves of the neurosecretory type. The corpus cardiacum of Leucophaea shows such synaptoid contacts between two types of neurosecretory neurons. The neural elements entering the corpora allata reveal intimate relationships between two or more peptidergic fibers of the same ultrastructural type. Their intriguing symmetrical (mirror-image-type) arrangement suggests mutual exchange of "information" presumably related to the control of the corpus allatum.

Most assuredly, the peptidergic neurons of insects present challenges for future exploration. Among the problems to be dealt with are the chemical characterization of the active principles they produce and their relationship with those already identified in higher vertebrates.

REFERENCES

Cassier, P., 1979, The corpora allata of insects, Internat. Rev. Cytol. 57:1-73.

Castel, M., 1977, Pseudopodia formation by neurosecretory granules, Cell Tiss. Res. 175:483-497.

Kopéc, S., 1917, Experiments on metamorphosis of insects, Bull. Acad. Sci. Cracovie, classe Sci. math. nat., Sér. B:57-60.

Scharrer, B., 1952, Neurosecretion. XI. The effects of nerve section on the intercerebralis-cardiacum-allatum system of the insect Leucophaea maderae, Biol. Bull. 102:261-272.

Scharrer, B., 1971, Histophysiological studies on the corpus allatum of Leucophaea maderae. V. Ultrastructure of sites of origin and release of a distinctive cellular product, Z. Zellforsch. 120:1-16.

Scharrer, B., 1978a, Histophysiological studies on the corpus allatum of Leucophaea maderae. VI. Ultrastructural characteristics in gonadectomized females, Cell Tiss. Res. 194:533-545.

Scharrer, B., 1978b, An evolutionary interpretation of the
    phenomenon of neurosecretion, The American Museum of
    Natural History, New York, pp. 1-17.
Scharrer, B. and Wurzelmann, S., 1974, Observations on synaptoid
    vesicles in insect neurons, Zool. Jahrbücher, Abt. Physiol.,
    Fischer-Verlag, Jena 78:387-396.
Scharrer, B. and Wurzelmann, S., 1977, Neurosecretion. XVI.
    Protrusions of bounding membranes of neurosecretory granules,
    Cell Tiss. Res. 184:79-85.
Scharrer, B. and Wurzelmann, S., 1978, Neurosecretion. XVII.
    Experimentally induced release of neurosecretory material by
    exocytosis in the insect Leucophaea maderae, Cell Tiss. Res.
    190:173-180.
Scharrer, E. and Scharrer, B., 1963, Neuroendocrinology, Columbia
    University Press, New York, 289 pp.

REGULATION OF CORPUS ALLATUM ACTIVITY IN LAST INSTAR <u>MANDUCA</u> <u>SEXTA</u> LARVAE

Govindan Bhaskaran

Institute of Developmental Biology
Texas A & M University
College Station, Texas

INTRODUCTION

The corpora allata (CA), like other endocrine glands, go through phases of activity and inactivity which correspond to well defined physiological changes occurring during post-embryonic development. This cyclic activity is not intrinsically programmed but is instead regulated by factors external to the gland. It is only logical that the secretory activity of the CA which produce juvenile hormone (JH), a hormone with profound effects on morphogenesis and reproduction, is finely tuned by environmental factors and the internal milieu via the central nervous system. Evidence suggests that the brain (and perhaps other ganglia) controls the activity of the CA by both humoral and nervous mechanisms during the stages preceding metamorphosis (Wigglesworth, 1954; 1970; Scharrer, 1958; Highnam, 1967; Doane, 1973).

A number of early studies involving transplantation of CA and severance of the nervi corpori allati (NCA) indicated that the brain hormonally stimulates the CA (Scharrer, 1958; Wigglesworth, 1964; 1970). This was supported by the cytological demonstration of neurosecretory material in the CA and the correlated change in activity of CA (Highnam, 1967; Cassier, 1979). Further confirmation was obtained by destruction of specific groups of NSC in the brain. Cautery of the lateral neurosecretory cells in fourth instar nymphs of <u>Locusta</u> <u>migratoria</u> (Girardie, 1965) and the NSC of the pars intercerebralis in <u>Rhodnius</u> <u>prolixus</u> (Baehr, 1976) gave rise to precocious adults and adultoids respectively. This suggests that the NSC are the source of an allatotropic hormone (ATH) in these insects. Recently it was found that transplantation of brains from normal or injured early last instar Galleria larvae into young last instar larvae caused supernumerary larval molting (Krishnakumaran, 1972; Granger and Sehnal, 1974; Pipa, 1976) whereas brains from older donors were ineffective (Sehnal and Granger, 1975).

Allatectomy of the hosts as well as cautery of the NSC in the pars intercerebralis of the implanted brain prevented supernumerary molting of the host larvae (Granger and Sehnal, 1974). These results suggest that brains from young donors activated the hosts' CA via secretion of an ATH whereas older brains lacked ATH (Krishnakumaran, 1972; Sehnal and Granger, 1975).

Neural stimulation of CA has been reported to occur in diapausing larvae of Diatraea grandiosella (Yin and Chippendale, 1979). The precocious appearance of imaginal characteristics after severance of NCA in the penultimate instar of Anisolabis maritima (Ozeki, 1962) is also indicative of neural activation.

Wigglesworth (1936) first surmised that the brain of Rhodnius inhibits the CA by a nervous mechanism, since fourth instar CA transplanted into fifth instar nymphs continued to remain active and induced additional nymphal molts in the hosts. Subsequently, numerous investigations confirmed that in some insects the brain exerts an inhibitory influence on the CA primarily by means of neural signals transmitted by the NCA. Sectioning of the NCA leads to reactivation of the CA in Leucophaea maderae (Scharrer, 1952; Lüscher and Engelmann, 1955; 1960), and Periplaneta americana (Fraser and Pipa, 1977). However, denervation has no effect on the CA in larvae of Bombyx (Bounhiol, 1957).

Evidence for neurohormonal inhibition of the CA has been found in a few insects. In Locusta migratoria the medial neurosecretory cells (M-NSC) appear to be the source of an inhibitory factor (Girardie, 1965). Williams (1976) postulated that in Manduca sexta the inhibitory factor (allatohibin) is transported via the NCA to the CA. The exact nature of this control mechanism is far from clear.

It is evident from this brief review that our understanding of the mechanism of regulation of CA activity during crucial periods of morphogenesis is very limited. The simplicity and economy of a dual control by means of ATH and a nervous inhibitory pathway are implicit in the tacit acceptance of this model by insect endocrinologists. However, any generalization must await the experimental demonstration of such a control mechanism in a wide variety of insects. In this context, our recent studies on the regulation of CA in Manduca sexta provide firm evidence for ATH and a two-step (a neurohormonal and neural) inhibition of CA in the last larval instar (Bhaskaran and Jones, 1980; Bhaskaran et al., 1980). In

the present paper, these results are discussed and CA inactivation
is related to other known mechanisms for reducing the hemolymph
JH titer prior to initiation of metamorphosis. A model for regulation of CA in larvae of Manduca sexta is also proposed with the hope
that it will stimulate further research on the regulation of CA
activity in this and other species of insects.

RESULTS

Developmental Characteristics and Synchronization of Fifth Instar Larvae

The growth and morphological changes taking place during the
fifth instar (last instar) and prior to pupation in Manduca sexta
have been previously recorded (Truman, 1972; Nijhout and Williams,
1974a; Jones et al., 1980). Ecdysis to the fifth instar occurs a
few hours before the onset of the scotophase, and two to three hours
after ecdysis larvae begin to feed. In a culture grown on a standard
diet, individual larvae show variation in rate of growth, time of
excretion of frosted frass (FF), exposure of dorsal vessel (EDV),
and purging of the gut; immediately thereafter larvae enter the
wandering stage (Truman, 1972; Nijhout and Williams, 1974a; Jones
et al., 1980a).

Using a 16:8 light-dark cycle at 27° and 60% RH, we are able to
achieve a high degree of synchronization of larval development with
a regimen which includes a short starvation period. Newly ecdysed
5th instar larvae are starved from the moment of emergence until
lights-on of the following day, after which they are given free
access to the diet. If at this point they are considered to be day
0 larvae, subsequent daily weighings show relatively low variance
(Fig. 1), and observations reveal that 80% excrete FF on day 3 and
EDV on day 4. They transform into prepupae (PP) on day 6 and ecdyse
(pupal ecdysis (PE)) on day 9 or 10 (Jones et al., 1980a). This
method has been used as the standard procedure for all experiments
hereafter described, except where otherwise stated.

The Starvation - Induced Supernumerary Larval Molt

Nijhout and Williams (1974a) observed that fifth instar larvae
starved indefinitely after reaching 3.0 - 4.0 g in weight trans-
formed into larval-pupal intermediates. Starvation had no effect
on larvae weighing more than 5.0 g but caused mortality in all larvae
less than 3.0 g in weight. Our own studies (Jones et al., 1980a)
showed that when newly-ecdysed fifth instar larvae were starved for
3 days and subsequently fed a regular diet, a majority molted into
sixth instars instead of undergoing pupation. Initial starvation
for only two days reduced the percentage of supernumerary molting
while starvation for 4 days or more sharply increased mortality.

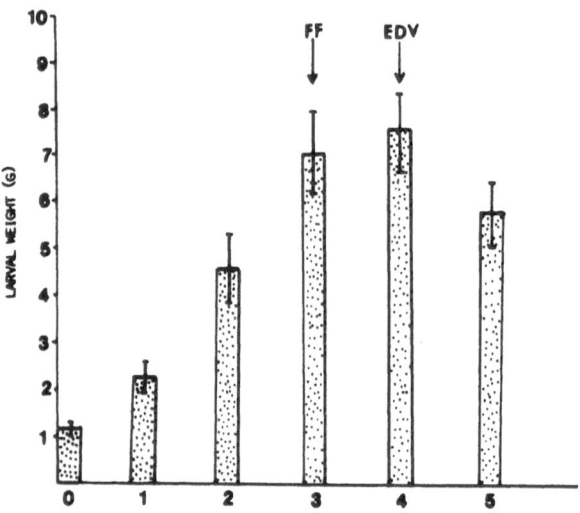

DAY OF THE FIFTH INSTAR

Fig. 1   Growth of fifth instar M. sexta larvae. Larvae which ecdysed
         between 19.00-22.00 hr (late photophase) were kept without
         food and water until lights-on of the following day (8.00
         hr) at which time they were weighed and fed regular diet.
         Larvae were weighed at the same hour on the following days
         until all larvae entered the wandering stage. A large
         majority (80%) of the larvae excreted frosted frass (FF)
         on day 3, exposed the dorsal vessel (EDV) on day 4 and became
         'wandering larvae'.

Such larvae either died before feeding or fed poorly and died within
a day or two. Therefore, in all experiments except those specifically noted, starved larvae were starved for three days before feeding.

     The sixth instar larvae showed no morphological abnormalities
and possessed normal mouth parts and prolegs with well developed
crochets. Most of these larvae died before pupation; a few grew
to about 13.0 g in weight and pupated, but several of them were
deformed. Recently, Cymborowski et al. (cited by Riddiford, 1980)
also reported the induction of supernumerary molting using similar
methods, except that under their laboratory conditions larvae had
to be starved for 5 days prior to feeding.

Endocrine Control of Starvation-Induced Supernumerary Molting

     The starvation-induced supernumerary larval molting could re-
sult from: (a) acceleration of a molt by precocious release of

ecdysone, or (b) elevation of hemolymph JH titer due either to
increased secretion by the CA or conservation of JH by suppression
of JH esterases. Evidence from our previous investigations
(Bhaskaran and Jones, 1980), which led to the conclusion that
starvation causes activation of the CA, is briefly summarized here.

Role of the CA in supernumerary molting

Larvae allatectomized prior to starvation failed to undergo a
supernumerary molt but pupated instead. Implantation of CA from 0-
day fifth instar donors into allatectomized larvae prior to starvation, restored the supernumerary molt (Table 1). For induction of
a supernumerary molt the CA were required until two days after feeding began. Larvae allatectomized three days after feeding molted
into sixth instars but exhibited varying degrees of black coloration.
The latter is characteristic of larvae deficient in JH during a
larval-larval molt (Truman et al., 1973).

Table 1. Effects of allatectomy (CA^-)[a]/medial neurosecretory cell
cautery $(M-NSC^-)$[a] with and without subsequent transplantation of CA, corpora cardiaca (CC) or M-NSC on the
starvation-induced supernumerary larval molt of $\underline{M}$. $\underline{sexta}$.

TREATMENT	No. larvae	No. Survived	(%)	% Survivors molted into:	
				6th instars	pupae
CA^-	50	38	(76)	0	100
CA^- and CA reimplanted	80	58	(73)	100	0
$M-NSC^-$	94	68	(72)	1.5	98.5
$M-NSC^-$ and brain[b] implanted	50	39	(78)	100	0
$M-NSC^-$ and one pair of CA[c] implanted	30	22	(73)	0	100
$M-NSC^-$ and one pair of CC[c] implanted	30	21	(70)	0	100

[a]All operations and transplantations were done on day 0 fifth instars prior to starvation.
[b]One brain each from donors of identical age.
[c]CA or CC from day 0, 3 day starved or 3 day starved and 1 day
fed donors (from Bhaskaran and Jones, 1980).

Role of the M-NSC

Cautery of the medial NSC (M-NSC) abolished the supernumerary molt (Table 1), but cautery of the lateral NSC (L-NSC) had no effect, since larvae subjected to the latter treatment molted into sixth instars. The hormonal role of the brain, and of group II M-NSC in particular, was confirmed by transplanting intact brains or brains with specific groups of NSC cauterized after removal from the donor, into M-NSC cauterized larvae prior to starvation. All group II M-NSC cauterized recipients wandered, whereas, those larvae which received intact brains or brains with other NSC regions destroyed, molted into sixth instars. The M-NSC were required only until one day after the initiation of feeding since destruction of these cells after this time did not prevent the supernumerary molt. We also determined whether the CA or CC could substitute for the M-NSC. Neither of these glands promoted a supernumerary molt when implanted into M-NSC-cauterized larvae (Table 1). These data demonstrate that both the CA and M-NSC are necessary for the induction of a supernumerary molt and suggest that starvation results in activation of the brain causing synthesis and/or release of an ATH.

ATH Secretion of Brains From Normal and Starved Larvae

The fact that M-NSC-cauterized-starved and fed larvae undergo pupation suggested to us the possibility of using such larvae as hosts for an ATH bioassay. The bioassay was developed on the basis of (a) our earlier findings that brains from starved day-3, but not normal day-3 larvae were able to induce a supernumerary molt when transplanted into M-NSC cauterized larvae on the day of feeding (Bhaskaran and Jones, 1980), and (b) the preliminary observation that the internal environment of fasting larvae may stimulate inactive brains but once feeding begins this effect declines. In determining the right conditions for the assay, it was necessary to avoid exposing the implanted test-brain to starvation stimulus and yet provide sufficient time for it to stimulate the CA so that the JH level would be high at the time of initiation of the molt. Therefore, we cauterized the M-NSC of day-0 fifth instar larvae, starved them for 3 days, fed them for 1 day and used them as recipients. Since the hosts had fed for 1 day after the starvation period we assumed that the internal environment would be sufficiently dissimilar from starving animals that it would not activate the implanted brain. This assay would, therefore, reveal whether the brain had intrinsic allatotropic activity at the time of transplantation.

Allatotropic Activity of Brain From Normal Fourth and Fifth Instars

Brains from newly ecdysed and 3-day old fourth instars, 0 and 3-4-day old fifth instars (either frosted frass or exposed dorsal vessel stage), and prepupae (2-3 days after exposure of dorsal vessel) were tested. Significant allatotropic activity was found only in the brains of newly ecdysed fourth instars. Brains from 3-day old fourths had slight activity and brains from all the other ages tested had no apparent activity (Table 2).

Starvation Induced Reactivation of Inactive Brains

The absence of allatotropic activity in brains from newly ecdysed fifth instars and its presence in brains from starved larvae suggest that starvation stimulates M-NSC to secrete allatotropin. In addition, activation does not require the presence of intact nerves, suggesting that metabolic changes in the starved animals trigger the process (Jones et al., 1980b). Can inactive brains from fifth instar larvae of any age be activated by exposure to starvation conditions?

Brains from newly ecdysed or 3-4-day old 5th instars or prepupae (2-3 days after exposure of dorsal vessel) were transplanted into M-NSC-cauterized hosts prior to starvation. By the sixth day after feeding all the recipients molted to 6th instar larvae (Table 3) demonstrating that inactive brains become active and secrete allatotropin in starved hosts.

In a similar experiment brains from 3-day-old fourth instars also promoted an extra larval molt in the recipients (Table 3).

Temporal Pattern of Activation of Brain in Starved Larvae

Since brains from newly ecdysed fifth instars did not possess ATH activity the time course of activation of the brain in starved larvae was examined. Brains from 1-day or 3-day starved larvae, and 1-day or 3-day fed larvae were bioassayed. The data summarized in Table 4 show that a majority of the recipients of 1-day starved larval brains molted to sixth instars and all those receiving brains from either 3-day starved or 3-day starved-1-day fed larvae underwent supernumerary molting. None of the hosts with brain implants from 3-day starved 3-day fed donors became extra larvae nor did the control group which received M-NSC cauterized brains from 3-day starved larvae. These results clearly indicate that activity of the M-NSC is initiated within 1 day of starvation and turned off between 1 and 3 days after feeding.

Table 2. Allatotropic activity of brains from normal fourth and
         fifth instar Manduca larva.

Implant	No. larvae	No. survived (%)	% survivors molted to	
			6th instars	pupae
Brain from newly ecdysed 4th instar larva	15	12 (80)	100	0
Brain from day-3 4th instar larva	15	11 (73)	18[*]	82
Brain from newly ecdysed 5th instar larva (from Table 1)	20	16 (80)	0	100
Brain from 3-4 day 5th instar larva	20	18 (90)	0	100
Brain from prepupa (2 days after exposure of dorsal vessel)	10	10 (100)	0	100

Allatotropic activity was determined using the following in vivo
assay. Medial neurosecretory cell-cauterized (M-NSC⁻) day-0
fifth instars were starved 3 days and fed 1 day. The test brains
were implanted in these M-NSC⁻-1 day-fed hosts. Induction of a
supernumerary larval molt (6th instar) indicates that the implanted brain had allatotropic activity. See the text for more
details.

---

[*] 1 black 6th instar and 1 larval-pupal intermediate.

Table 3. Starvation-induced stimulation of allatotropic activity
          in implanted larval brains of Manduca.

Implant	No. larvae	No. survived (%)	% survivors molted to 6th instar	pupae
Brain from newly ecdysed 5th instar larva	50	39 (78)	100	0
Brain from 3-4-day 5th instar larva	10	7 (70)	100	0
Brain from prepupa (2 days after exposure of dorsal vessel)	20	18 (90)	100	0
Brain from 3-day, 4th instar larva	32	22 (69)	100	0
Control (no implant)	27	26 (96)	4	96

Test brains were implanted into M-NSC cauterized day-0 fifth
instars which were then starved 3 days and fed normal diet.
Induction of a supernumerary molt indicates that the implanted
brains were stimulated to secrete allatotropin (see Table 2
for a comparison).

Table 4. Allatotropic activity of brain at different periods during
          starvation and feeding of fifth instar <u>Manduca</u> larvae[a].

Implant	No. Larvae	No. Survived (%)	% survivors molted to 6th instar	pupae
Brain of newly ecdysed 5th instar	20	16 (80)	0	100
Brain of day-1 starved larva	20	18 (90)	78	22
Brain of day-3 starved larva	22	16 (73)	100	0
Brain of day-1 fed larva	19	17 (90)	100	0
Brain of day-3 fed larva	6	6 (100)	0	100
M-NSC⁻ brain from day-3 starved larva (control)	10	6 (60)	0	100

[a]For details of assay refer to Table 2.

Transplantation of Active Brains or Brain-CC-CA Complexes Into Normal 5th Instar Larvae

Young last instar larvae of Galleria mellonella can be made
to undergo extra larval molts by implantation of brains from injured larvae (Krishnakumaran, 1972) or brains from early last instar
larval donors (Sehnal and Granger, 1975; Pipa, 1977). Our preceding
observations that starvation stimulated the brain to secrete ATH
and that brains from 3-day starved or 3-day starved-1-day fed larvae
appeared to possess maximum allatotropic activity in our in vivo
assay, prompted us to test whether implantation of active brains
into normal early last instar Manduca larvae can cause supernumerary
molting. Brains from either 3-day starved or 3-day starved-1-day
fed larvae were implanted into (a) newly ecdysed and (b) 'frosted
frass' (during the preceding night larvae first excreted frosted
frass) fifth instars. The data (Table 5) show that none of the
hosts molted into extra larvae. In fact, these hosts exposed the
dorsal vessel in synchrony with operated control larvae from the
same group. The resulting pupae were normal, i.e., they did not re-
tain any larval characteristics. Therefore, either the implanted
brains failed to secrete allatotropin in normal hosts or the host
CA were not responsive to it. Further, when CA alone or brain-CC-
CA complexes from starved animals were transplanted into newly
ecdysed or 2-day fifth instar hosts all the surviving recipients
pupated normally without any significant delay (Table 5).

Transplantation of Active CA into M-NSC-Cauterized, Starved Larvae

The previous data indicate that CA activity declines in the
absence of allatotropin. Therefore, an additional experiment was
done to test whether active CA maintain their activity in M-NSC-
cauterized, 3-day starved-1-day fed larvae which were used to
assay allatotropic activity of brains. Since CA taken from 3-day
starved or 3-day-starved-1-day fed fifth instars secreted more JH
in vitro than CA from any other stage (see later section), they
were implanted into the M-NSC-cauterized assay animals. If the
implanted CA remained active the host larvae could be expected to
undergo supernumerary molting. Of the 20 hosts, 13 survived and
all of them pupated. Evidently, in the absence of allatotropin the
active CA were unable to maintain their activity and secrete a
sufficient amount of JH to induce supernumerary molting.

JH Synthesis in vitro by CA From Normal and Starved Fifth Instar Larvae

The in vivo studies described in the preceding sections show
that in starved larvae ATH stimulated increased secretory activity
of the CA. We sought to confirm these observations by testing the

Table 5. Results of transplantation of active brains and/or CA
          into normal fifth instar <u>Manduca</u> larvae (fed on standard
          diet from the time of ecdysis to 5th).

Implant	No. Larvae	No. Surviving (%)	% survivors molted to	
			6th instars	pupae
Host - Newly ecdysed fifth				
Brain of 3-day starved or 3-day starved, 1-day fed 5th	30	27 (90)	0	100
CA from newly ecdysed 5th	15	15 (100)	0	100
CA from 3-day starved or 3-day starved, 1-day fed 5th	20	18 (90)	0	100
Brain-CC-CA complex from 3-day starved 5th	25	24 (96)	0	100
Host - 2-3 day old fifth				
Brain of 3-day starved or 3-day starved, 1-day fed 5th	45	43 (96)	0	100
CA from 3-day starved or 3-day starved, 1-day fed 5th	24	24 (100)	0	100
Brain-CC-CA complex from 3-day starved 5th	25	23 (92)	0	100

physiological status of the CA in vitro (Judy et al., 1973; Dahm
et al., 1976). Preliminary results summarized in Table 6
(Bhaskaran and Shirk, unpublished) indicate that in normal larvae
the CA decline in activity from the time of ecdysis. By day 1 they
no longer produce JH-I and secrete JH-II in much smaller amounts,
whereas JH-III synthesis shows an increase. CA from 3-4 day larvae
(FF, EDV) do not synthesize detectable amounts of any JH. The
changing pattern of synthesis and loss of activity parallel that
of adult female M. sexta CA at successive time intervals during
prolonged in vitro culture (Dahm et al., 1976) and may indicate a
preferential loss of enzymes for JH-I and II synthesis during the
stages of inactivation. The data also support our conclusion
that the CA are inactivated in two stages (Bhaskaran et al., 1980),
a differential reduction in biosynthesis occurring by day 1 and
shut down taking place by day 3.

Surprisingly, our experiments showed that starvation did not
immediately increase the activity of the CA. On the contrary, CA
from larvae 2 days after initiation of starvation were less active
than day-0 CA. This is puzzling because brains from 1-day starved
larvae contained substantial ATH activity, although that does not
prove that ATH was being secreted in these larvae. In any event,
by day-3 the CA had high activity and CA from 1-day fed larvae
had even higher activity. This, in conjunction with the hemolymph
JH titer profile of starved larvae (Riddiford, 1980), provides
strong evidence in support of our conclusion that supernumerary
molting of starved larvae is caused by increased JH synthesis by
the CA.

CA from 3-day starved larvae which were M-NSC-cauterized prior
to starvation synthesized only small quantities of JH, indicating
that CA maintain only a low level of activity in the absence of ATH.
The preliminary data thus support our interpretation that M-NSC
secrete ATH.

Relationship Between Hemolymph Trehalose Levels and Supernumerary Molting

What is the mechanism whereby starvation stimulates the
activity of the ATH center in the brain?

Since transplanted brains devoid of nerve connections become
activated in starved larvae we can infer that changes in the internal
milieu trigger the brain to produce ATH. Dahlman's (1973; 1975)
studies on 5th instar Manduca sexta larvae showed that a drastic
reduction in hemolymph trehalose, the major blood sugar, occurred
within 24 hours of fasting, whereas none of the other blood properties such as protein content, hemocyte count, sp. gravity, etc.,
was significantly influenced by starvation. Thus, it was pertinent

Table 6. In vitro activity of M. sexta fifth instar larval CA.

Treatment of Larvae	Age of Larvae	dpm/day x gland pair		
		JH I	JH II	JH III
Normal	<12 hr	117	677	161
Normal	1 d	nil	119	314
Normal	3-4 d	nil	nil	nil
Starved	2 d	nil	116	52
Starved	3 d	116	1833	678
Starved and fed	1 d	121	2657	672
M-NSC⁻ starved	3 d	18	99	43

CA were cultured in Grace's medium without methionine (Grand Island Biol. Co., New York), supplemented with 1% BSA (fraction V) and 33 μCi/ml of [^{3}H-methyl]-L-methionine (Schwarz-Mann, New York, 3.7 Ci/mmol). After incubation, the media were extracted and processed as in Dahm et al. (1976).

to ask whether the fall in trehalose content is the primary cue to which the brain responds, a question which could be tested by provisioning larvae with various dietary factors during the starvation period and noting the nature of the molt that occurred after resumption of feeding on a regular diet. Our results demonstrated that a majority of larvae provided with sugars molted into pupae; larvae fed a diet containing all factors (proteins, vitamins, cholesterol, fatty acids, mineral mixtures) except sugar, molted to sixth instars (Jones et al., 1980b). Measuring the hemolymph trehalose and protein levels after various dietary manipulations it was found that the concentration of hemolymph trehalose was very low (<15 μM as compared to 30-82 μM in normally fed larvae) in larvae provided with agar only, or a diet lacking carbohydrates. Such larvae were destined to undergo a supernumerary molt. The

concentration of trehalose was high (>68 µM) in larvae provided
with sucrose, and intermediate (∿25-60 µM) for fructose-or mannose-
fed larvae. About 90% of the sucrose-fed larvae, 58% of those fed
fructose and 32% of the mannose-fed larvae were destined to pupate
(Jones et al., 1980b). The strong correlation between hemolymph
trehalose levels and supernumerary molting suggests that the ATH
center is activated when the trehalose titer is drastically reduced.

Responses of CA From Normal 5th Instar Larvae-A Two Step Mechanism of Inhibition

Neurohormonal Inhibition; The First Step of Inactivation

The classical theory of insect metamorphosis postulates that
hemolymph JH titer is reduced significantly during the final larval
instar leading to initiation of a metamorphic molt (Wigglesworth,
1954; 1970; Piepho, 1951; Schneiderman and Gilbert, 1964; Doane,
1973). Inactivation of CA, increase in hemolymph JHE, dilution of
JH by increase in blood volume, all of these may contribute to
lower JH titer toward the end of the feeding period (Akamatsu et
al., 1975; Gilbert et al., 1978). In the last larval instar of
M. sexta the decline in JH titer has been attributed to a cessation
of ATH secretion and the release of an inhibitory factor, allatohibin, via the allatal nerves to the CA (Williams, 1976). Our finding
that ATH activity increased in starved larvae suggested to us the
possibility of using the allatectomized, starved larva as a bioassay
system to evaluate the responsiveness of CA from larvae at different
periods during the fifth instar and from pharate pupae. When CA
from these stages were transplanted into newly ecdysed, allatectomized fifth instar hosts which were starved and then fed we were
surprised to find that only CA from 0-day fifth instar larvae were
capable of inducing a supernumerary larval molt in the hosts. CA
from 1-day or older larvae and from pharate pupae failed to promote
a larval molt and all the recipients pupated (Bhaskaran et al., 1980).
Obviously the responsiveness of CA to ATH had changed by day 1.
CA remained in this non-responsive state until at least one to two
days prior to pupal ecdysis which was the last stage tested in this
study.

Further studies showed that the change in CA was independent
of neural connections and was controlled by the group II M-NSC in
the brain (Bhaskaran et al., 1980). This suggests that group II
M-NSC secrete a blood-borne factor (allatohibin) which lowers or
eliminates the response of CA to ATH (secreted by the same cell
group under starvation conditions). Allatohibin activity was not
present in the hemolymph of larvae on day 2 or thereafter (Table
7).

Table 7. Determination of the period during which allatohibin is present in the hemolymph of fifth instar <u>Manduca sexta</u> larvae[*].

Age of the test host at the time of CA-		Source of CA	# Surviving assay larvae	% Survivors molted to	
Implantation	Removal for assay			6th instars	pupae
Day 0	Day 1	Implant	14	0	100
		Test-Host	7	0	100
Day 2	Day 3	Implant	15	100	0
		Test-Host	8	0	100
Day 3	Day 4	Implant	17	100	0
		Test-Host	9	0	100

[*]CA were excised from 0 day fifth instars and implanted into fifth instar test-hosts of different ages. One day later the implanted CA and the test-host CA were removed and reimplanted seperately into CA⁻ day 0 fifth instars (assay larvae) which were then starved for 3 days and fed. The occurrence of a supernumerary molt (6th instar) denotes that the CA were responsive to allatotropin (from Bhaskaran et al., 1980).

Neural Inhibition; The Second Step in CA Inactivation

The decrease in CA activity detected by day 1, using our bio-assay, apparently conflicts with the suggestion of Nijhout and Williams (1974b) that inactivation of CA occurs only after larvae reach 5.0 g in weight (i.e., during the night of day 2). However, our findings are indirectly supported by the sharp fall in hemolymph JH titer observed on day 1 by Fain and Riddiford (1975). Since the titer curve shows a plateau between days 1 and 3 and a further re-duction of JH concentration to undetectable levels on day 3, it is logical to infer that a second level of inhibition occurs at this time, leading to the inactivation of the CA as suggested by Nijhout and Williams (1974b). The verification of this second inhibitory step required a different bioassay which could discriminate between

less active and inactive CA. Based on the observations that (a)
fourth instar larvae allatectomized one day after ecdysis undergo
precocious pupation whereas a majority of larvae allatectomized on
day 2 molted into black (indicating low JH levels) fifth instars
(Kiguchi and Riddiford, 1978; Bhaskaran et al., 1980) and (b) the
JH titer begins to drop rapidly by day 2 of the fourth instar (Fain
and Riddiford, 1975), we hypothesized that less active, but not
inactive, fifth instar CA would, perhaps, support a larval molt if
transplanted into allatectomized, day-1 fourth instar larvae. The
results of this experiment demonstrated that CA from day-0 and
day-1 fifth instars cause a larval molt in the 4th instar assay
hosts, whereas CA from day 3 and older larvae are unable to do so
(Fig. 2B). Evidently, the CA of day-3 larvae are much less active
than day-1 larvae, indicating that day-1 CA still secrete JH,
although not enough to induce a supernumerary molt in the 5th
instar assay, and that a second level of inhibition occurs by day 3
(larvae excrete FF during the preceding night). This second
inhibition is neurally mediated since severance of allatal nerves
prior to day 3 prevented the inhibition. Once the neural inhibition
occurred, sectioning of the nerves did not reverse the inhibition
since the CA remained inactive when transplanted into allatectomized
day-1 fourth instar larvae (Bhaskaran et al., 1980).

When fifth instar CA were assayed in allatectomized 0-day
fourth instar larvae the results were identical to those obtained
with the starved fifth instar assay; that is, 0-day fifth instar

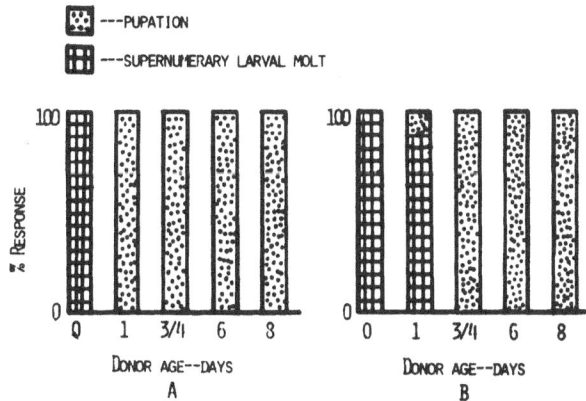

Fig. 2   Effect of implantation of CA from fifth instar larvae into
         (A) allatectomized 0-day fourth instars (B) allatectomized
         1-day fourth instars. The ordinate shows the percentage
         of surviving fourth instar larvae which molted into fifth
         instar larvae or pupated prematurely (adapted from
         Bhaskaran et al., 1980).

CA caused a larval molt whereas CA from all other age groups were inactive (Fig. 2A). Thus, the two fourth instar assays provide enough resolution to discriminate between the three levels of activity - high, low and none (Bhaskaran et al., 1980). In conclusion, highly active CA were able to promote a larval molt in allatectomized fourth instars of any age, less active CA only in day-1 and older fourth instar hosts and inactive glands in none of the hosts tested.

The Reversibility of the Two Inhibitory States

From the foregoing data it is evident that the neurohormonal inhibition was not reversible even when the CA were exposed to ATH for a prolonged period. For example, day-1 CA transplanted into allatectomized, day-0 fifth instars which were subsequently starved and fed (and hence CA were subjected to prolonged exposure to high levels of ATH) did not cause a larval molt. CA from older stages behaved in the same manner. We then tested glands cultured in vitro for their responsiveness to ATH. Responsive CA (from day-0 instar) cultured for 1-3 days promoted a larval molt when transplanted into 0-day fourth instar assay hosts. Hence they had not lost their responsiveness to ATH. CA which were exposed to allatohibin (day-1 fifth instar CA), did not regain the capacity to become stimulated by ATH after culture for three days in vitro. Such glands were not reactivated by ATH in vivo, as determined by the failure of assay hosts to undergo a larval molt.

The above data do not reveal whether CA will be neurally inhibited if they are not first exposed to allatohibin; and if they are only neurally inhibited, whether the inhibition can be reversed by ATH. Experiments designed to answer these questions consisted of cauterizing the M-NSC in day-0 fifth instar larvae and then assaying the CA on days 1 and 3, using either allatectomized, starved fifth instars or allatectomized, day-0 or day-1 fourth instars. Preliminary data indicate that cautery of M-NSC prevents neurohormonal inhibition but does not prevent neural inhibition. More importantly, CA subjected only to neural inhibition can be reactivated by ATH. It is interesting to note that the source of neural inhibition is in a region of the brain other than that in which the M-NSC are located.

DISCUSSION

Several mechanisms may be involved in the regulation of hemolymph JH titer in larval stages of insects (Akamatsu et al., 1975; Williams, 1976; Gilbert et al., 1978). The data presented here suggest that regulation at the level of the CA is a major factor in determining the rise and fall in JH levels in the hemolymph of the fourth and fifth instar M. sexta. The activity of the CA is

controlled by both activating and inhibiting stimuli from the
brain (Bhaskaran and Jones, 1980; Bhaskaran et al., 1980).

The Endocrine Basis of Starvation-Induced Supernumerary Molting

Several observations supporting the above conclusions were made
possible by the discovery of starvation-induced supernumerary molting
in fifth instar larvae (Jones et al., 1980; Cymborowski et al.,
cited by Riddiford, 1980). Investigations into the endocrine basis
of the supernumerary molt provided the first clear evidence for
ATH in M. sexta (Bhaskaran and Jones, 1980). Starvation causes a
significant elevation in hemolymph JH concentration (Riddiford,
1980) primarily due to increased JH synthesis by the CA (Table 6).
The increased activity of the CA is not a direct response to changes
in the internal milieu but in fact appears to be stimulated by an
ATH from the brain. Available evidence suggests that the group II
M-NSC are the source of the ATH (Bhaskaran and Jones, 1980) but
which among the many cells found in this region actually secrete
ATH remains unknown. Our preliminary studies indicate that the
group of 8 cells (4 on each side) in the anteromedial region are
not the ATH secreting cells since their stainability with PAF does
not differ with various physiological changes occurring during
normal development or after starvation and feeding. However, we
cannot exclude the possibility that the hormone content may vary
without corresponding changes in the stainable, carrier protein
content (Meola and Lea, 1971; Mahon and Nair, 1975).

The data presented here do not support the interpretation that
the brain promotes supernumerary molting by acceleration of a molt.
Besides the fact that L-NSC and not the M-NSC secrete PTTH in M.
sexta (Gibbs and Riddiford, 1977; Agui et al., 1979), the timing of
PTTH release in the various experimental groups had no bearing on
the nature of the ensuing molt. These aspects are discussed in
detail elsewhere (Bhaskaran and Jones, 1980). Finally, since there
was no premature increase in JHE levels in M-NSC cauterized larvae,
these cells do not appear to regulate JH levels in starved larvae
by suppression of JHE release by the fat body (Seligman et al.,
in preparation).

It is evident that prolonged starvation stimulates the M-NSC
to secrete ATH, since brains from 3-day-starved, and 3-day starved-
1-day fed animals exhibited the highest levels of ATH activity.
CA from these larvae secreted more JH in vitro than CA from day-0
fifth instars and much more than CA of M-NSC cauterized-3-day starved
larvae (Table 6). Therefore, our preliminary data complement the
observations of Cymborowski et al. (cited in Riddiford, 1980), that
hemolymph JH titers rise markedly during starvation and decline
gradually after feeding. Furthermore, Nijhout (1975b) observed a
5-fold increase in hemolymph JH titer at the end of 24 hours of

starvation of 3.0–3.5 g larvae (fifth instar). Larvae starved at
these weights eventually formed larval-pupal intermediates whereas
allatectomy abolished this effect. Starvation induces supernumerary
molting in other Lepidoptera and presumably in all these cases the
primary neuroendocrine response is activation of the ATH center.
However, in some insects starvation has the opposite effect, that
of causing precocious metamorphosis, which may be either due to
lack of stimulation or active inhibition of CA.

In the context of regulation of CA during normal development
our data demonstrate that the decline in hemolymph JH titer during
the later part of the fourth instar and during the first day after
ecdysis to the fifth instar (Fain and Riddiford, 1975) is caused
by a gradual decrease in CA activity. This conclusion is supported
by the following observations; (a) the presence in brain of high
ATH activity in early fourth instar and very low activity in late
fourth and day 0 fifth instar larvae (b) the absence of significant JH esterase activity in hemolymph of fourth and early fifth
instars (Weirich et al., 1973; Sanburg et al., 1975) and (c) the
presence of large amounts of JH binding proteins, which protect
bound JH from degradation by general esterases (Sanburg et al.,
1975) during these stages (Goodman and Gilbert, 1978). Furthermore, continuous stimulation by allatotropin is required to maintain the CA at a high level of activity as shown by (a) the inability of CA to induce a supernumerary molt in M-NSC-cauterized
larvae (Bhaskaran and Jones, 1980), (b) the failure of active
CA from starved larvae to restore supernumerary molting in the M-
NSC-cauterized, 3 day starved and 1 day fed assay larvae, and (c)
the relatively low level of JH production in vitro by CA from M-NSC
cauterized starved larvae. Such dependence of CA function on ATH
has been demonstrated in the penultimate larval instar of the
earwig Anisolabis maritima (Ozeki, 1965), in larvae of Bombyx mori
(Morohoshi and Shimada, 1974), in some species of adult locusts
(Strong, 1965; McCaffery, 1976; Lazarovici and Pener, 1978) and in
adult Diploptera punctata (Stay and Tobe, 1978). In diapausing
larvae of the borer Diatraea grandiosella, continuous neural
stimulation maintains the CA in an active state (Yin and Chippendale,
1979). On the other hand, in many insects removal of neural inhibition is sufficient to reactivate and sustain CA activity (Wigglesworth, 1970; Doane, 1973; Novak, 1975).

Our failure to detect allatotropic activity in brains of newly
ecdysed fifth instars (<12 hr old) needs to be explained, since
hemolymph JH level is high at this time (Fain and Riddiford, 1975).
The question can only be resolved with more detailed assays of
brains removed at intervals between the time of head capsule slip-
page and ecdysis to the fifth instar. It is possible that a transient activation of the ATH center occurs in pharate fifth instars
and is turned-off immediately after ecdysis. Only under exceptional

conditions is the allatotropin center reactivated during the fifth
instar, as for example, after prolonged starvation (Bhaskaran and
Jones, 1980). The pattern of allatotropic activity of the brain
in the last instar larva of M. sexta distinctly differs from that in
G. mellonella. In the latter, the brain is reported to be active
during the first few days of the last larval instar (Granger and
Sehnal, 1974).

It is interesting to note that inactive brains from fourth
and fifth instars of any age can be reactivated by the starvation
stimulus. Even brains from wandering and prepupal animals could
induce a supernumerary molt in M-NSC-cauterized larvae if implanted
prior to starvation. The positive response of implanted brains also
demonstrates that intact nerve connections are not required. Hence
the brain must be responding to some changes in the hemolymph that
are induced by starvation. Preliminary evidence indicates that
the primary factor is the hemolymph trehalose concentration (Jones
et al., 1980). Circumstantial evidence suggests that supernumerary
molting caused by parasitization (Beckage and Riddiford, 1978) may
also be due to low hemolymph trehalose levels (Dahlman, 1975)
causing stimulation of the allatotropin center.

In striking contrast to the situation in Galleria larvae
(Krishnakumaran, 1972; Granger and Sehnal, 1974; Pipa, 1976; 1977),
active brains implanted into young last instar Manduca larvae show
no allatotropic activity. We can infer from this that the internal
milieu is not suitable for sustained allatotropin secretion and that,
as a consequence, the transplanted brains stop producing allatotropin.
On the other hand, even if the implanted brains secrete allatotropin
thereby stimulating the host CA to secrete JH, the high JH esterase
activity in the hemolymph by day 4 (Weirich et al., 1973; Vince and
Gilbert, 1977) can effectively eliminate the extra amount of JH.
This may also explain the absence of any JH-effect in such animals
even after transplantation of one or more pairs of CA or brain
CC-CA complexes. Therefore, the importance of JH esterase in
eliminating hemolymph JH (Akamatsu et al., 1975) cannot be ignored.

Inactivation of CA in the Fifth (Last) Instar Larva

Inactivation of CA in the last larval instar of holometabolous
insects or in the penultimate larval instar of hemimetabolous insects
has been shown by transplantation of CA into suitable assay hosts
or by hemolymph JH titer determinations. In Bombyx mori inactivation
occurs midway through the last instar and shortly before the termination of the phagoperiod (Fukuda, 1944). Likewise, CA of last
instar larvae of Hyalophora cecropia become inactive toward the end
of the feeding period (Williams, 1961). Hemolymph JH titer determinations carried out on several lepidopterans (Varjas et al., 1976;
Yagi, 1976; Hsiao and Hsiao, 1977; Mauchamp et al., 1979; Yin and

Chippendale, 1979) reveal similarities in the JH profile and tend
to corroborate the conclusions drawn from the transplantation
studies. In M. sexta, hemolymph JH level decreases in two stages.
The first reduction occurs within a day after ecdysis, and JH titer
levels off until a day before the wandering stage, when the second
drastic reduction lowers JH to undetectable levels (Nijhout and
Williams, 1974b; Fain and Riddiford, 1975; Judy, personal communication). Besides direct inactivation of CA, the large increase in
hemolymph volume and the rapid increase in JHE levels (Gilbert
et al., 1978) contribute to lowering the JH titer.

In most instances, inactivation of the CA is believed to be
mediated by nervous stimuli whether it occurs in the last larval
instar or in the adult female (Scharrer, 1958; Wigglesworth, 1970;
Doane, 1973; Novak, 1975). In M. sexta larvae Williams (1976)
postulated that the release of a neurohormone via the allatal
nerves inactivates the CA shortly before the wandering stage.
Results presented here show that the first decrease in JH level,
occurring by day 1, can be partly attributed to the absence of ATH.
In addition, the data summarized here and discussed in detail
elsewhere (Bhaskaran et al., 1980), demonstrate that the CA lose
their responsiveness to ATH by day 1. This change in CA responsiveness is induced by a neurohormonal factor secreted by the M-NSC.
This factor disappears from the hemolymph by day 2 and is never
again released prior to pupation. We have adopted the term allatohibin coined by Williams (1976) to identify this factor (our interpretation of the postulated allatohibin of Williams (1976), is stated
in a later section of the discussion). Glands exposed to allatohibin are not totally inactivated but continue to secrete JH at a
low rate. This is inferred from the plateau in JH titer between
days 1-3 and from the evidence that they will cause a larval molt
when implanted into allatectomized day-1 fourth instar larvae. Thus
allatohibin causes a step-down in ATH responsiveness of the CA,
and these glands cannot again increase their JH synthetic rate to
the level necessary to induce a larval molt in the assay larvae.
This suggests that allatohibin may be acting on a rate-limiting step
in the biosynthetic pathway.

The neural inhibition which occurs prior to the appearance of
FF on day 3 produces a complete shut-down of the biosynthetic
activity. The timing of the neural inhibition reported here coin-
cides with that of the inhibition postulated earlier by Nijhout and
Williams (1974b) and Williams (1976). It also requires the presence
of intact nerves. This in itself does not preclude a neurohormonal
factor since it is known that some neurohormonal substances are
normally transported via the axons directly into target tissues
(Maddrell and Nordmann, 1979).

PTTH As The Axonally Transported CA-Inhibiting Neurohormone

On the basis of circumstantial evidence it would appear that the second inhibition is actually caused by PTTH which is transported axonally into the CA and that direct release of PTTH into the CA is imperative for PTTH to exert its inhibitory effect. The rationale for this is as follows: First, there is a remarkable coincidence in timing between the initial release of PTTH and the occurrence of the neural inhibition. We have found that the CA undergo neural inhibition on the night of day 2, and a recent estimate indicates that PTTH release also occurs at this time (Riddiford, 1980). Second, the CA exhibit PTTH activity in bioassays (Gibbs and Riddiford, 1977) and are now believed to be the sites of release of PTTH (Morohoshi and Shimada, 1975; Agui et al., 1979). Third, the decline in JH titer during the fourth instar (Fain and Riddiford, 1975) begins at approximately the same time as the release of PTTH (Fain and Riddiford, 1976), and CA from older fourth instar larvae cultured in vitro secrete more JH when cultured alone than when cultured as brain-CC-CA complexes with intact NCA (Granger et al., see this volume). Finally, Morohoshi and Shimada (1976) found that in Bombyx mori PTTH is released from the CA and that when a large amount of PTTH is present in the CA the latter become inactivated. Our own data show that neural inhibition of CA does not originate in the M-NSC, which is also the case with PTTH (secreted by lateral NSC group III). There is also reason to believe that in older fourth instars the brain inhibits CA by means of a nervous stimulus (Granger et al., see this volume), an inhibition which is reversed later during the pharate fifth instar, as may be deduced from the JH titer curve. The axon-mediated inhibition of CA in older fifth instars is also reversible if it occurs without prior exposure to allatohibin. With all of these evidences, the question arises as to why circulating PTTH does not inhibit CA in nerve sectioned larvae? The answer may lie in the sensitivity of CA to PTTH. Perhaps even the highest concentration of PTTH normally present in the blood is insufficient to inactivate the CA. Alternatively, one may postulate the existence of a permeability barrier in the connective tissue sheath surrounding the CA.

The use of PTTH as a CA-inactivating factor by the larva would be economical and timely in the light of speculation that the PTTH center and/or the PG are inhibited by JH (Williams, 1976; Riddiford, 1980). Transport of PTTH axonally to CA and its release from the CA into the hemolymph would rapidly inactivate the CA prior to (and permitting) the accumulation of the critical concentration of PTTH in the hemolymph required to fully activate the PG.

A Model For Regulation of CA in Larvae

The reversible PTTH-neural inhibition in combination with ATH is probably used in earlier stages of larval development for modulation of CA activity. In the final instar the attainment of a critical head capsule size with the attendant stretch stimulus (Nijhout, 1975a), and initiation of feeding with resulting elevation of hemolymph trehalose level (Jones et al., 1980) signal the time for turning-off the ATH center and activating the allatohibin center (Fig. 3A). The neurohormonal inhibition not only ensures that the CA is irreversibly limited in its capacity to synthesize JH as a prelude to metamorphosis but may also be a mechanism for maintenance of CA activity at a low level so that under adverse nutritional

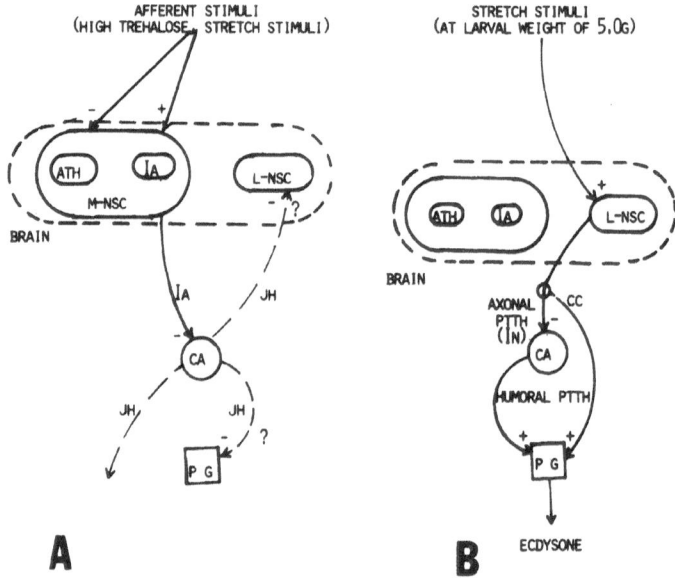

Fig. 3   Schematic diagrams illustrating the postulated role of the brain-neurosecretory centers in regulation of the CA activity in fifth instar M. sexta larvae. (A) Shortly after ecdysis, the ATH center is turned off. Within 24 hours after initiation of feeding the allatohibin center (I_A) is turned on. Allatohibin either depresses or eliminates the capacity of the CA to respond to ATH. The glands continue to produce low amounts of JH until day 3. (B) On day 3, the L-NSC (PTTH center) are activated and PTTH which is transported axonally into the CA causes complete inactivation of the CA. This represents the nervous inhibition (I_N).

conditions the phagoperiod is extended, enabling the animal to accumulate sufficient energy resources for metamorphosis. Glands exposed to allatohibin continue to secrete JH at a low level, but, during this period, the brain gradually becomes competent to secrete PTTH, since the low level of hemolymph JH is insufficient to block the PTTH center (Nijhout and Williams, 1974b). When larvae attain a critical size, stretch-stimuli trigger the axonally conducted release of PTTH into the CA causing complete inactivation of the glands (Fig. 3B). The reduction and elimination of JH from the blood and tissues is also facilitated by JHE. This, again, is a fool-proof mechanism for ensuring the absence of JH during metamorphosis. The need for a stable inhibition during premetamorphic and metamorphic stages is eminently clear, and it is reasonable to assume that only after exposure to the metamorphic environment will the inhibitory states be reversed. Noteworthy is the fact that in Bombyx mori the CA regain activity only during the pharate adult stage (Fukuda, 1962). Although many aspects of this model are at present speculative, our recent studies (Bhaskaran and Jones, 1980; Bhaskaran et al., 1980; Jones et al., 1980) have provided a basis for further analysis of the mechanisms regulating CA activity in M. sexta larvae.

ACKNOWLEDGEMENTS

     I thank Grace Jones, Davy Jones, Paul Shirk and Morris Seligman for their contributions to the work discussed here; Larry Keeley and Günter Weirich for helpful comments on the manuscript and H. A. Röller for his interest and support. The investigations were supported by grants from NSF (PCM 77-25417) and by Organized Research, Texas A & M University.

REFERENCES

Agui, N., Granger, N. A., Gilbert, L. I., and Bollenbacher, W. E., 1979, Cellular localization of the insect prothoracicotropic hormone: In vitro assay ot a single neurosecretory cell, Proc. Natl. Acad. Sci. U.S. 76:5694-5698.

Akamatsu, Y., Dunn, P. E., Kezdy, F. J., Kramer, K. J., Law, J. H., Reibstein, D., and Sanburg, L. L., 1975, Biochemical aspects of juvenile hormone action in insects, in: "Control Mechanisms in Development", R. Meints and E. Davies, eds., Plenum Press, New York, pp. 123-149.

Baehr, J. C., 1976, Étude   du contrôle neuro-endocrine du fonctionnement du corpus allatum chez les larves du quatrième stade de Rhodnius prolixus, J. Insect Physiol. 22:73-82.

Beckage, N. E., and Riddiford, L. M. 1978, Developmental interactions between the tobacco hornworm Manduca sexta and its braconid parasite Apanteles congregatus, Ent. Exp. Appl. 23:139-151.

Bhaskaran, G., and Jones, G., 1980, Neuroendocrine regulation of
    corpus allatum activity in Manduca sexta. The endocrine basis
    of starvation induced supernumerary larval molt, J. Insect
    Physiol. 26:in press.

Bhaskaran, G., Jones, G., and Jones, D., 1980, Neuroendocrine regu-
    lation of corpus allatum activity in Manduca sexta: Sequential
    neurohormonal and nervous inhibition in the last larval instar,
    in preparation.

Bounhiol, J. J., 1957, La métamorphose se produit, chez Bombyx mori,
    après supression, au dernier stade larvaire, des relations
    nerveuse entre cérébroides et corps allates, ceux-ci restant.
    longtemps imprégnés de neurosecretion, C. R. Acad. Sci. 245:
    1087-1089.

Cassier, P., 1979, The corpora allata of insects, Int. Rev. Cytol.
    57:1-73.

Dahlman, D. L., 1973, Starvation of the tobacco hornworm, Manduca
    sexta. I. Changes in hemolymph characteristics of 5th-stage
    larvae, Ann. Entomol. Soc. Am. 66:1023-1029.

Dahlman, D. L., 1975, Trehalose and glucose levels in hemolymph of
    diet-reared, tobacco leaf-reared and parasitized tobacco
    hornworm larvae, Comp. Biochem. Physiol. 50A:165-167.

Dahm, K. H., Bhaskaran, G., Peter, M. G., Shirk, P. D., Seshan,
    K. R. and Röller, H., 1976, On the identity of the juvenile
    hormone in insects, in: "Juvenile Hormones", L. I. Gilbert, ed.,
    Plenum Press, New York, pp. 19-47.

Doane, W. W., 1973, Role of hormones in insect development, in:
    "Developmental Systems: Insects", S. J. Counce and C. H.
    Waddington, eds., Academic Press, New York, Vol. 2, pp. 291-
    497.

Fain, M. J., and Riddiford, L. M., 1975, Juvenile hormone titers
    in the hemolymph during late larval development of the tobacco
    hornworm, Manduca sexta (L.), Biol. Bull. Woods Hole 149:506-521.

Fain, M. J., and Riddiford, L. M., 1976, Reassessment of the critical
    periods for prothoracicotropic hormone and juvenile hormone
    secretion in the larval molt of the tobacco hornworm, Manduca
    sexta, Gen. Comp. Endocrinol. 30:131:141.

Fraser, J., and Pipa, R., 1977, Corpus allatum regulation during
    the metamorphosis of Periplaneta americana: axon pathways,
    J. Insect Physiol. 23:975-984.

Fukuda, S., 1944, The hormonal mechanism of larval molting and
    metamorphosis in the silkworm, J. Fac. Sci. Tokyo Univ.
    Sec. IV 6:477-532.

Fukuda, S., 1962, Secretion of juvenile hormone by the corpora
    allata in pupae and moths of the silkworm, Bombyx mori,
    Annot. Zool. Japan 35:199-212.

Gibbs, D., and Riddiford, L. M., 1977, Prothoracicotropic hormone
    in Manduca sexta: Localization by a larval assay, J. Exp.
    Biol. 66:255-266.

Gilbert, L. I., Goodman, W., and Granger, N. A., 1978, Regulation of
    juvenile hormone titer in Lepidoptera, in: "Comparative
    Endocrinology", P. J. Gaillard and H. H. Boer, eds., Elsevier/
    North Holland, Amsterdam, pp. 471-486.
Girardie, A., 1965, Contribution à l'etude du contrôle de l'activité
    des corpora allata par la pars intercerebralis chez Locusta
    migratoria (L.), C. R. Acad. Sc. Paris 261:4876-4878.
Goodman, W., and Gilbert, L. I., 1978, The hemolymph titer of
    juvenile hormone binding protein and binding sites during the
    fourth larval instar of Manduca sexta, Gen. Comp. Endocrinol.
    35:27-34.
Granger, N. A., and Sehnal, F., 1974, Regulation of larval corpora
    allata in Galleria mellonella, Nature 251:415-417.
Highnam, K. C., 1967, Insect hormones, J. Endocr. 39:123-150.
Hsiao, T. H., and Hsiao, C., 1977, Simultaneous determination of
    molting and juvenile hormone titers of the greater wax moth,
    J. Insect Physiol. 23:89-93.
Jones, D., Jones, G., and Bhaskaran, G., 1980a, Induction of super-
    numerary molting by starvation in Manduca sexta larvae,
    Ent. Exp. Appl. 28:in press.
Jones, D., Jones, G., and Bhaskaran, G., 1980b, Dietary sugars,
    hemolymph trehalose levels and supernumerary molting of
    Manduca sexta larvae, in preparation.
Judy, K. J., Schooley, D. A., Dunham, L. L., Hall, M. S., Bergot,
    B. J., and Siddall, J. G., 1973, Isolation, structure and
    absolute configuration of a new natural insect juvenile hormone
    from Manduca sexta, Proc. Natl. Acad. Sci. U.S.A. 70:1509-1513.
Kiguchi, K., and Riddiford, L. M., 1978, A role of juvenile hormone
    in pupal development of the tobacco hornworm, Manduca sexta,
    J. Insect Physiol. 24:673-680.
Krishnakumaran, A., 1972, Injury induced molting in Galleria
    mellonella larvae, Biol. Bull. Woods Hole 142:281-292.
Lazarovici, P., and Pener, M. P., 1978, The relations of the pars
    intercerebralis, corpora allata, and juvenile hormone to
    oocyte development and oviposition in the African migratory
    locust, Gen. Comp. Endocrinol. 35: 375-386.
Lüscher, M., and Engelmann, F., 1955, Über die Steuerung der Corpora
    allata-Funktion bei der Schabe Leucophaea maderae, Rev. Suisse
    Zool. 62:649-657.
Lüscher, M., and Engelmann, F., 1960, Histologische und experi-
    mentelle Untersuchungen über die Auslösung der Metamorphose
    bei Leucophaea maderae, J. Insect Physiol. 5:240-258.
Mauchamp, B., Lafont, R., and Jourdain, D., 1979, Mass fragmento-
    graphic analysis of juvenile hormone I levels during the last
    larval instar of Pieris brassicae, J. Insect Physiol. 25:545-
    550.
Maddrell, S. H. P. and Nordmann, J. J., 1979, "Neurosecretion",
    John Wiley & Sons, New York.

Mahon, D. C., and Nair, K. K., 1975, A comparison of aldehyde fuchsin
    and alcian blue staining of neurosecretory material in
    Oncopeltus fasciatus, Cell. Tiss. Res. 161:477-484.
Meola, R., and Lea, A. O., 1971, Independence of paraldehyde fuchsin
    staining of the corpus cardicaun and the presence of the neuro-
    secretory hormone required for egg development in the mosquito,
    Gen. Comp. Endocrinol. 16:105-111.
McCaffery, A. R., 1976, Effects of electrocoagulation of cerebral
    neurosecretory cells and implantation of corpora allata on
    oocyte development in Locusta migratoria, J. Insect Physiol. 22:
    1081-1092.
Morohoshi, S., and Shimada, J., 1974, The control of growth and
    development in Bombyx mori. XXI. Function of the brain by
    strains in the activity of corpora allata of the fifth instar
    larvae, Proc. Jap. Acad. 50:155-160.
Morohoshi, S., and Shimada, J., 1975, The control of growth and
    development in Bombyx mori. XXVII. Release of the brain
    hormone from the corpora allata through nerve axons from brain
    neurosecretory cells, Proc. Japan Acad. 51(10):744-747.
Morohoshi, S., and Shimada, J., 1976, The control of growth and
    development in Bombyx mori. XXXIV. Inhibition of corpus
    allatum activity by brain hormone, Proc. Japan Acad. 52(8):
    442-445.
Nijhout, H. F., 1975a, A threshold size for metamorphosis in the
    tobacco hornworm, Manduca sexta (L.), Biol. Bull. Woods Hole
    149:214-225.
Nijhout, H. F., 1975b, Dynamics of juvenile hormone action in
    larvae of the tobacco hornworm, Manduca sexta, Biol. Bull. Woods
    Hole 149:568-579.
Nijhout, H. F., and Williams, C. M., 1974a, Control of moulting and
    metamorphosis in the tobacco hornworm, Manduca sexta (L.):
    Growth of the last-instar larva and the decision to pupate,
    J. Exp. Biol. 61:481-491.
Nijhout, H. F., and Williams, C. M., 1974b, Control of moulting
    and metamorphosis in the tobacco hornworm, Manduca sexta:
    cessation of juvenile hormone secretion as a trigger for
    pupation, J. Exp. Biol. 61:493-501.
Novak, V. J. A., 1975, "Insect Hormones", Chapman and Hall Ltd.,
    London.
Ozeki, K., 1962, Studies on the secretion of the juvenile hormone
    in the earwig, Anisolabis maritima. Sci. Pap. Coll. Gen. Educ.
    Univ. Tokyo 12(1):65-72.
Ozeki, K., 1965, Studies on the function of the corpus allatum during
    the last nymphal stage in the earwig, Anisolabis maritima, Sci.
    Pap. Coll. Gen. Educ. Univ. Tokyo, 15:149-156.
Piepho, H., 1951, Über die Lenkung der Insektenmetamorphose durch
    Hormone, Verh. dtsch. Zool. Ges., Wihelmshaven pg. 62-75.

Pipa, R. L., 1976, Supernumerary instars produced by chilled wax
    moth larvae: Endocrine mechanisms J. Insect Physiol. 22:1641-
    1648.
Pipa, R. L., 1977, Do the brains of wax moth larvae secrete an alla-
    totropic hormone?, J. Insect Physiol. 23:103-108.
Riddiford, L. M., 1980, Interaction of ecdysteroids and juvenile
    hormone in the regulation of larval growth and metamorphosis
    of the tobacco hornworm, in "Progress in Ecdysone Research",
    J. A. Hoffman, ed., North-Holland, Amsterdam, in press.
Sanburg, L. L., Kramer, K. J., Kézdy, F. J., and Law, J. H.,1975,
    Juvenile hormone-specific esterases in the hemolymph of the
    tobacco hornworm, Manduca sexta, J. Insect Physiol. 21:873-888.
Scharrer, B., 1952, Neurosecretion. XI. The effects of nerve
    section on the intercerebralis-cardiacum-allatum system of the
    insect Leucophaea maderae, Biol. Bull. Woods Hole 102:261-272.
Scharrer, B., 1958, Neuro-endocrine mechanisms in insects, Proc.
    Second International Symposium. Neurosecretion, Springer-
    Verlag, Berlin- Göttingen-Heidelberg, pp. 79-84.
Sehnal, F. and Granger, N. A., 1975, Control of corpora allata
    function in larvae of Galleria mellonella, Biol. Bull. Woods
    Hole 148:106-116.
Schneiderman, H. A., and Gilbert, L. I., 1964, Control of growth
    and development in insects, Science 143:325-333.
Stay, B., and Tobe, S. S., 1978, Control of juvenile hormone bio-
    synthesis during the reproductive cycle of a viviparous cock-
    roach II. Effects of unilateral allatectomy, implantation of
    supernumerary corpora allata and ovariectomy, Gen. Comp.
    Endocrinol. 34:276-286.
Strong, L., 1965, The relationship between the brain, corpora allata
    and oocyte growth in the Central American locust, Schistocerca
    sp. I. The cerebral neurosecretory system, the corpora allata
    and oocyte growth, J. Insect Physiol. 11:135-146.
Truman, J. W., 1972, Physiology of insect rhythms.   Circadian
    organization of the endocrine events underlying the moulting
    cycle of larval tobacco hornworms, J. Exp. Biol. 57:805-820.
Truman, J. W., Riddiford, L. M., and Safranek, L., 1973, Hormonal
    control of cuticle coloration in the tobacco hornworm, Manduca
    sexta: basis of an ultrasensitive bioassay for juvenile hor-
    mone, J. Insect Physiol. 19:195-203.
Varjas, J., Paguia, P., and DeWilde, J., 1976, Juvenile hormone
    titers in penultimate and last instar of Pieris brassicae and
    Barathra brassicae, in relation to the effect of juvenoid
    application, Experientia 32:249-251.

Vince, R. K., and Gilbert, L. I., 1977, Juvenile hormone esterase activity in precisely timed last larval instar larvae and pharate pupae of Manduca sexta, Insect Biochem. 7:115-120.

Weirich, G., Wren, J., and Siddall, J. B., 1973, Developmental changes of the juvenile hormone esterase activity in hemolymph of the tobacco hornworm, Manduca sexta, Insect Biochem. 3:397-407.

Wigglesworth, V. B., 1936, The function of the corpus allatum in the growth and reproduction of Rhodnius prolixus (Hemiptera), Quart. J. Micr. Sci. 79:91-121.

Wigglesworth, V. B., 1954, "The Physiology of Insect Metamorphosis", Cambridge University Press, England.

Wigglesworth, V. B., 1964, The hormonal regulation of growth and reproduction in insects, in: "Advances in Insect Physiology", Academic Press, New York, Vol. 2, pp. 247-336.

Wigglesworth, V. B., 1970, "Insect Hormones", W. H. Freeman and Company, San Francisco.

Williams, C. M., 1961, The juvenile hormone II. Its role in the endocrine control of moulting, pupation and adult development in the Cecropia silkworm, Biol. Bull. Woods Hole 121:572-585.

Williams, C. M., 1976, Juvenile hormone . . . In retrospect and in prospect, in: "The Juvenile Hormones", L. I. Gilbert, ed., Plenum Press, New York, pp. 1-14.

Yagi, S., 1976, The role of juvenile hormone in diapause and phase variation in some lepidopterous insects, in: "The Juvenile Hormones", L. I. Gilbert, ed., Plenum Press, New York, pp. 288-300.

Yin, C. M., and Chippendale, G. M., 1979, Diapause of the southwestern corn borer, Diatraea grandiosella: further evidence showing juvenile hormone to be the regulator, J. Insect Physiol. 25:513-523.

AN IN VITRO APPROACH FOR INVESTIGATING THE REGULATION OF THE

CORPORA ALLATA DURING LARVAL-PUPAL METAMORPHOSIS

N. A. Granger, W. E. Bollenbacher and L. I. Gilbert

Department of Biological Sciences
Northwestern University
Evanston, Illinois 60201

INTRODUCTION

According to the now classical scheme for the hormonal control
of insect metamorphosis, the type of molt an insect undergoes is
determined by the concentration of juvenile hormone (JH) at a
critical period before the molt (see Gilbert, 1964). Generally, it
is believed that larval development occurs in the presence of high
concentrations of JH, and that in the presence of intermediate and
negligible concentrations of the hormone, pupal and adult development occur, respectively. Determinations of JH hemolymph titers
during the postembryonic development of a wide variety of holometabolous insects have tended to support this idea. Logically,
the mechanisms involved in the regulation of the JH titer would
involve modulation of its synthesis, its transport in the hemolymph,
catabolism and excretion.

Although little is known about JH excretion, and particularly
about its relationship to fluctuations in the JH hemolymph titer
(see Gilbert et al., 1980), information on the transport and
catabolism of JH in the hemolymph has expanded significantly in the
last five years, primarily due to the discovery in the hemolymph
of binding proteins for JH and of JH-specific esterases (Gilbert
et al., 1978). In one insect, Manduca sexta, the tobacco hornworm,
it has been proposed that the temporally precise fluctuations in
hemolymph esterase activity account for the decline in the JH titer
necessary for larval-pupal metamorphosis (Akamatsu et al., 1975).
However, JH titer data for the last larval instar of this insect
indicates that the concentration of the hormone begins to decrease
prior to the increase in JH esterase activity. In addition,

fluctuations in the JH titer during the penultimate larval instar apparently occur in the absence of these enzymes (Gilbert et al., 1978). Thus, mechanisms other than catabolism may contribute significantly to the modulation of the JH titer. The most obvious and probably the most important point at which regulation of the JH titer could occur is at the level of the gland, i.e., synthesis and/or release of the hormone by the CA.

The majority of the data on the control of JH biosynthesis by the CA are from studies on gland activity during the reproductive cycle of hemimetabolous insects (Doane, 1973). From these studies, several generalizations can be made about the types of control mechanisms regulating adult CA activity. First, a neurohumoral factor apparently activates the gland. This activation is possibly direct, in that the factor may be released from neurosecretory cell axons which terminate in the CA. Second, inactivation of the CA apparently is a nervous phenomenon, exerted via the non-neurosecretory axons which also innervate the glands.

These generalizations are based principally upon the results of studies employing an in situ approach, one from which it is difficult to make a distinction between the effect of nervous innervation and that of the direct release of neurosecretory factors, since both reach the gland via the same nerve bundle. In addition, biological assays are usually employed to assess putative CA regulation. Such assays are by nature indirect, CA activity being inferred rather than directly quantified. Recently, an in vitro approach has been taken to this problem in which JH synthesis by the glands is directly quantified, thus bypassing many of the shortcomings of the exclusively in situ approach (Tobe and Pratt, 1975; Tobe et al., 1977; Stay and Tobe, 1980). The data from studies utilizing this in vitro protocol have provided the most compelling evidence for control mechanisms for JH synthesis by the adult CA. In direct contrast to the many studies conducted on the regulation of adult CA, very few have dealt with the control of these glands in immature stages. In this chapter, the studies on the regulation of the larval CA will be reviewed and the development of a new in vitro approach to this problem will be discussed. In addition, preliminary data generated with this in vitro system suggesting the existence of cerebral regulatory factors for the CA will be presented.

REGULATION OF THE LARVAL CA

Nearly fifty years ago, Wigglesworth (1934; 1936) suggested that the nervous connection of the CA to the brain via the corpora cardiaca (CC) provided a structural basis for the control of these glands by the brain. Some of the earliest studies which supported this suggestion investigated the effect of brain extirpation on the activity of the larval CA. Removal of the brain from a nymph of the earwig, Anisolabis maritima, was found to inactivate the CA (Ozeki, 1959), while the same manipulation in larvae of the silkworm, Bombyx mori, resulted in gland activation (Fukuda, 1962). Humoral control of the CA was first suggested when it was found that inactive CA from last instar nymphs of Anisolabis were activated when cultured in penultimate instar nymphs (Ozeki, 1962) and that the activity of transplanted CA in Bombyx larvae often corresponded to the developmental stage of the host (Fukuda, 1962). The source of the humoral factor(s) affecting larval CA activity was assumed to be the brain, and certain neurosecretory cells in the brains of Locusta migratoria cinerascens nymphs were subsequently implicated as the source of both stimulatory (allatotropins) and inhibitory (allatohibins) factors for gland activity (Girardie, 1965; 1967).

Recently, a study of the control of the CA in larvae of the wax moth, Galleriamellonella, initiated by the observation (Sehnal, 1966; Pipa, 1971) that the implantation of brains into freshly ecdysed last instar larvae elicited supernumerary larval molts, further supported the concept of humoral control of the CA. Implanted brains did not have an effect on the pupal development of alla- tectomized larvae, suggesting that the brains stimulated the host CA to produce the extra larval molt. The response of host larvae with intact CA to implanted brains was dose dependent, i.e., in- creasing numbers of supernumerary larvae were obtained with in- creasing numbers of brains (Fig. 1), and the number of brains necessary to elicit the expected response, that is, an extra larval molt, varied according to the developmental stage of the brain. Specifically, for brains from the penultimate and last larval instar, stimulatory activity was highest during the first part of each instar (Fig. 1). From these data, it appeared that a neurohormonal factor from the implanted brain, an allatotropin, activated the CA. Furthermore, if the pars intercerebralis, a region of the brain containing a large heterogeneous group of neurosecretory cells, was cauterized before implantation, significantly fewer extra larval molts resulted, suggesting that this area of the brain was the source of the factor (Granger and Sehnal, 1974; Sehnal and Granger, 1975).

Although the results of these studies were straightforward, the conclusions drawn from the data could have been interpreted differently, i.e., prothoracicotropic hormone from the implanted brains may have activated the prothoracic glands precociously at a time when the JH titer was still high, thus eliciting a super- numerary larval molt (Pipa, 1977). However, three additional lines of evidence support the conclusion that an allatotropin was the neurohemal factor released by the brains. First, a precocious stimulation of the prothoracic glands by implanted brains was not evident from the results of the original experiments. The few larvae which developed into normal pupae after brain implantation,

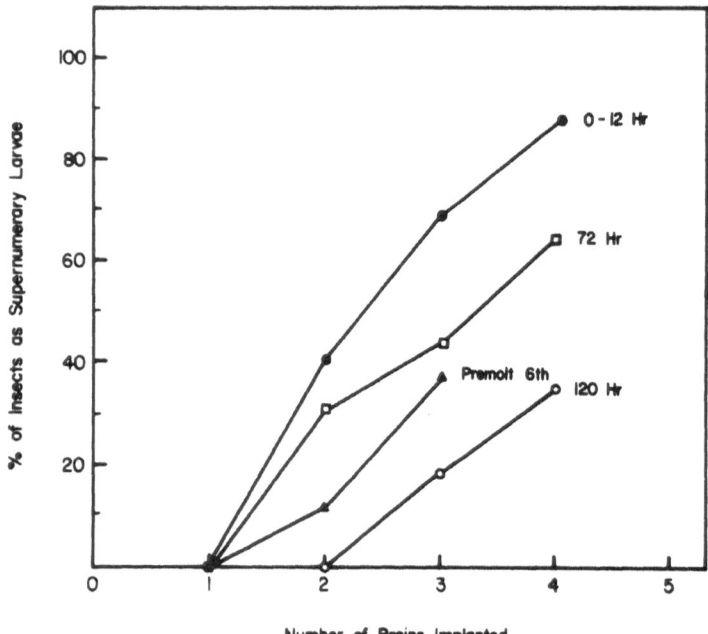

Fig.1   Percent supernumerary larval development in <u>Galleria</u>
        <u>mellonella</u> as a function of the number of brains implanted
        into freshly ecdysed last instar larvae. Larval hosts were
        0-12 hr post ecdysis, and brains were taken from penultimate
        and last instar larvae. Premolt 6th, sixth (penultimate)
        instar larva immediately prior to the last larval molt;
        0-12 hr, seventh (last) instar larva 0-12 hr after ecdysis;
        72 hr, last instar larva 72 hr after ecdysis; 120 hr, last
        instar larva 120 hr after ecdysis. (Granger, unpublished)

as well as allatectomized larvae implanted with as many as six
brains, pupated in the same length of time as sham-operated animals
(Sehnal and Granger, 1975).   Second, stimulation of a molt by brain
implantation into debrained, allatectomized larvae (neck-ligated)
required approximately two weeks.   Third, developmental changes in
cerebral allatotropic activity did not parallel fluctuations in
the prothoracicotropic activity of the <u>Galleria</u> brain during the
last two larval instars (Malá et al., 1977).

   In addition to humoral control, nervous regulation of the
larval CA may also exist.   In late last instar <u>Galleria</u> larvae, the
CA become refractory to the allatotropic activity of implanted
brains, and thus presumably to that of the <u>in situ</u> brain.   Nerve

severence studies suggested that nervous inhibition of the glands
is exerted via the nerves connecting the CA to the brain and sub-
esophageal ganglion (Sehnal and Granger, 1975). Although the brain
may be the main source of the inhibition, the integrity of the
nervous connections between the brain and the rest of the nervous
system is required for the inhibitory effect (Sehnal and Granger,
unpublished results). In Manduca sexta, both nervous and neuroendocrine control of the larval CA also appears to exist. Allo-
metry may have an important role in this regulation since when the
larva attains a critical weight (size), the CA are inactivated
(Nijhout and Williams, 1974a;b). The inactivation of the CA in
the last larval instar appears to occur as the result of curtailed
secretion of an allatotropin by the brain and the simultaneous
secretion of a cerebral allatohibin (Williams, 1976). It has been
suggested that this turnover in the neurohormonal control of the CA
is triggered by the release of a factor from an unknown source in
the abdomen. Evidence supporting the existence of this abdominal
factor derives from the observation that larval brain-CC-CA com-
lexes possessing active CA synthesize less JH in vitro when cultured
with hemolymph taken from early in the last instar (days 2-4).
Direct proof for the existence of allatotropins and allatohibins,
as well as of the abdominal factor, in the control of CA activity
in Manduca is still lacking, but may soon be forthcoming due to
the development of new methods and experimental designs with which
to approach the problem (see subsequent discussion and Bhaskaran,
this volume).

Clearly, the dearth of information on the control of the larval
CA is a result of the type of assay employed. These assays have
been mainly biological and, as previously stated, are indirect by
definition. They are also frequently cumbersome, non-specific, and
subject to interpretive variability. These various problems inherent
in an in vivo approach to quantifying CA activity, i.e., rates of
JH synthesis, can be avoided with an in vitro protocol. The advantages of an in vitro approach are several: 1) synthesis of JH
in vitro can be measured directly, precisely and reproducibly; 2)
the effect of potentially complex endocrine interrelationships is
eliminated; 3) non-specific effects on CA activity are minimized
in vitro, maximizing the potential for resolution of specific
regulatory factors for gland activity; and 4) donor selection for
CA maximally sensitive to putative allatotropins and allatohibins
is possible.

Of several methods now available for quantifying JH synthesis
by CA in vitro, the JH radioimmunoassay (RIA) is best suited for
in vitro studies of the regulation of CA. The RIA is characterized
by a high degree of specificity, is rapid and easy to perform, and
requires minimum equipment. This is in marked contrast to current
analytical chemical methods employed to quantify JH. In addition,

with the availability of RIA's for the three JH homologs, there
exists the capacity to measure simultaneously the synthesis of the
three JH homologs. Such simultaneous measurements are not easily
made with the radioenzymological assay for JH (Tobe and Pratt, 1975).
Also, the RIA measures total JH synthesis, while the radioenzymological assay quantifies JH synthesis by measuring the rate of a
terminal step in the biosynthetic pathway. This is an important
difference, in that changes in rate of methylation of the JH molecule may not accurately reflect actual changes in gland activity,
although variations in gland activity found thus far with the
radioenzymological assay do correlate with overt JH-mediated events.

Only one in vitro system has been developed for culture of
the CA in which a JH RIA is employed for quantifying rates of
hormone synthesis (Granger et al., 1979). This system has been
used to study the JH biosynthetic activity of the larval CA of
Manduca sexta and provides the basis on which studies of the regulation of the larval glands have been initiated. The remainder of
the chapter will deal specifically with the development of this in
vitro assay for CA activity and its application to investigations of
CA activity during larval-pupal metamorphosis and of the possible
neuroendocrine regulation of this activity.

DEVELOPMENT OF THE IN VITRO ASSAY FOR CA ACTIVITY

The successful development of an in vitro culture system for
studying the regulation of the larval CA depended on the establishment of culture conditions that would ensure both maintenance of
gland activity and accurate measurement of JH biosynthesis. The
latter was a critical factor to consider in light of the unusual
physical properties of JH which make it difficult to keep this
molecule quantitatively in solution. Marks medium 19AB was ultimately selected because of the previous successful maintenance of
both Galleria brains and CA in vitro in this medium and because
loss of JH from this medium was minimal up to 6 hr and less than
30% after 48 hr. In addition, polyethylene glycol-coated glass
culture vessels were used to ensure that JH would remain in solution
by minimizing nonspecific adsorption of the hormone to the culture
vessel.

Once the appropriate in vitro conditions for the culture of
the CA had been established, it was then necessary to adapt the JH
I RIA for the direct measurement of JH in culture medium. It was
critical that the conditions for maximum RIA sensitivity be defined
since previous studies have shown that JH biosynthesis by the
Manduca larval CA in vitro is fairly low (Dahm et al., 1976). The
protocol for the JH I RIA is described in Fig. 2. The K_d for the
JH I antibody in this assay was 1.58×10^{-9} M, a value which suggested a lower limit of sensitivity of approximately 50 pg JH I.

This, in fact, is what was obtained. The ligand specificity of the JH I RIA was of particular practical importance since all three homologs of JH appear to be synthesized in vitro by Manduca larval CA. This meant that it might be possible to quantify differentially all three homologs or specifically measure only one. A determination of the cross reactivity of the JH I antibody with the JH homologs, as well as with analogs and metabolites, revealed that the JH I RIA has a high degree of both homolog and isomeric specificity (Table 1) (Granger et al., 1979). Other than JH I, and to a lesser degree JH II, the JH I acid was the only compound which cross reacted significantly. The equivalency of the cross reactivity of the JH I acid to that of JH I was not unexpected since the hapten (JH I) to which this antibody was generated had been conjugated via its ester function. The cross reactivity of the JH acid proved to be an attribute of the JH I RIA in that esteratic conversion of newly synthesized JH I to JH I acid in vitro would not affect an accurate measurement of the total amount of JH biosynthesized.

Once the culture conditions for the CA and the protocol for the JH RIA had been established, a time course of JH synthesis and release by the CA was determined in order to demonstrate the validity of the approach and its potential for probing the existence of putative allatotropins and allatohibins.

For the initial kinetics of synthesis, CA from wandering stage larvae (day 5) were chosen because these larvae can be conveniently and precisely staged and because their CA are large and easy to extirpate. Most importantly, preliminary screening of gland activity revealed that JH synthesis by day 5 CA in vitro can be quantified within 1 hr of culture, and that synthesis by as few as two pairs can be reproducibly determined. Data from a kinetics study using six pairs of CA revealed that synthesis was linear for 4-6 hr, with an evident plateau at $\sim$ 12 hr (Fig. 3). Although it appeared from the kinetics data that the JH in the culture medium was a product of synthesis, it was also possible that these data in part or totally represented release of hormone stored in the gland. To show that synthesis was occurring rather than release, aqueous and organic solvent extracts of CA were assayed for JH I RIA-active material. Significant levels of JH RIA activity were not present in these extracts, demonstrating that the RIA activity in the culture medium resulted from actual synthesis of the hormone. In addition, this result revealed that storage and release of JH are probably not points at which CA activity is controlled. These

Assay conditions: 1) Unlabelled ligand

                     2) Culture medium 19AB (100 μl stock
                        or 100 μl from cultures of corpora
                        allata)

                     3) Labelled ligand (approximately
                        6.5×10^3 cpm ^{3}H-JH I in 100 μl
                        phosphate buffer pH 7.2, containing
                        0.1% BSA and 0.1% sodium azide)

                     4) JH I antiserum (0.04% antiserum in
                        100 μl 0.2% rabbit IgG in phosphate
                        buffer containing BSA and sodium
                        azide)

                                    |

            Incubation: 11.5 hr at 4°C

                                    |

      Termination: 300 μl 100% $(NH_4)_2SO_4$; 20 min at 4°C

                                    |

      Centrifugation: 2000 x g for 10 min at 4°C

                                    |
        _____|_____

Supernatant (discard)   Pellet

                                        |
                                        | Washes: 2 x 600 μl
                                        | with 50% $(NH_4)_2SO_4$
                                        |
                                        | Centrifugation: 2000 x g;
                                        | 10 min at 4°C
                                        |
                        _____|_____

           Supernatant (discard)   Pellet

                                            LSC: 1) 25 μl H_2O
                                                 2) 575 μl Aquasol

        Fig.2   Protocol for the JH I radioimmunoassay.

Table 1. Cross reactivity of JH homologs and analogs in the modified RIA for JH I.

HORMONE/ANALOG		% CROSS REACTIVITY[1]
(structure) COOCH₃	JH O	37.0
(structure) COOCH₃	JH I	100
(structure) COOCH₃	JH II	12.6
(structure) COOCH₃	JH III	1.7
(structure) COOH	JH I ACID	100
(structure) COOCH₃	JH I DIOL	0.5[2]
(structure) E,E,cis 17%	JH I MIXED ISOMERS	37.2
(structure) COOH	LINOLEIC ACID	0
(structure) COOCH₃	METHYL LINOLEATE	0

1. % Cross reactivity represents the number of nanograms of JH I necessary to compete off 50% of the labelled ligand divided by the number of nanograms of the homolog or analog necessary for 50% competition, x 100. [10-³H]-JH I was used as the labelled ligand.

2. Competition study done with phosphate buffer instead of 19AB. Virtually identical cross reactivities were observed for all homologs and analogs with the phosphate buffer RIA, except for methyl linoleate (0.6%) (see text) (from Granger et al., 1979).

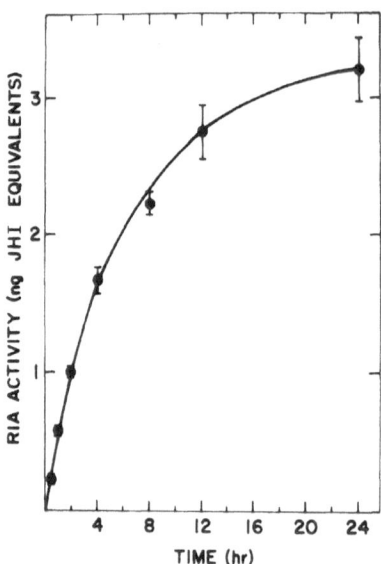

Fig. 3   In vitro time course of JH synthesis by CA from day 5
          last instar larvae of Manduca sexta. Each datum point
          (+ SEM) is the mean of four cultures, each containing six
          pairs of CA (from Granger et al., 1979).

findings were consistent with those from previous studies on adult
Orthopteran CA in vitro in which JH was apparently released by
diffusion immediately following its synthesis.

    With the demonstration of synthesis and release of an ap-
parent JH in the in vitro system, it became necessary to identify
chemically the RIA active material synthesized by the day 5 CA.
High pressure liquid chromatography of organic solvent extracts of
multiple CA cultures coupled with RIA analysis of the chromatographi-
cally identified fractions revealed that JH I RIA activity represented both JH I and JH II in a ratio of approximately 1:4. JH III
was not detected, but this was due to the fact that in the JH I
RIA, JH III does not cross react. RIA of the same culture medium
using a JH III antibody revealed that $\sim$ 1 ng JH III is synthesized
per pair of day 5 CA (Granger, unpublished). Since $\sim$ 0.5 ng JH I
is synthesized in vitro by a pair of day 5 CA, the ratio of JH I:
JH II:JH III synthesis by these CA is approximately 1:4:2.

    With the establishment of a viable in vitro protocol for
assessing rates of JH synthesis by larval CA, the assay system could
then be utilized for an investigation of the control of CA activity.

DEVELOPMENTAL CHANGES IN JH BIOSYNTHESIS IN VITRO

A necessary antecedent to the proposed investigation of CA regulation was a developmental study of CA activity during larval-pupal metamorphosis, since glands with particular biosynthetic activities were needed for the demonstration of putative regulatory factors. Inactive CA would theoretically be maximally sensitive to allatotropins and active CA to allqtohibins. The experimental protocol was designed to generate time courses of JH synthesis for both isolated CA and brain-CC-CA complexes from penultimate and last instar larvae and early pupae and then to determine relative rates of synthesis at a time when all rates were linear. The kinetics data revealed that during the last two instars and the early pupal period, the rates of JH synthesis by the CA in brain-CC-CA complexes was linear for at least 6 hr, and the rates varied depending on the stage of the donor (Fig. 4). This difference in rates alone was suggestive of a modulation in CA activity during development. The developmental variation in the rate data is more clearly illustrated in Figure 5, where JH synthesis in 6 hr is compared for each stage. Synthetic activity was relatively high at the beginning of each larval instar and decreased as the instar progressed. A second increase in CA activity was observed in the last instar at ∿ day 7, and during the early pupal period, JH I synthesis was not detectable.

When compared to changes in the CA activity of brain-CC-CA complexes, fluctuations in the JH biosynthetic activity of CA alone were temporally the same; however, quantitatively they were not as dramatic (Fig. 6). These absolute differences in the rates of JH synthesis by isolated CA in comparison to brain-CC-CA complexes argue for a direct involvement of the brain in modulating CA activity, an involvement that can affect the glands in both a positive and negative way. Examples of this exist at head cap slip stage of the fourth instar when the isolated CA synthesize approximately three times the JH as CA complexed with the brain, and at day 7 of the fifth instar when isolated CA synthesize only 75% that synthesized by CA in a complex. A synthesis ratio over time of JH produced by two complexes to that by two pairs of CA clearly demonstrates that the brain has the capacity to affect CA activity both positively and negatively (Fig. 7).

Overall, these developmental data on CA activity have provided the information requisite to an in vitro investigation of the humoral control of the CA. To demonstrate the existence of an allatotropin, CA from day 3 of the fourth instar or day 4 of the fifth would be ideal since the rate of JH synthesis by these glands is low. To demonstrate an allatohibin, CA from either days 0-2 or day 7 of the fifth instar could be used because their JH biosynthetic activity is the highest.

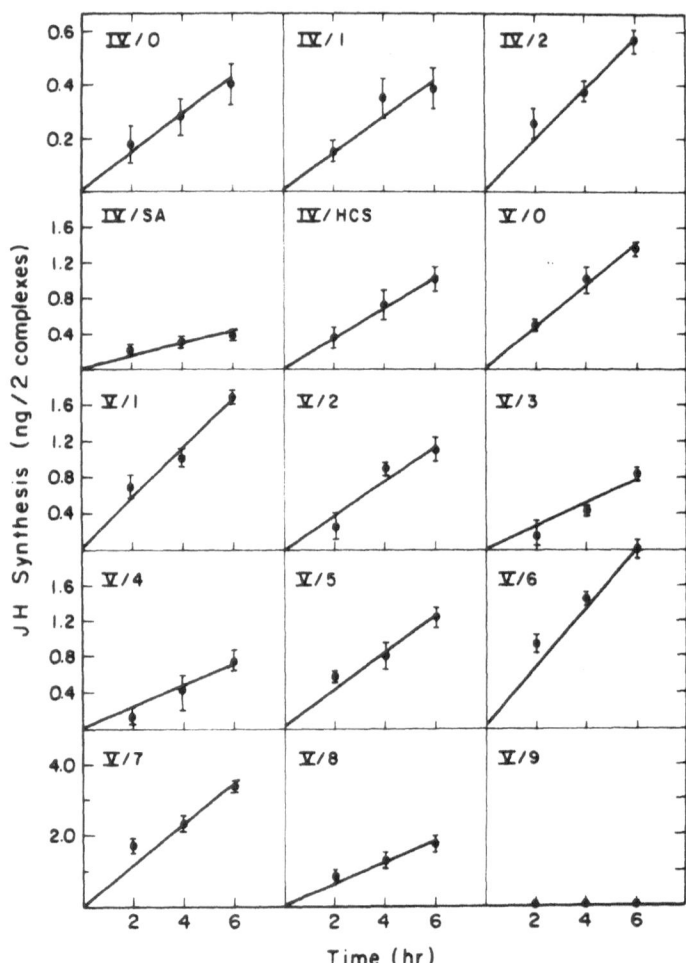

Fig.4   In vitro time course of JH synthesis by Manduca sexta
        brain—CC—CA complexes during the penultimate and last larval
        instars. JH synthesis is expressed as ng JH I RIA equiva-
        lents. Stages used were days 0-2, fourth instar (IV/0-2):
        spiracle apolysis, fourth instar (IV/SA); head cap slip,
        fourth instar (IV/HCS); and days 0-9, fifth instar (V/0-9).
        Each datum point is the mean of six cultures (± SD), each
        containing four (fourth instar) or two (fifth instar)
        complexes.

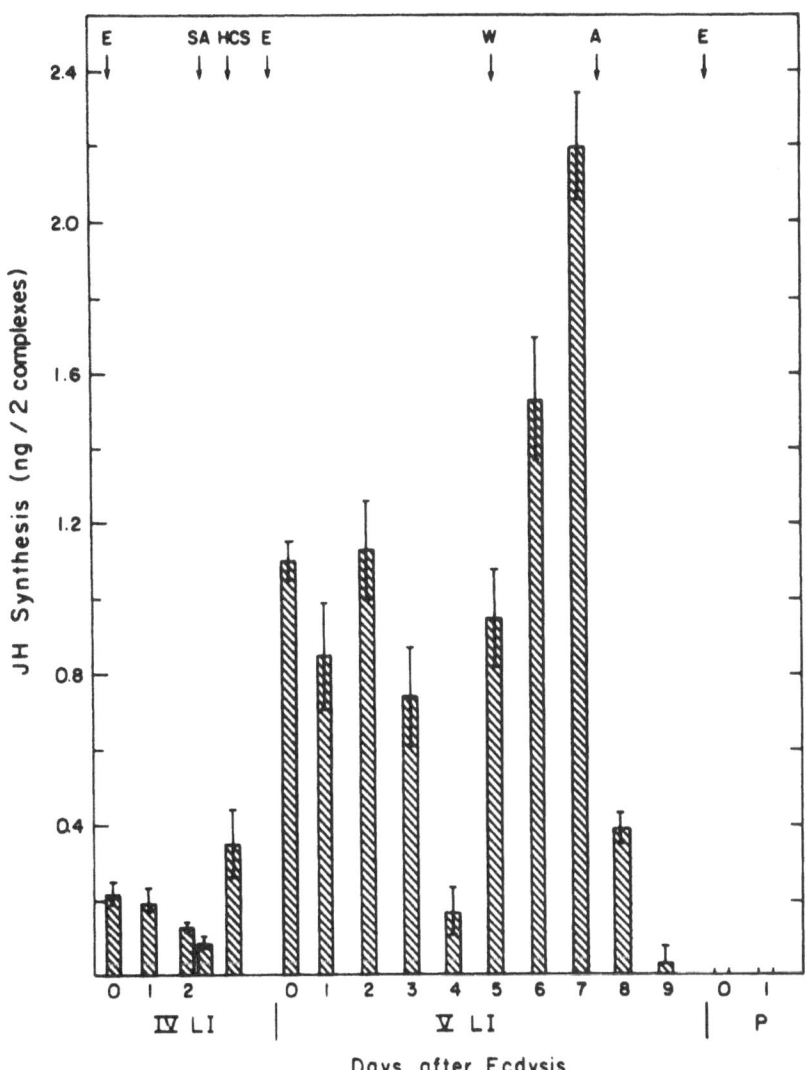

Fig.5   In vitro JH biosynthetic activity of Manduca sexta brain-
CC-CA complexes during the penultimate and last larval
instars and early pupal period. JH synthesis is expressed
as ng JH I RIA equivalents synthesized in 6 hr. IV LI,
fourth larval instar; V LI, fifth larval instar; P, pupa;
E, ecdysis; SA, spiracle apolysis; HCS, head cap slip; W,
wandering stage; A, apolysis. Each datum point is the mean
of six cultures (± SD), each containing four complexes
(fourth instar) or two complexes (fifth instar, early
pupal period).

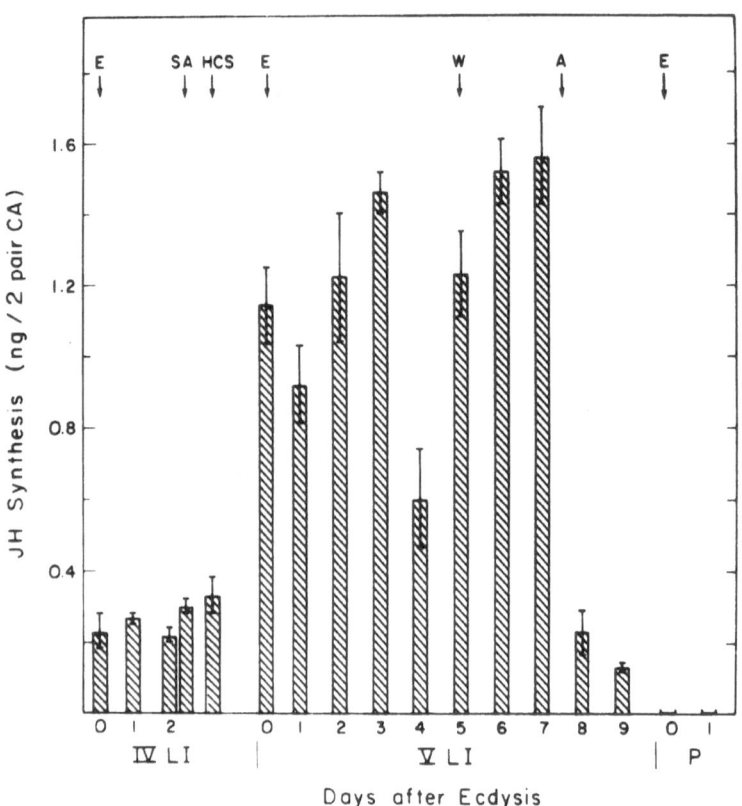

Fig. 6   In vitro biosynthetic activity of isolated Manduca sexta
CA from the penultimate and last larval instars and the early
pupal period. JH synthesis is expressed in ng JH I RIA
equivalents synthesized in 6 hr. IV LI, V LI, P, E, SA,
HCS, W, and A as in Fig. 5. Each datum point ($\pm$ SD) is
the mean of six cultures each containing four pairs of
CA (fourth instar) or two pairs of CA (fifth instar, early
pupal period).

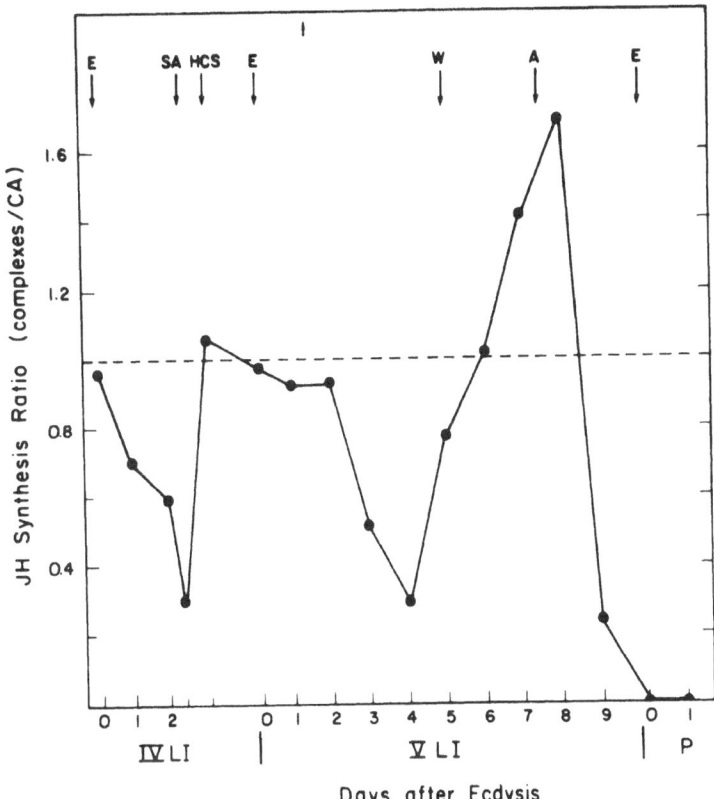

Fig. 7   Ratios of the synthesis of JH by brain-CC-CA complexes to
          that by isolated CA during the penultimate and last larval
          instars and early pupal period. IV LI, V LI, P, E, SA,
          HCS, W, and A as in Fig. 5.

RELATIONSHIP OF CA ACTIVITY IN VITRO AND THE JH TITER IN VIVO

        In addition to providing information critical for a study of
the control of the CA, the developmental profile of CA activity
in vitro (Figs. 5, 6) demonstrates that the modulations in CA
activity contribute to the hemolymph titer of JH in Manduca during
this same period. When relative synthesis by brain-CC-CA from
the fourth and fifth larval instars is compared to a JH titer derived from the combined data of several partial titers (Riddiford
and Truman, 1978), it is immediately evident that there is general
agreement between the two sets of data (Fig. 8). Although there
appears to be one major temporal discrepancy late in the last instar, this can be easily explained if the conditions under which
the larvae were reared are considered. In another lepidopteran,

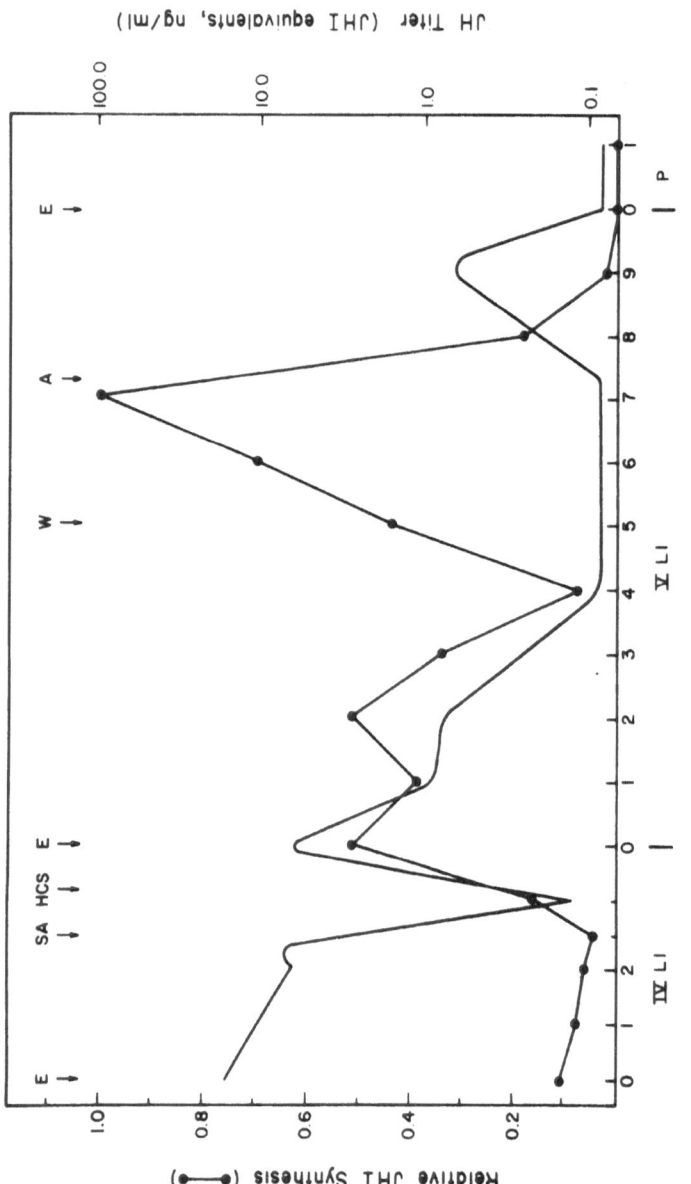

Fig. 8   Comparison of in vitro JH biosynthetic activity of Manduca sexta brain–CC–CA complexes from the penultimate and last larval instars and early pupal period to the JH hemolymph titer. JH synthesis by complexes (●—●) is expressed relative to the value for day 7 of the fifth instar. JH titer data (——) is expressed as ng JH I per ml hemolymph. JH hemolymph titer from Riddiford and Truman (1978). IV LI, V LI, P, E, SA, NCS, W, and A as in Fig. 5.

it has been observed that the peak in the JH titer at the end of the
last instar occurs 24-48 hr later in larvae reared under short-day
(diapausing) conditions than in larvae reared under long-day conditions (Yagi, 1976). Since the Manduca JH hemolymph titer was determined in larvae reared under short-day conditions, while the CA
activity profile was obtained with larvae reared under long-day
conditions, the peak in CA biosynthetic activity around day 7 and
the peak in the JH hemolymph titer around day 9 may in fact be
coincidental. Although changes in the JH titer and CA activity
in vitro are temporally similar for the fourth and early fifth larval
instars, there are quantitative discrepancies in the relative CA
activities during these two instars in comparison to the JH titer.
For example, relative synthesis by CA at the beginning of the fourth
instar is not equivalent to that at the beginning of the fifth,
although the JH hemolymph titers are the same at these times. However, if CA activity is considered relative to larval size and thus
to hemolymph volume, the quantitative differences in CA activity
become more apparent than real. To demonstrate this, a relative JH
titer was extrapolated from the in vitro JH synthesis data and the
hemolymph volumes for each of five developmental stages (Table 2).
The resulting peaks at the beginning of the fourth and fifth instars
were approximately the same in this relative titer as were the peaks
in the actual JH hemolymph titer (Fig. 8). Furthermore, the relative JH titer on day 5 was actually quite low, which is consistent
with the low level of JH in the hemolymph at this time. Thus, it
appears that in addition to modulated gland activity and catabolism
of JH, growth of the animal, as represented by its increasing
hemolymph volume, can contribute in a passive way to the regulation
of the JH titer.

NEUROENDOCRINE CONTROL OF THE CA

Initial investigations to demonstrate the involvement of neuroendocrine mechanisms in the regulation of CA activity have utilized
the established in vitro system for Manduca larval CA and yielded
quite promising data. The results thus far suggest the existence
of both allatotropins and allatohibins that specifically affect
the synthesis of individual JH homologs.

Studies directed at probing the interaction of the brain and
CA from specific stages of development were first conducted to
determine if JH synthesis by a given CA could be either positively or
negatively affected by co-culture with brain-CC complexes from the
same and different stages of development. An example of this
experimental design is presented in Table 3, where the effect of day
0 fifth instar brains on the relatively inactive day 4 CA and of
day 4 brains on active day 0 CA was determined. For these studies,
both JH I and JH III synthesis was monitored with specific RIA's
for each hormone (Granger, unpublished). When the rate of synthesis

Table 2. Comparison of in vitro rates of JH synthesis by CA from
          Manduca sexta larvae at different developmental stages
          in relation to their hemolymph volumes.

Stage[a]	JH Synthesis[b]		Relative JH Titer[c]
	per gland pair	per ml hemolymph	
IV/1	0.18 + 0.02	1.05	1.0
IV/SA	0.06 + 0.02	0.16	0.16
Pharate V	0.33 + 0.01	1.03	0.99
V/1	0.38 + 0.01	0.46	0.49
V/5	0.40 + 0.03	0.15	0.14

[a]Stages used were day 1 of the fourth instar (IV/1); day 3 of
the fourth instar, early spiracle apolysis (IV/SA); day 3 of the
fourth instar, pharate fifth instar (pharate V); day 1 of the
fifth instar (V/1); day 5 of the fifth instar (V/5).

[b]ng JH I eqivalents ($\pm$ SEM) after 6 hr in vitro; values per ml
hemolymph were obtained by dividing the rate data per gland
pair by the hemolymph volume of the animals.

[c]The relative titer data was derived by normalizing the values
per ml hemolymph to the value for day 1 of the fourth instar.
Catabolic and excretory variables were deliberately not considered so that the extrapolated titer would truly reflect the
rates of JH synthesis in relationship to the size and hemolymph volume of the larva (Table after Granger et al., 1979).

by day 4 CA co-cultured with day 0 brains was compared to the
rate by day 4 CA cultured alone (synthesis ratio), there was a
2.5-fold increase in the rate of synthesis of JH I, but no increase
in the synthesis of JH III. The result is consistent with the
hypothesis that a day 0 brain would possess a high level of an allatotropin to which the relatively inactive day 4 CA might respond.
Conversely, the day 4 brains appear to inhibit the synthesis of JH
I by the day 0 glands by $\sim$ 50%, indicating the involvement of an
allatohibin, which might be predicted on the basis of the titer of

Table 3. Stimulation and inhibition of JH I synthesis _in vitro_
          by CA co-cultured with brain-corpora cardiaca complexes.

Culture System[a]	JH I RIA Equivalents[b]	JH III RIA Equivalents[b]
V/0 CA	1.00 $\pm$ 0.05	1.43 $\pm$ 0.03
V/0 CA + V/0 BRCC	0.75 $\pm$ 0.04	2.01 $\pm$ 0.09
V/4 CA	0.14 $\pm$ 0.06	1.15 $\pm$ 0.03
V/4 CA + V/4 BRCC	0.24 $\pm$ 0.05	1.67 $\pm$ 0.10
V/4 CA + V/0 BRCC	0.36 $\pm$ 0.11	1.46 $\pm$ 0.23
V/0 CA + V/4 BRCC	0.46 $\pm$ 0.03	1.57 $\pm$ 0.18

Synthesis Ratios

$\dfrac{\text{V/4 CA + V/0 BRCC}}{\text{V/4 CA}}$	2.57	1.27
$\dfrac{\text{V/0 CA + V/4 BRCC}}{\text{V/0 CA}}$	0.46	1.10

[a]V/0, day of ecdysis to the fifth instar; V/4, day 4, fifth instar;
BRCC, 2 brain-corpora cardiaca complexes; CA, 2 pair corpora
allata.

[b]ng $\pm$ SEM.

JH biosynthesis by CA and brain-CC-CA complexes. Of considerable
importance was the observation that JH III synthesis was not affected in these cultures, a result which suggests a considerable
degree of specificity in the activation and inhibition of JH I
synthesis. This result alone argues for the existence of allatotropins and allatohibins specific for a single JH homolog.

     The same approach was taken utilizing glands and brains from
day 7 fifth instar larvae and freshly ecdysed pupae (Table 4). In
this case, the brain never appeared to affect JH I synthesis, while

Table. 4.   Stimulation of JH III biosynthesis in vitro by CA co-
cultured with brain-corpora cardiaca complexes.

Culture System[a]	JH I RIA Equivalents[b]	JH III RIA Equivalents[b]
V/7 CA	2.05 $\pm$ 0.05	4.70 $\pm$ 0.09
V/7 CA + V/7 BRCC	1.99 $\pm$ 0.09	3.63 $\pm$ 0.13
P/O CA	0	0
P/O CA + P/O BRCC	0.04 $\pm$ 0.03	0.96 $\pm$ 0.07
P/O CA + V/7 BRCC	0.09 $\pm$ 0.05	0.27 $\pm$ 0.01
V/7 CA + P/O BRCC	2.03 $\pm$ 0.05	10.86 $\pm$ 0.23

Synthesis Ratios

$\dfrac{\text{V/7CA + P/O BRCC}}{\text{V/7 CA}}$	0.99	2.99
$\dfrac{\text{P/O CA + V/7 BRCC}}{\text{P/O CA}}$	0	0

[a]V/7, day 7, fifth instar; P/O, day of ecdysis to pupa; BRCC, 2
brain-corpora cardiaca complexes; CA, 2 pair corpora allata.

[b]ng $\pm$ SEM.

a slight stimulation of JH III synthesis by the pupal glands may
have occurred in the presence of day 7 brains. However, when day
7 larval CA were co-cultured with pupal brains, a significant
three-fold stimulation of JH III synthesis resulted. As in the
previous study, a stimulatory effect on the synthesis of an individual JH homolog was observed, which strongly suggests that specific
neurohumoral factors are involved in the modulation of CA activity.

     Although these data must be considered preliminary, they
serve to emphasize the overall potential of this in vitro approach
for answering questions about the existence of neuroendocrine
mechanisms for the control of CA activity. In fact, based on these

initial findings, it appears that this system may permit a demonstration of neurohormonal modulation of the synthesis of the individual JH homologs at specific stages of development.

SUMMARY

The present approach for elucidating the regulation of the larval CA has been designed to maximize the possibility of demonstrating the existence of regulation and the means by which it is exerted. With this approach, differential rates of JH synthesis were measured in cultures of CA with and without the brain, strongly suggesting that regulatory factors stimulating (allatotropins) and inhibiting (allatohibins) CA activity do indeed exist, and preliminary evidence has been obtained suggesting the existence of allatotropins specific for JH I and JH III and an allatohibin specific for JH I. Thus, it is anticipated that this in vitro approach will provide the means by which to investigate the endocrinology of CA regulation.

ACKNOWLEDGEMENTS

This research was supported by grants 1 RO1 NS 14816-01 from the National Institutes of Health, PCM-76-23291 and PCM-76-23201-A01 from the National Science Foundation and a grant from the Whitehall Foundation. The authors would like to thank Sheryl Niemiec for her expert technical assistance and for the photography, and Nancy Grousnick for her secretarial assistance.

REFERENCES

Akamatsu, Y., Dunn, P. E., Kezdy, F. J., Kramer, K. J., Law, J. H., Reibstein, D., and Sanburg, L. L., 1975, Biochemical aspects of juvenile hormone action in insects, in: "Control Mechanisms and Development", R. H. Meintz and E. Davies, ed., Plenum Press, New York, pp. 123-149.

Dahm, K. H., Bhaskaran, G., Peter, M. G., Shirk, P. D., Seshan, K. R., and Röller, H., 1976, On the identity of the juvenile hormone in insects, in: "The Juvenile Hormones", L. I. Gilbert, ed., Plenum Press, New York, pp. 19-47.

Doane, W. W., 1973, Role of hormones in insect development, in: "Developmental Systems: Insects", S. J. Counce and C. H. Waddington, ed., Academic Press, New York, pp. 291-497.

Fukuda, S., 1962, Secretion of juvenile hormone by the corpora allata in pupae and moths of the silkworm, Bombyx mori, Annot. Zool. Japan 35:199-212.

Gilbert, L. I., 1964, Physiology of growth and development: Endocrine aspects, in: "Physiology of Insecta", M. Rockstein, ed., Academic Press, New York, 1:149-225.

Gilbert, L. I., Bollenbacher, W. E., and Granger, N. A., 1980, Insect endocrinology: Regulation of endocrine glands,

hormone titer and hormone metabolism, Ann. Rev. Physiol.
42:493-510.

Gilbert, L. I., Goodman, W., and Granger, N. A., 1978, Regulation
of juvenile hormone titer in Lepidoptera, in: "Comparative
Endocrinology", P. J. Gaillard and H. H. Boer, ed., Elsevier/
North Holland, Amsterdam, pp. 471-486.

Girardie, A., 1965, Contribution à l'étude du controle de l'activité
des corpora allata par la pars intercerebralis chez Locusta
migratoria (L.), C. R. Acad. Sci. Paris 261:4876-4878.

Girardie, A., 1967, Controle neuro-hormonal de la métamorphose et
de la pigmentation chez Locusta migratoria cinerascens, Bull.
Biol. Fr. Belg. 101:79-114.

Granger, N. A., Bollenbacher, W. E., Vince, R., Gilbert, L. I.,
Baehr, J. C., and Dray, F., 1979, In vitro biosynthesis of
juvenile hormone by the larval corpora allata of Manduca sexta:
Quantification by radioimmunoassay, Mol. Cell. Endocrinol. 16:
1-17.

Granger, N. A., and Sehnal, F., 1974, Regulation of larval corpora
allata in Galleria mellonella, Nature 251:415-417.

Malá, J., Granger, N. A., and Sehnal, F., 1977, Control of prothoracic gland activity in larvae of Galleria mellonella, J.
Insect Physiol. 23:309-316.

Nijhout, H. F., and Williams, C. M., 1974a, Control of moulting
and metamorphosis in the tobacco hornworm, Manduca sexta
(L.): Growth of the last-instar larva and the decision to
pupate, J. Exp. Biol. 61:481-491.

Nijhout, H. F., and Williams, C. M., 1974b, Control of moulting
and metamorphosis in the tobacco hornworm, Manduca sexta (L.):
Cessation of juvenile hormone secretion as a trigger for
pupation, J. Exp. Biol. 61:493-501.

Ozeki, K., 1959, Further studies on the effects of the corpus allatum
hormone on the development of the genital organs in males of
the earwig, Anisolabis maritima, Sci. Pap. Coll. Gen. Educ.
Univ. Tokyo 9:127-134.

Ozeki, K., 1962, Studies on the secretion of the juvenile hormone
in the earwig, Anisolabis maritima, Sci. Pap. Coll. Gen. Educ.
Univ. Tokyo 12:65-72.

Pipa, R., 1971, Neuroendocrine involvement in the delayed pupation
of space-deprived Galleria mellonella (Lepidoptera), J. Insect
Physiol. 17:2441-2450.

Pipa, R. L., 1977, Do the brains of wax moth larvae secrete an
allatotropic hormone?, J. Insect Physiol. 23:103-107.

Riddiford, L. M., and Truman, J. W., 1978, Biochemistry of insect
hormones and insect growth regulators, in: "Biochemistry of
Insects", M. Rockstein, ed., Academic Press, New York, pp.
308-357.

Sehnal, F., 1966, Kritisches Studium der Bionomie und Biometrik der
in verschiedenen Lebens bedingunger gezüchteten Wachsmotte,
Galleriamellonella L. (Lepidoptera), Z. Wiss. Zool. 174:53-82.

Sehnal, F., and Granger, N. A., 1975, Control of corpora allata
    function in larvae of Galleria mellonella, Biol. Bull. 148:
    106-116.
Tobe, S. S., Chapman, C. S., and Pratt, G. E., 1977, Decay in
    juvenile hormone biosynthesis by insect corpus allatum after
    nerve transection, Nature 268:728-730.
Tobe, S. S., and Pratt, G. E., 1975, Corpus allatum activity in
    vitro during ovarian maturation in the desert locust,
    Schistocerca gregaria, J. Exp. Biol. 62:611-627.
Tobe, S. S., and Stay, B., 1980, Control of juvenile hormone bio-
    synthesis during the reproductive cycle of a viviparous cock-
    roach. III. Effects of denervation and age on compensation
    with unilateral allatectomy and supernumerary corpora allata,
    Gen. Comp. Endocrinol. 40:89-98.
Wigglesworth, V. B., 1934, The physiology of ecdysis in Rhodnius
    prolixus (Hemiptera). II. Factors controlling moulting and
    "metamorphosis", Quart. J. micr. Sci. 77:191-222.
Wigglesworth, V. B., 1936, The function of the corpus allatum in the
    growth and reproduction of Rhodnius prolixus (Hemiptera),
    Quart. J. micr. Sci. 79:91-121.
Williams, C. M., 1976, Juvenile hormone . . . In retrospect and
    in prospect, in: "The Juvenile Hormones", L. I. Gilbert, ed.,
    Plenum Press, New York, pp. 1-14.
Yagi, S., 1976, The role of juvenile hormone in diapause and
    phase variation in some lepidopterous insects, in: "The
    Juvenile Hormones", L. I. Gilbert, ed., Plenum Press, New
    York, pp. 288-300.

A MINIMAL MODEL OF METAMORPHOSIS: FAT BODY COMPETENCE TO RESPOND

TO JUVENILE HORMONE

Joseph G. Kunkel

Zoology Department
University of Massachusetts
Amherst, Massachusetts 01003

ARGUMENTS FOR A MINIMAL MODEL

The ontogeny of larval forms of invertebrates and vertebrates has fascinated embryologists and evolutionary biologists for centuries (Gould, 1977). We are now at a stage when appropriate choices of models of metazoan developmental phenomena may allow us to understand them on a mechanistic and molecular level. Success in this venture may depend on the ultimate complexity of the model chosen to study and for this reason potential models of minimal complexity should be sought.

The scientific literature is replete with examples in which concentrating on a simpler system affords rapid progress in a subject. Neurospora and later E. coli revolutionized the study of genetics and biochemistry. The further search by geneticists for simpler systems led to the use of viruses and ultimately to the simplest of all replicating organisms, the RNA-coliphage. For some strains of this phage, which has only 3 genes, science is in the enviable position in which it can claim an almost total inventory of an organism's molecules and behavior.

While microbial models for metamorphosis do exist, i.e., bacterial sporulation (Losick, 1973), or cellular slime mold morphogenesis (Bonner, 1967), one must eventually ask the general question of whether slime mold, amphibian and insect metamorphosis are based on the same mechanism. For this reason it is useful to explore minimal models of metamorphosis at a number of levels of complexity in a variety of organisms.

The phenomenon of metamorphosis connotes a change of form associated with a change in life style. In cellular slime molds the change is from a unicellular to a multicellular form, brought on by starvation. In amphibians the change usually accompanies the transition from water to land and is most dramatically represented by the metamorphosis of tadpole to frog, which involves both morphological and physiological adaptations to a new ecology. Changes in feeding, excretory physiology and respiratory physiology are among the dramatic shifts. Sexual maturation, the transition to a reproductively competent life style, is not accompanied by radical morphological changes in vertebrates. Changes which do occur at puberty in the higher vertebrates, birds and mammals, usually involve development of secondary sexual characters such as mammary glands or changes in plumage, developmental changes which do not have the dramatic qualities which have been labeled metamorphosis. Interestingly the evolution of metamorphosis in both insects and amphibians may have occurred at the same time, in the Devonian era, 350-400 million years ago, as a response to the severely fluctuating ecological conditions (Wigglesworth et al., 1963; Wigglesworth, 1976).

Among insects the phenomenon of metamorphosis is associated with maturation from a larval to adult form concomitant with a cessation of further molting, and is restricted to the higher, winged, species. The evolution of the metamorphic process is linked to the origins of flight but is shrouded in obscurity due to a poor fossil record (Wooten, 1976). The adults of the earliest known winged insects, from the Carboniferous period, already possessed two pairs of perfect wings (Snodgrass, 1952). At that stage of evolution the wings were the only exoskeletal structures other than genitalia which differentiated adults from larvae (Clarke, 1973). Since, at that time, there already existed two radically different ways of articulating the wing to the body, it is entirely possible that there was more than one independent origin of flight and, likewise, of metamorphosis. However, arguments to the contrary abound (Wigglesworth et al., 1963).

Due to the confusion of the fossil history, if a unified theory of insect metamorphosis is to be established, we must deal with the comparative morphology, physiology, and development of extant forms. The exact correspondence of developmental stages in ametabolous, hemimetabolous and holometabolous types of insect development is as yet unresolved (Hinton and Mackerras, 1970). Indeed the resolution of the correspondence of metamorphosis and stages in the different groups may come from more detailed knowledge of the regulation of the mechanism(s) of metamorphosis in higher insects and how these mechanisms relate to the primitive regulatory system found in the closest relatives of the Pterygotes among the Apterygotes, the Thysanura (Snodgrass, 1952).

In apterygote insects little or no external metamorphic changes are noticeable in the larval to adult transition. The basic ecology and physiology of the maturing animal does not change, and if an observer did not notice the reproductive behavior and egg production, it would be hard to tell that an important transition had occurred. Silverfish (Thysanura) development is a typical example of this "ametabolous" type of maturation, Fig. 1a. At the transition from last larval to first adult stage some changes in scale pattern do occur, and these small differences may have been amplified by natural selection into the types of differences which preadapted a Thysanuraform ancestor to give rise to a metamorphic, winged line of descent (Wigglesworth et al., 1963).

In modern day Thysanurans the pattern of molting cycles of the larval phase is continued in the adult, but each female adult molting cycle is preceded by a yolk deposition and ovulation cycle (Watson, 1964). Ovulation and molting alternate in a mutually exclusive pattern for the remainder of the life of the female sex of this ametabolous insect. This alternating pattern in the adult is under neuroendocrine control and could serve as a model of a primitive control system from which all the higher insect molting and reproductive control systems are derived.

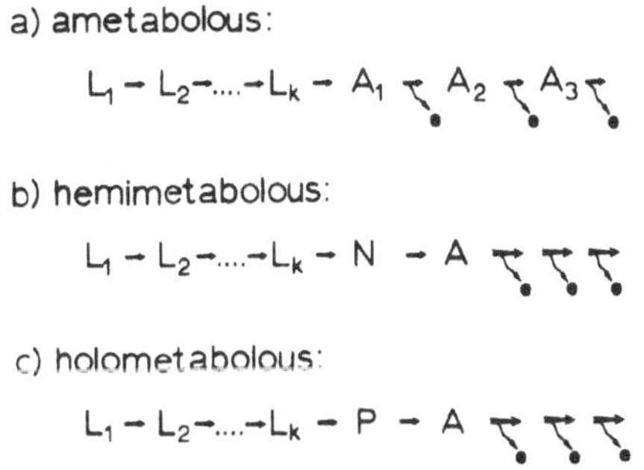

a) ametabolous:

$$L_1 - L_2 - \dots - L_k - A_1 \quad A_2 \quad A_3$$

b) hemimetabolous:

$$L_1 - L_2 - \dots - L_k - N - A$$

c) holometabolous:

$$L_1 - L_2 - \dots - L_k - P - A$$

Fig.1   Three major types of insect development: a) Ametabolous development typified by the silverfish, b) Hemimetabolous development typified by the cockroach; in some hemimetabola such as termites and grasshoppers there may be more than one stage with wing pads that could be called nymphal; c) Holometabolous development typified by flies, moths and butterflies, and beetles. Abbreviations: 'A', adult stage; 'L', larval stage; 'N', nymphal stage (=last instar larva); 'P', pupal stage; 'e', eggs.

At the other extreme the holometabolous insects undergo dramatic changes in ecology, physiology and morphology at the larval to adult transition, Fig. 1c. Their so-called "complete" metamorphosis is so extreme that an intermediate stage, the pupa, is required to accommodate the replacement of larval structures by adult structures (Hinton, 1963). In the Diptera the changes from maggot to fly are such that few functional larval tissues survive in the adult. The fly tissues are derived from nests of cells, imaginal discs, which do not participate in forming larval structures (Poodry, 1979). In these extreme cases it is inappropriate to talk about a particular tissue in terms of its larval to adult transition. Metamorphosis is accomplished by the programmed death of one set of tissues, the larval set, and growth and terminal differentiation of a second set, that of the adult. In those holometabolous groups in which a tissue does function continuously in larva, pupa and adult, such as in the lepidopteran abdominal epidermis (Willis, 1969 and this volume), the external morphology, ultrastructure and biochemistry of the tissue undergo numerous changes. As adults the majority of holometabolous insects still produce eggs in batches in a cyclical fashion; most with the involvement of juvenile hormone, JH, and some as suggested by various investigators, with the involvement of both JH and ecdysone (for reviews see Hagedorn and Kunkel, 1979; Engelmann, 1979).

These holometabolous metamorphic systems, though they serve well to demonstrate and elucidate the neuroendocrine controls and phenomenology of metamorphosis, are too complex to allow an understanding of the mechanism of metamorphosis at the biochemical level. The basic problem as I see it is to separate those biochemical processes which are part of the larval, pupal or adult phenotypes from those which are important for the transition between these stages. In holometabolous development too many of the observed phenomena are likely to be attributes of a stage phenotype rather than a part of the ontogenetic mechanism of transition. Since the larval to pupal and pupal to adult molting cycles are one of a kind, i.e., occurring only once in the normal development of an individual, it is additionally hard to distinguish the stage specific molting physiology from the metamorphic process.

On the other hand the hemimetabolous type of metamorphosis, Fig. 1b, as exemplified by the cockroach, is, in many respects, a minimal model of metamorphosis. The ecology of the larva and adult is often quite similar, and most of the functional larval tissues continue to be functional in the adult. The metamorphosis is often described as 'incomplete' or 'gradual'. A transition stage between immature and adult, the nymph, is distinguished by some investigators in specific groups of hemimetabolous insects. Although certain schools eschew the use of the term 'nymph', I find it useful. The nymph, as opposed to a larva in general, has developed wing pads and

in some groups, including the cockroaches, is diagnostic of the fact
that metamorphosis will occur with the next molt. I will subsequently refer to the last instar larva as a nymph, while retaining
the more traditional term 'penultimate larva' for the next to last
larval instar. The nymph stage has been given special meaning and
formalized with relation to the caste system in the termites,
immunologically close relatives of the cockroach (Kunkel and Lawler,
1974). The termite nymph has already undergone some metamorphic
changes and is restricted in its developmental potencies; it is capable of becoming a functional reproductive but not a soldier (Wilson,
1975).

Since the majority of tissues in the hemimetabola are continuously functional through the larval, nymphal and adult stages, one
can ask in real terms how particular tissues are changed in function
through metamorphosis. If the hemimetabolous and holometabolous
mechanisms of metamorphosis are basically the same, then by choosing
the simpler hemimetabolous model we have enriched for the details of
the metamorphic mechanism rather than the details of the changed
stages.

Development vs. Physiology

Even after establishing a hemimetabolous minimal model of
metamorphosis, there remains the problem of separating the important
events of metamorphosis from the physiology of the molting process.
Growth and development of arthropods are inextricably tied to the
molting process, during which the old cuticle is replaced by a
fresh, usually larger, and, if metamorphosis is occurring, geo-
metrically different exoskeleton. The problem of deciding what is
molting and what is metamorphosis is most clearly explained by discussing the epidermis. Even during isometric growth, which is
predominant during larval - larval molt cycles, Fig. 2a, the
cockroach epidermis undergoes distinct differentiative and prolifera-
tive phases of mitosis (Kunkel, 1975a). The differentiation of new
sensory and glandular organules in the expanding epidermis occurs
during the intermolt phase of the stadium. Proliferation of the
generalized epidermal cells, contributing to the expansion, occurs
at the beginning of the molting phase of the stadium but prior to
apolysis. In the later part of the molting phase, after apolysis,
the epidermal cells go through their program of digestion of old
cuticle and deposition of new cuticle.

The intermolt phase during which the determinative cell divi-
sions for new structures occur, can be expanded by various developmental processes including regeneration, Fig. 2b (Kunkel, 1975a;
1977), and metamorphosis, Fig. 2c. The process of metamorphosis
involves changes in all phases of the stadium. However, in the
cockroach the length of the molting phase remains relatively

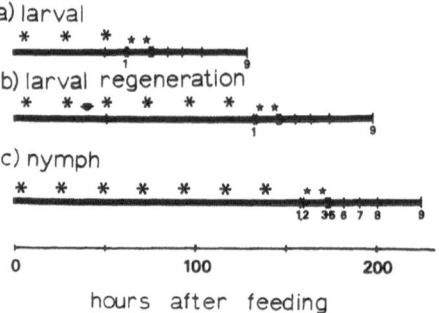

Fig.2   Timing of events during molt-intermolt cycles of the German
         cockroach, <u>Blattella</u> <u>germanica</u>. Periods of proliferative
         (stars) and differentiative (asterisks) cell divisions are
         illustrated for a) premetamorphic IV instar larvae, b) IV
         instar larvae regenerating a leg autotomized at 40 hours
         after feeding (arrow), c) metamorphosing nymphs. Key to the
         molting cycle events: 1) brain critical period, 2) regen-
         eration critical period, 3) prothoracic gland critical
         period, 4) apolysis, 5) end of epidermal mitosis, 6)
         bristle growth initiation, 7) bristle growth termination,
         8) muscle attachment release and 9) ecdysis.

constant despite any type of lengthening of the entire stadium,
seemingly due to a stereotyped program of epidermal molting behavior
initiated in an all-or-none fashion by the brain-prothoracic gland
endocrine axis.

     Thus, the allometric changes which occur on a biochemical and
morphological level during metamorphosis must be viewed against the
backdrop of the cyclical molting physiology (Kunkel, 1975a; 1975b;
Duhamel and Kunkel, 1978) that dominates insect growth and development.

The Fat Body as a Metamorphosing Tissue

     The epidermis, responsible for the synthesis and secretion of
the exoskeleton, has received most attention in studies of metamorphosis. Unfortunately, metamorphosis of this tissue is usually
evaluated by its secretion of a cuticle at the end of the molting
phase, and local changes in shape and coloration of this structure
are difficult to quantify. The fat body, another system which
undergoes metamorphic changes may be better tissue in which to study
this phenomenon. The fat body is the source of the majority of
serum proteins (Wyatt and Pan, 1978) including the JH binding
protein (Nowock et al., 1975), storage proteins, such as calli-
phorin (Thomson, 1975) and cockroach larval specific protein (Kunkel

and Lawler, 1974), and the vitellogenins (Pan et al., 1969). The
secretion of these proteins by the fat body constitutes a phenotype
which can be continuously monitored in its transition from a larval
to an adult pattern. The metamorphosis of the fat body in the
holometabola, like epidermal metamorphosis, is dramatic, involving
changes in tissue identity, morphology and physiology, which make
comparisons of larval and adult fat bodies difficult to interpret.
On the other hand, the fat body of the cockroaches and other
hemimetabolous groups remains morphologically intact throughout
metamorphosis. While the epidermis is structurally complex and its
metamorphosis involves a variety of localized changes, the fat body
is made up of only a few cell types and has the potential of reacting
in a uniform manner to the forces of metamorphic change: a biochemically tractable problem.

The Fat Body as a Minimal Tissue Model of Metamorphosis

It is convenient to assess a major aspect of the in vivo synthetic activity of the fat body by monitoring the concentration of
serum proteins in the hemolymph. All of the major serum proteins
of the cockroach are large, with molecular diameters greater than
100 Å (Kunkel and Pan, 1976). In the majority of the larval and
adult stages there is very little turnover of this group of proteins.
As a result, it is possible to follow the increases in concentration
and labeling of the major serum proteins seen during each molting
cycle (Duhamel and Kunkel, 1978; Duhamel, 1977) and interpret them
as accumulations of newly synthesized and secreted protein.

About 24 hours prior to each ecdysis in B. germanica, all of
the accumulated large serum proteins are cleared from the circulation in a precipitous fashion. Even injected foreign proteins such
as horse ferritin and E. coli β-galactosidase, which do not turn
over at an appreciable rate during the first three quarters of each
stadium, disappear rapidly as ecdysis approaches (Duhamel, 1977;
Duhamel and Kunkel, in preparation). In addition, the protein
storage granules of the fat body also disappear as ecdysis approaches, and do not reappear until about 48 hours after feeding in
the next stadium (Kunkel, 1975a). At present, the best guess as
to where these serum and fat body protein resources are going at
this time is into the production of the new cuticle. The analogous
(and possibly homologous) storage proteins in the holometabola
(Thomson, 1975; Wyatt and Pan, 1978) contribute to the general
tissue remodeling which occurs at metamorphosis.

Upon this background of cyclical synthesis of serum proteins
by the cockroach fat body there occurs a metamorphosis of the
capacity to synthesize and secrete the larval serum protein, LSP,
and vitellogenin, Vg. The ability of the fat body to secrete LSP
into the serum disappears at the metamorphic molt (Kunkel and

Lawler, 1974), while Vg appears for the first time shortly after the
first feeding following metamorphosis. This reciprocal change in
secretion of two major serum proteins represents the major change
of secretory behavior expressed by the cockroach fat body due to
metamorphosis. If the process by which LSP secretion is turned off
and Vg secretion turned on could be elucidated, we might have a
better understanding of metamorphosis in general. Since we know
that the synthesis and secretion of Vg in the cockroach is under
the control of juvenile hormone (Engelmann and Friedel, 1974), it
is also appropriate to ask when and how the fat body becomes
competent to respond to this hormone.

The fat body of hemimetabolous insects is an ideal minimal
tissue model of metamorphosis. A major larval gene is turned off
and a major adult gene turned on while the gross morphological structure is retained intact.

ESTABLISHMENT OF THE COCKROACH MODEL

Heterogeneous Metamorphic Rates of the Cockroach

Despite the fact that feeding can be used to synchronize each
molting cycle of a number of species of cockroach (Kunkel, 1966;
1977), the development from egg to adult is not entirely predictable.
Cockroaches have been observed to undergo a variable number of molting cycles before reaching the adult stage. The number of instars
in the cockroach life cycle has been shown to be affected by various
physiological stresses (e.g., antennal amputation (Pohley, 1962),
leg amputation (O'Farrell and Stock, 1956)) and ecological factors
(e.g., nutrition and colony density (Wharton et al., 1968)), as
well as being different in each sex (Kunkel, unpublished). In
cultures in which these extrinsic variables have been minimized or
controlled (Kunkel, in preparation), there still remains a substantial variability in the instar of metamorphosis. Genetic
selection canalizing for a particular instar of metamorphosis under
a particular set of conditions has provided strains in which larvae
can be counted on, with high reliability, to become adults at a
particular instar, Fig. 3. An example of the progression of such
a genetic selection program is given in Fig. 4. Some strains are
rapidly canalized to metamorphose at the particular instar selected
for; some strains are more reluctant to cooperate and persist in
spreading out the instar of metamorphosis among two or three
consecutive instars.

It seems clear that the instar of metamorphosis is a polygenic
trait in the cockroach with a good deal of environmental input involved in its expression. Even in highly inbred lines the proportion
of individuals metamorphosing at a particular instar, though relatively constant for one set of conditions, can be varied by changing
parameters such as the size of the culture container, the amount

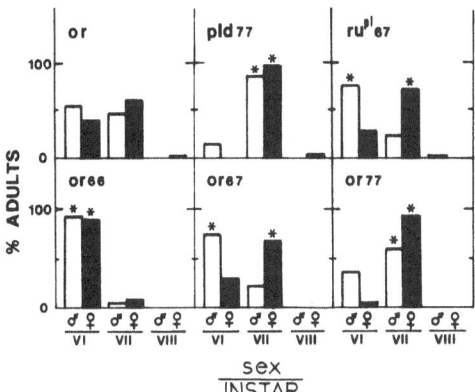

Fig.3   Instar canalized strains of Blattella germanica compared to
        an unselected strain, ro. The instars genetically selected
        for in each strain are indicated by asterisks. The three
        strains derived from or illustrate the results of eight
        generations of selection for the indicated adults and illu-
        strate in strain or67 the potential of sexual dimorphism in
        instar number that is not apparent in the parent strain.

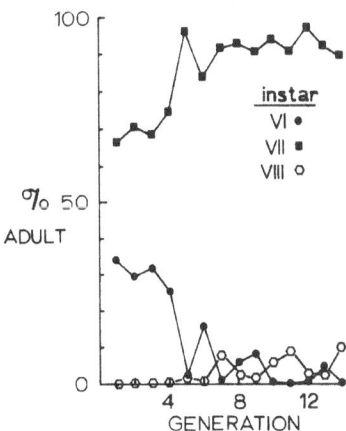

Fig.4   Canalization for seventh instar adults of Blattella germanica.
        The original strain was offspring of a cross between the or
        bearing strain and a New York strain (Kunkel, 1966). At
        each generation only the VII instar adults were allowed to
        mate and contribute to the next generation.

of food and water available, or the temperature. Clearly our ability
to control the instar of metamorphosis by manipulating culture conditions can be a valuable asset in elucidating the neurohormonal
controls of the metamorphic process (cf. Nijhout and Williams, 1974).
Using uniform, optimum culture conditions combined with canalized
strains with reproducible metamorphic patterns, it is now possible
to approach the study of the metamorphic process itself. Our ability
to assume a future schedule of development for an individual animal
is crucial to this study of the early phases of metamorphosis.

Normal Pattern of Vitellogenesis in Blattella germanica

Vitellogenesis in adult B. germanica is under the extrinsic
control of feeding (Kunkel, 1966; 1973) and the intrinsic control
of JH (Kunkel, 1973; Kunkel and Pan, 1977). Vg appears in the
serum of normal adult females between 12 and 24 hours after feeding
commences, Fig. 5, reaches a peak concentration in the serum by
48 hours and starts declining after ovulation during day 6.
Vitellin deposition in the terminal oocytes starts in earnest about
36 hours after feeding, and during the next 5 days the oocytes
continue to grow, removing the Vg from the serum by specific adsorb-
tive pinocytosis (Anderson, 1970; Kunkel and Pan, 1976). In order
to observe the true Vg secretory capability of the fat body, one
may examine secretion in ovariectomized females, Fig. 6, in which
the Vg accumulates in the serum and is, initially, only slowly

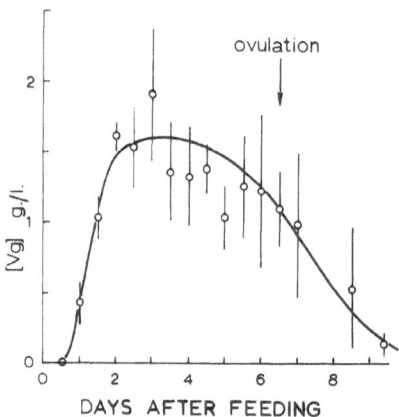

DAYS AFTER FEEDING

Fig.5 Normal Blattella germanica female vitellogenin (Vg) titer.
       Animals were starved for one week after the metamorphic
       ecdysis then fed at 30° C. Vg was measured by quantitative
       immunoelectrophoresis using purified Vg as a standard. The
       mean of five observations (+ standard error of the mean)
       is plotted.

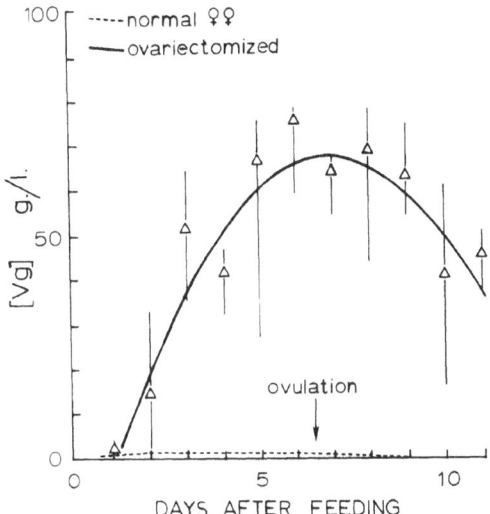

Fig.6   Ovariectomized <u>Blattella</u> <u>germanica</u> serum vitellogenin (Vg)
        titer. Animals were ovariectomized in the nymphal stage,
        allowed to metamorphose to the adult, starved for one week
        after the metamorphic ecdysis then fed at 30° C. Vg was
        measured as in Fig. 5. The mean and range of five observa-
        tions (= approximate 95% confidence interval of the mean)
        are plotted. The titer for normal females is included for
        comparison.

degraded. The Vg in these animals approaches 8% of the serum by
weight and after the time at which they would normally ovulate
there is a decrease in concentration. However, their blood volumes
are increasing in such an unpredictable fashion at this time that
it is difficult to interpret changes in the Vg titer. Vg is
synthesized indefinitely in these ovariectomized animals, as has
been described previously in other species (Engelmann, 1978).
Additional pathology due to the ovariectomized condition has been
discovered in the corpora allata (Scharrer, this volume).

Metamorphosis of Vitellogenic Competence

     The vitellogenic response of the adult animal to JH is
interesting in a biochemical and physiological sense as an example
of hormonal induction of a specific macromolecule. Recent reviews
cover this rapidly developing area (Engelmann, 1979; Hagedorn and
Kunkel, 1979). However, the more interesting, but rarely examined
developmental phenomenon is how the larva, which initially does not
produce Vg in response to JH, metamorphoses to the vitellogenically

competent state. This can be timed by injecting immature cockroaches
with JH at different stages and asking when they become capable of
producing Vg (Fig. 7). The nymph (i.e., last instar larva), early
in its stadium accumulates less Vg in its serum within 24 hours after
JH injection than does the late nymph. Late penultimate instar
larvae (actually pharate nymphs) secrete even less Vg at equivalent
JH doses and earlier stages secrete none. These results may be
taken to suggest a gradual development of competence to respond to
JH during the nymphal stage (Kunkel and Pan, 1977), but a closer
look at the kinetics of the vitellogenic response gives a slightly
more complex picture (Fig. 8). When the secretion of Vg into serum
is followed at finer time intervals during the nymphal instar it
is seen that there is a 16 hour lag in Vg appearance after JH
injection which is independent of the age of the injected nymph.
This nymphal lag is almost three times longer than the six hour lag
in Vg secretion seen when adult females carrying an ootheca are
injected with JH.

This difference in lag time may be a reflection of the difference between a primary and secondary response to a hormone, as·
has been described for the estrogen induction of ovalbumin in
chicken oviduct (Schimke et al., 1973). However, attempts to decrease the lag in Vg secretion in response to a JH injection late
in the nymphal instar by application of submaximal doses early in
the instar were unsuccessful. Irrespective of the meaning of the
latter result, the action involved in establishing the difference
in lag time between nymph and adult may represent a significant

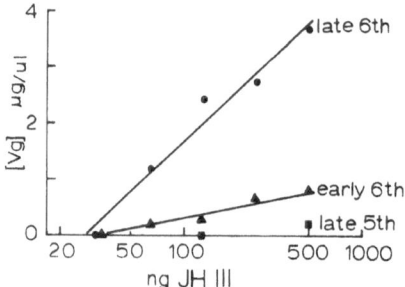

Fig.7   Juvenile Hormone (JH) induced serum vitellogenin (Vg)
        titers in larval female Blattella germanica. JH III was in-
        jected in various doses into early and late VI instar nymphs
        and late V instar larvae (= pharate nymphs). Vg was measured
        as in Fig. 5. Each point represents the mean for eight
        animals.

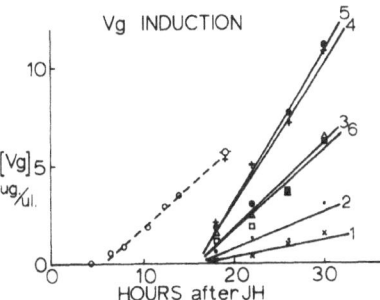

Fig.8   Kinetics of serum vitellogenin (Vg) titer increase in re-
        sponse to juvenile hormone injection. One microgram of JH
        III was injected into <u>Blattella germanica</u> nymphs on days one
        through six after feeding (solid lines labeled with day of
        injection) and into adult females carrying oothecae eleven
        days post ovulation (dashed line). Each point represents
        the mean of six animals. Analysis of variance of the nymphal
        data suggests that there are no significant non-linear com-
        ponents to the Vg accumulation data over the time span ob-
        served and that each curve extrapolates back to 16 hours after
        JH injection. The corresponding lag in Vg appearance for the
        adult females is approximately 6 hours.

portion of the metamorphosis of the fat body. The short time span
involved in the change of a primary to a secondary response, and
the lack of substantial gross morphological changes in the fat
body during the primary response, may make this fat body phenomenon
an important model system for the examination of the development
of a hormonal response.

<u>Intrusion of Molting Physiology on Metamorphosis</u>

        The induced rate of secretion of Vg into the serum increases
steadily during the nymphal instar until day 5 (Fig. 9), after
which the response to injected JH declines. The early increase in
response parallels the increasing capacity of the fat body to secrete all the serum proteins (Duhamel and Kunkel, 1978). At first
the decline in JH sensitivity after day 5 was disconcerting because
it appeared to be a reversal in the metamorphic process. However,
on further analysis, it proved to be another of the properties of
the fat body which is sensitive to the stage of the molting process.
Day five in these animals is the day when the molting cycle is
initiated by the joint action of the brain and the prothoracic glands
(Kunkel, 1975a). It is also the time at which the fat body is responding to the molting hormones by an accelerated increase in
secretion of serum proteins (Kunkel, 1975b; Duhamel and Kunkel,

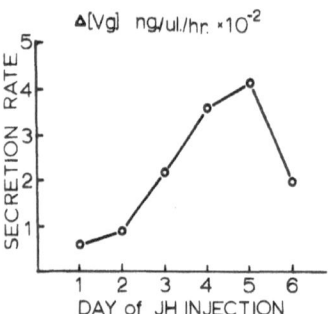

Fig. 9   Secretion rates of vitellogenin (Vg) into the serum of larvae
         injected with JH on different days of their nymphal stadium.
         The slopes of the lines in Fig. 8 are plotted against the day
         of JH injection.

1978). The physiology of the fat body during this later phase of
the stadium is dominated by the molting process. To test directly
whether the molting hormone, ecdysone, has an inhibitory effect on
JH induction of Vg secretion, nymphs which had been feeding for
four days were injected with JH and then injected one hour later with
ecdysterone (Fig. 10). The ecdysterone injection, which contained
sufficient hormone to induce a normal molting cycle (Kunkel, 1975b;
1977), clearly inhibited the JH response by 50%. In further studies
on another species, Periplaneta americana, it has been possible to
totally extinguish the JH vitellogenic response in nymphs with
ecdysterone (Storella and Kunkel, in preparation).

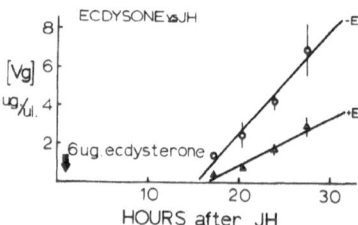

Fig. 10 Ecdysterone inhibition of JH induced secretion of vitello-
         genin (Vg) in VI instar larvae of Blattella germanica. A
         single dose of 6 µg ecdysterone was injected one hour after
         JH injection on day four of the nymphal instar. Vg was
         measured as in Fig. 5. Each point represents the mean (+
         standard error of the mean) for five animals.

JH Antagonism of the Molting Process

The antagonistic effect of ecdysteroids on the JH response, which has been observed in both hemi- and holometabolous orders of insects (see review by Engelmann, 1979), is paralleled in the cockroach by an antagonism of JH to the molting process. Figure 11 illustrates the response of nymphs to injections of 1 µg of JH-I on days 1 through 9 after last instar feeding begins. The animals pictured are the adults which resulted from molting under the influence of JH. Each column of animals in the figure represents an injection-day group and the rows illustrate the relative timing of ecdysis. Day 1 and day 5 through 9 injected animals did not delay molting at all, but day 2 through 4 injected groups were progressively delayed in their ecdysis times. Also, day 1 through day 4 animals were minimally and unevenly juvenilized by the JH injection, while day 5 through 9 animals were juvenilized in a status quo sense. For example, in day 5 animals, adult features such as male tergal glands which undergo most of their formative mitoses during the expanded metamorphic intermolt phase (Fig. 2c), appear normal: their morphological status is determined prior to JH injection. However, features such as the wings, which undergo extensive mitoses at the beginning of the molting phase, are severely stunted by the juvenilizing effects of a JH injection at a time when their mitoses are scheduled to occur. By day 7 the wings and all other structures appear to have been determined since JH injections at this time or later have no gross juvenilizing effects save for cuticle coloration. Despite the adult character of the exoskeletal structures which the 7 day nymphal epidermal cells are poised to secrete, JH injected at this stage is able to induce the cells to lay down a larval melanic pattern of coloration in all the secreted structures. This stepwise metamorphosis of the epidermal cells, which can be visualized by the effect of JH injections at various times during the nymphal instar, is encouraging in that it may also be possible to see such a series of stages in the developing fat body.

Although the groups which delay their molting cycles do not show a uniform juvenilization, the animals of groups 5 through 9 do respond consistently. The ability of JH to uniformly affect the epidermis may extend to internal tissues such as the fat body. For instance, all of 15 nymphs injected with 0.5 µg of JH on day 5 after feeding were juvenilized to the same extent (Fig. 12). When they were fed as adults they produced no eggs, and, on dissection two weeks later, their ovaries showed no signs of yolk deposition and no Vg in the serum. The JH injection had produced a uniform batch of sterile females. It is not known where the physiological lesions are which prevent a normal reproductive cycle in these particular animals, but it is obvious that this type of preparation may be a valuable aid in our experimental dissection of the metamorphic process. Indeed, this kind of stepwise control in a morphologically

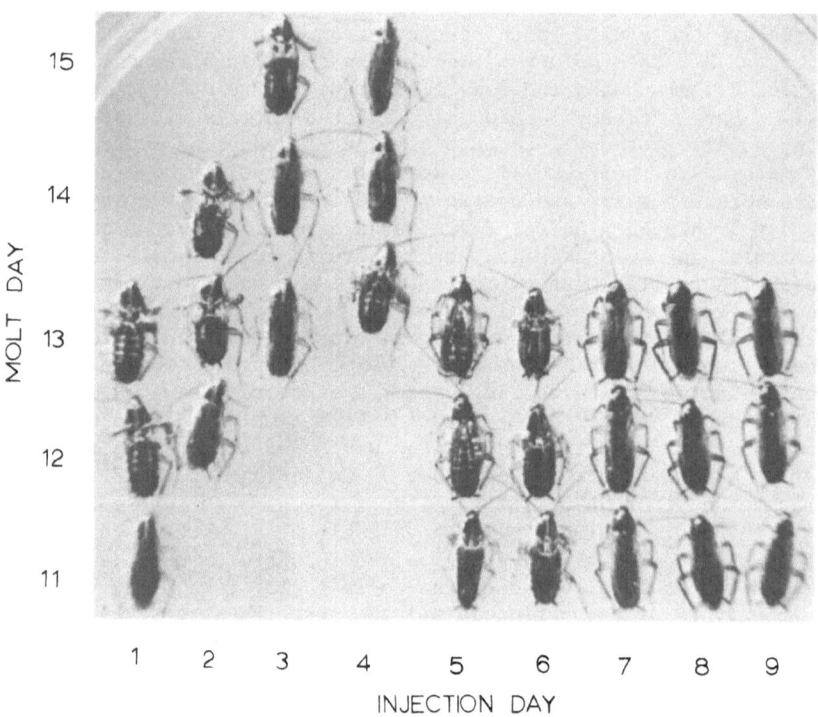

Fig.11   Molting and juvenilization record for groups of fifteen
         Blattella germanica nymphs injected with 1 μg JH-I on days
         one through nine (columns 1-9) of their metamorphic stadium.
         The rows labeled 11 through 15 are the days after feeding
         during which the animals in that row underwent their meta-
         morphic ecdysis.   The animals depicted in each column are
         representative of the fifteen animals injected on that day.

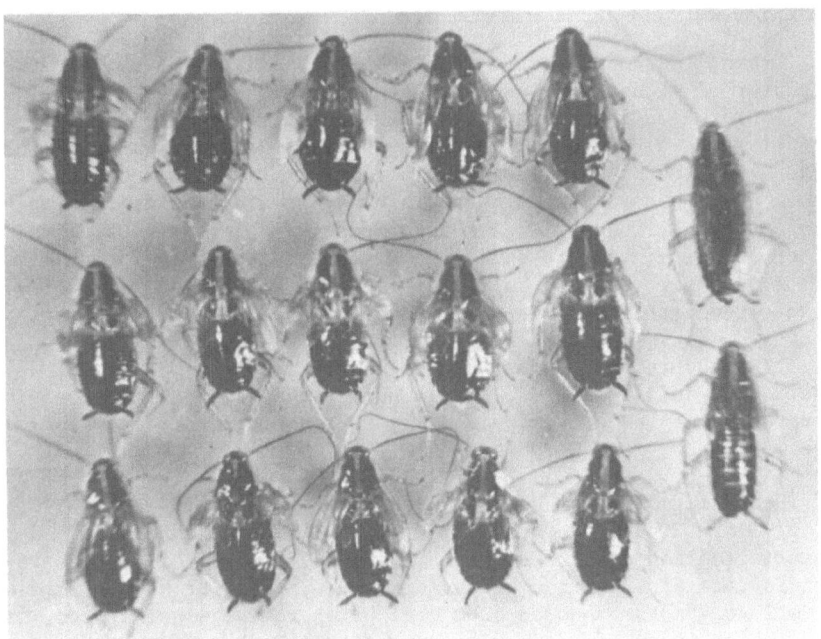

Fig.12   Juvenilized adult <u>Blattella</u> <u>germanica</u> resulting from nymphs
          injected with 0.5 µg of JH-I on day five after feeding.
          Two normal adult females, one with its wings clipped to
          bare its abdomen, are included for comparison at the right
          of the juvenilized adults.

unchanging structure such as the fat body, which is responsible for
Vg production may permit us to better understand metamorphosis at
the molecular level.

      What causes the JH induced molting delay in injection groups 2
through 4 of Fig. 11 is open to speculation at this point. It may
correspond to a similar phenomenon observed in last instar larval
<u>Manduca sexta</u> in which JH in hypothesized to inhibit the release
by the brain of prothoracotrophic hormone (Nijhout and Williams,
1974). A similar ability of JH to inhibit the initiation of a
molting cycle has been observed in another lepidopteran, <u>Spodoptera</u>
<u>littoralini</u> (Cymborowski and Stolarz, 1979). From an evolutionary
point of view, the mutual antagonisms of ecdysone on the JH induced
vitellogenic response and JH on the molting response may both be
relics of the primitive regulatory system in the apterygotes, in
which molting and reproduction alternate in a mutually exclusive
way (Fig. 1a).

Degrees of Metamorphosis at the Last Larval Instar

     As mentioned above, even in highly inbred lines of B. germanica
the instar of metamorphosis is not a determined feature. In one
inbred strain (23 generations of brother-sister mating), under a
particular set of culture conditions, 50% of the females are nymphs
in the V instar and the remainder in the VI instar. All of these
animals are close to isogenic yet when nymphal females of the two
instars are challenged with a maximal dose of JH at various times
in their stadium, they respond with different levels of Vg in their
serum (Fig. 13). The two fold difference in vitellogenic capacity
of these genetically (perhaps) but superficially identical nymphs
reflects some as yet unexplained relationship between instar
and metamorphosis of the fat body. The two fold difference in
secretion is not likely to be explained by the minor differences in
blood volume or fat body mass and may represent another example of a
stage in fat body metamorphosis.

     The maximum rate of secretion observed for the VI instar nymphs
(650ng/μl/hr, 4 days after feeding), approaches the rate observed
in ovariectomized adult females (Fig. 6),   730 ng/μl/hr.   With respect to rate of secretion, the VI instar nymphal fat body is close
to completely metamorphosed five days prior to the imaginal ecdysis.
The V instar nymph, five days prior to its imaginal ecdysis, is
only capable of secreting at half that rate. The corresponding
adults from these two types of nymphs produce approximately the same

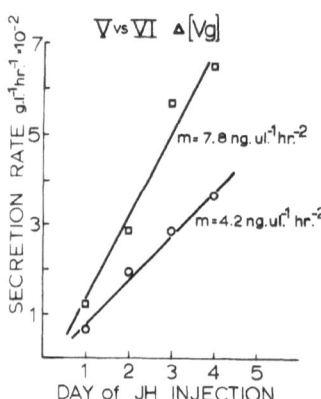

Fig.13   Instar dependent competence to secrete vitellogenin (Vg) in
         Blattella germanica nymphs. Each point represents the slope
         of a JH induced accumulation of Vg in the serum of a group of
         nymphs injected with JH on the day indicated on the ordinate.
         The V instar nymphs' secretion rates (circles) are at each
         time of the molting cycle about half the secretion rate of
         comparable VI instar nymphs (squares).

weight of eggs, so the eventual capacity of the fat bodies of the
two types of nymphs is not in question. This kind of investigation,
which contrasts genetically uniform but developmentally different or
experimentally altered animals, will, hopefully, provide a biochemical explanation for these developmental phenomena.

The above results also sound a cautionary note to anyone studying the metamorphosis of the fat body in hemimetabolous insects and
particularly, cockroaches. In order to obtain consistent results
one must know more than the animal's apparent developmental stage
(i.e., penultimate larva or nymph); one must know the number of
larval instars that have preceded the metamorphic instar. Efforts
to predict the developmental stages of cockroaches using simple
structure:size ratios are successful in identifying the early instars, but fail particularly with respect to distinguishing the
critical penultimate larval stage and discriminating between V and
VI instar nymphs (Tanaka and Hasegawa, 1979). The only alternative
to the uncertainty is to employ some form of genetic canalization
and synchronous culture technique (Kunkel, 1966; 1977), in which
the developmental and chronological ages of the animals are known
precisely.

PROSPECTS AND ALTERNATIVES

An argument for a minimal complexity model of metamorphosis has
been presented suggesting that such a model would allow easier access
to the mechanism of metamorphosis and avoid the confusion of the
excessive number of differences between the larval and adult physiological phenotypes. At the moment it may seem far fetched to
suggest that the advantages posed by the cockroach, Blattella
germanica, or any other hemimetabolous insect for that matter, could
make up for the deficit in experience with its biology compared with
that of the holometabolous models of development, Drosophila, and
the giant silk moths. Surely, the exquisitely developed genetics
and emerging molecular biology of Drosophila makes that organism
difficult to compete with.

However, a respectable genetics is developing for B. germanica
(Ross and Cochran, 1975). Inbred strains of this species are available, as described above. This species in particular compares
favorably with the laboratory mouse in generation time, ten weeks
for the cockroach at 30° C versus nine weeks for the mouse. Com-
bining the capacity to do genetics with the ability to synchronize
the molting cycles of the cockroach results in an experimental
animal well suited for studying the process of metamorphosis.

Two major gene products show dramatic changes during metamorphosis in Blattella: the LSP gene is shut off, and the Vg gene
is turned on. This reciprocal change epitomizes the phenomenon of

metamorphosis. The fact that the Vg gene is actually first
inducible in the pharate nymph by injections of exogenous JH, can
be construed as support of the concept of a stepwise metamorphic
process in the hemimetabola. However, it might also be evidence
for independence of the two gross metamorphic products, i.e.,
reproductive competence, and flight associated cessation of molting.
Metamorphosis may have evolved as a gradual accretion of independently controlled changes, each capturing JH titer as its signal for
activation.

Clearly, we must learn more about the metamorphic process, and
the modulation of Vg production is a significant activity which can
be followed in detail. Two steps in the turning on of vitellogenic
competence have been described, a difference in lag time between
the larval and adult Vg responses to JH, and a difference in maximum inducible Vg secretion rate for different instar nymphs.

The difference in lag time between nymphs and adults has a
number of possible explanations. The extra lag in the nymphal fat
body response to JH may represent the time necessary to develop the
enzymatic and ultrastructural machinery for processing the pre- and
provitellogenin molecules. Currently we are working on characterizing the nonprotein portion of Vg, including its phosphate and
oligosaccharide (Kunkel et al., 1978, and in preparation). The
typical branched oligosaccharide of Vg seems to be a uniform 10 or
11 mannose residues long, attached through a chitobiosyl linkage
to an asparagine residue on the protein backbone. The phosphate
is attached relatively close to, but not on, the oligosaccharide.
Presumably, specific glycosyl transferases and kinases would have
to be in existence prior to the processing of Vg before its secretion into the hemolymph. These may need to be induced by a
primary response to JH similar to that seen in nymphs.

The difference in maximal secretion rate by V versus VI instar
nymphs could have two non trivial explanations: either each individual fat body cell gradually develops its competence to respond
vitellogenically to JH and is present at different stages of its
development in the above two types of nymphs, or fat body cells as
a population are only gradually recruited to be able to respond to
JH and the two types of nymphs represent recruitments of different
proportions of the total population of fat body cells. Either of
the alternatives would be interesting if confirmed.

When the mechanism of metamorphosis of the expression of the
Vg gene is worked out, it will become possible to ask whether the
turning off of the larval gene for LSP is metamorphically tied to,
or merely a concomitant of the gain of vitellogenic competence.

ACKNOWLEDGEMENTS

I am indebted to Judy Willis for discussions of material
contained in this chapter and to John Storella for criticisms of an
earlier draft. The original research reported in this chapter
was supported in part by grants from the National Institutes of
Health, AI11269, RR07048 and the National Science Foundation,
PCM 79-03653.

BIBLIOGRAPHY

Anderson, E., 1970, Two types of coated vesicles in oocyte develop-
    ment, J. Microsc. 8:721.
Bonner, J. T., 1967, "The Cellular Slime Molds" 2nd ed., Princeton
    U. Press, Princeton.
Clarke, K. U., 1973, "The Biology of the Arthropoda," Edward Arnold
    Ltd., London.
Cymborowski, B., and Stolarz, G., 1979, The role of juvenile hormone
    during larval-pupal transformation of Spodoptera littoralini:
    Switchover in the sensitivity of the prothoracic glands, J.
    Insect Physiol. 25:939.
Duhamel, R. C., 1977, "The Hemolymph Proteins of the Cockroach
    Blatta orientalis: Changes During the Molting Cycle and in
    Adults," Dissertation, University of Massachusetts, Amherst.
Duhamel, R. C., and Kunkel, J. G., 1978, A molting rhythm for serum
    proteins of the cockroach, Blatta orientalis, Comp. Biochem.
    Physiol. 60B:333.
Engelmann, F., 1978, Synthesis of vitellogenin after long term
    ovariectomy in a cockroach, Insect Biochem. 8:149.
Engelmann, F., 1979, Insect vitellogenin: identification, bio-
    synthesis and role in vitellogenesis, Adv. Insect Physiol.
    14:49.
Engelmann, F., and Friedel, T., 1974, Insect yolk protein precursor:
    A juvenile hormone induced phosphoprotein, Life Sci. 14:587.
Gould, S. J., 1977, "Ontogeny and Phylogeny," Harvard U. Press,
    Cambridge.
Hagedorn, H. H., and Kunkel, J. G., 1979, Vitellogenin and vitellin
    in insects, Ann. Rev. Entomol. 24:475.
Hinton, H. E., 1963, The origin and function of the pupal stage,
    Proc. R. Ent. Soc. Lond. A 38:77.
Hinton, H. E., and Mackerras, I. M., 1970, Reproduction and meta-
    morphosis, in: "Insects of Australia," CSIRO.
Kunkel, J. G., 1966, Development and the availability of food in
    the German Cockroach, Blattella germanica (L.), J. Insect
    Physiol. 12:227.
Kunkel, J. G., 1973, Gonadotrophic effect of juvenile hormone in
    Blattella germanica: a rapid simple quantitative bioassay,
    J. Insect Physiol. 19:1285.

Kunkel, J. G., 1975a, Cockroach molting. I. Temporal organization of events during molting cycle of Blattella germanica (L.), Biol. Bull. 148:259.

Kunkel, J. G., 1975b, Larval-specific protein in the order Dictyoptera II. Antagonistic effects of ecdysone and regeneration on LSP concentration in the hemolymph of the oriental cockroach, Blatta orientalis, Comp. Biochem. Physiol. 51B:177.

Kunkel, J. G., 1977, Cockroach molting II. The nature of regeneration induced delay of molting hormone secretion, Biol. Bull, 153:145.

Kunkel, J. G., and Lawler, D. M., 1974, Larval-specific protein in the order Dictyoptera. I. Immunologic characterization in larval Blattella germanica and cross-reaction throughout the order, Comp. Biochem. Physiol. 47B:697.

Kunkel, J. G., and Pan, M. L., 1976, Selectivity of yolk protein uptake: Comparison of vitellogenesis of two insects, J. Insect Physiol. 22:809.

Kunkel, J. G., and Pan, M. L., 1977, Ovary and vitellogenin synthesis in two insects, Am. Zool. 17:914.

Kunkel, J. G., Ethier, D. B., and Nordin, J. H., 1978, Carbohydrate structure and immunological properties of Blattella vitellin, Fed. Proc. 37:948.

Losick, R., 1973, The question of gene regulation in sporulating bacteria, in: "Genetic Mechanisms of Development," F. H. Ruddle, ed., Academic Press, New York.

Nijhout, H. F., and Williams, C. M., 1974, Control of moulting and metamorphosis in the tobacco hornworm, Manduca sexta (L.): Cessation of juvenile hormone secretion as a trigger for pupation, J. Exp. Biol. 61:493.

Nowock, J., Goodman, W., Bollenbacher, W. E., and Gilbert, L. I., 1975, Synthesis of juvenile hormone binding proteins by the fat body of Manduca sexta, Gen. Comp. Endocr. 27:230.

O'Farrell, A. F., Stock, A., and Morgan, J., 1956, Regeneration and the molting cycle in Blattella germanica. IV. Single and repeated regeneration and metamorphosis, Austr. J. Biol. Sci. 9:406.

Pan, M. L., Bell, W. J., and Telfer, W. H., 1969, Vitellogenic blood protein synthesis by insect fat body, Science 164:393.

Pohley, J. H., 1962, Untersuchungen uber die Veranderung der Metamorphoserate durch Antennenamputation bei Periplaneta americana. Roux Archiv fur Entwicklungsmechanik 153:492.

Poodry, C. A., 1979, Imaginal discs: morphology and development, in: "Genetics and Biology of Drosophila", 2nd edition, in preparation.

Ross, M. A., and Cochran, D., 1975, The german cockroach, Blattella germanica, in: "Handbook of Genetics," Vol. 3, R. C. King, ed., Plenum Press, New York.

Schimke, R. T., Palmiter, R. D., Palacios, R., Rhoads, R. E.,
    McKnight, S., Sullivan, D., and Summers, M., 1973, Estrogen
    regulation of ovalbumin mRNA content and utilization, in:
    "Genetic Mechanisms of Development," F. H. Ruddle, ed., Academic
    Press, New York.
Snodgrass, R. E., 1952, "A Textbook of Arthropod Anatomy," Hafner,
    New York.
Tanaka, A., and Hasegawa, A., 1979, Nymphal development of the German
    cockroach, Blattella germanica Linne (Blattaria: Blattellidae),
    with special reference to instar determination and intra-instar
    staging, Kontyu 47:225.
Thomson, J. A., 1975, Major patterns of gene activity during develop-
    ment in holometabolous insects, Adv. Insect Physiol. 11:321.
Watson, J. A. L., 1964, Moulting and reproduction in the adult
    firebrat, Thermobia domestica (Packard) (Thysanura, Lepismati-
    dae) II. The reproductive cycles, J. Insect Physiol. 10:399.
Wharton, D. R. A., Lola, J. E., and Wharton, M. L., 1968, Growth
    factors and population density in the American cockroach,
    J. Insect Physiol. 14:637.
Willis, J. H., 1969, The programming of differentiation and its
    control by juvenile hormone in saturniids, J. Embryol. Exp.
    Morph. 22:27.
Wilson, E. O., 1975, "Sociobiology: The New Synthesis," Harvard
    U. Press, Cambridge.
Wigglesworth, V. B., 1976, The evolution of insect flight, in:
    "Insect Flight," R. C. Rainey, ed., Blackwell Scientific Pub.,
    Oxford.
Wigglesworth, V. B., Hinton H. E., Johnson, C. C., and Leston, D.,
    1963, The origin of flight in insects, Proc. R. Ent. Soc.
    Lond. (C) 28:23.
Wooton, R. J., 1976, The fossil record and insect flight, in:
    "Insect Flight," R. C. Rainey, ed., Blackwell Scientific
    Pub., Oxford.
Wyatt, G. T., and Pan, M. L., 1978, Insect plasma proteins, Ann.
    Rev. Biochem. 47:779.

THE PHYSIOLOGY OF PUPAL DIAPAUSE IN FLESH FLIES

David L. Denlinger

Department of Entomology
Ohio State University
1735 Neil Avenue
Columbus, Ohio 43210

INTRODUCTION

While visiting the University of Paris in 1963, Gottfried
Fraenkel was attracted by a thriving colony of Sarcophaga argyrostoma
cultured in the Laboratoire de Zoologie. He decided to transport
some of these flesh flies back to Urbana, but since his journey included several more visits along the way he tried to slow down their
rate of development. It was December. The days were short, the
temperature cold. Each night he diligently placed the flies on the
cool windowsills of his hotel room. When he arrived in Urbana, a
few flies emerged immediately, but the majority remained as undif-
ferentiated pupae and emerged only after several months. Professor
Fraenkel thus inadvertently discovered diapause in flesh fly pupae
and, along with Catherine Hsiao, published two papers (Fraenkel and
Hsaio, 1968a, 1968b), that preliminarily described its environmental
and endocrine regulation. Although Roubaud had experimented with a
few diapausing pupae of S. argyrostoma as early as 1922 and diapause
was also reported for Pseudosarcophaga affinis (Coppel et al., 1959;
House, 1967), the Fraenkel and Hsiao papers provided the real impetus
for the over 35 publications on flesh fly diapause that have since
appeared.

The potential for developmental arrest in the pupal stage of
flesh flies appears to be widespread; thus far it has been docum-
mented in 22 species representing a wide diversity of geographic
regions (Table 1). The arrest invariably occurs in the young
phaenerocephalic pupa. Adult antennal discs usually are not yet
visible, cells of the fat body are still intact and the hemolymph

Table 1. Reports from various geographic areas documenting pupal diapause in the Sarcophagidae.

Geographic Origin	Species*	Reference
North America	Sarcophaga bullata	Fraenkel and Hsiao, 1968a
	S. crassipalpis	Denlinger, 1971
	Pseudosarcophaga affinis	Coppel et al., 1959
South America	S. ruficornis	Denlinger, 1979
Europe	S. argyrostoma	Roubaud, 1922
	S. scoparia	Zdarek and Denlinger, 1975
	S. haemorrhoidalis	Vinogradova and Zinovjeva, 1972
	S. semenovi	Vinogradova and Zinovjeva, 1972
	S. similis	Vinogradova and Zinovjeva, 1972
	Bellieria melanura	Vinogradova and Zinovjeva, 1972
	Ravinia striata	Vinogradova and Zinovjeva, 1972
	Boettcherisca septentrionalis	Vinogradova, 1976
	Wohlfahrtia magnifica	Ternovoy, 1978
Africa	S. par	Denlinger, 1979
	S. inzi	Denlinger, 1979
	S. exuberans	Denlinger, 1979
	S. monospila	Denlinger, 1979
	S. haemorrhoidalis	Denlinger, 1979
	Poecilometopa spilogaster	Denlinger, 1979
	P. punctipennis	Denlinger, 1979
Japan	S. peregrina	Ohtaki and Takahashi, 1972
Australia	Tricholioproctia impatiens	Roberts and Warren, 1975

*Nomenclature conforms to Stone et al., 1965.

is clear (Fraenkel and Hsiao, 1968b). Occasional individuals of
Sarcophaga bullata and S. argyrostoma diapause in a slightly later
developmental stage, with the antennal discs already visible through
the cuticle (Fraenkel and Hsiao, 1968a), and in a European species
Sarcophaga scoparia, antennal discs are readily visible during
diapause in nearly all individuals (Zdarek and Denlinger, 1975).

The metabolic economy of diapause is readily discernible
from Fig. 1. Oxygen consumption drops to about 10% of the lowest
rate observed in nondiapausing flies (Denlinger et al., 1972) allowing pupae to survive on limited nutrient reserves for a prolonged period.

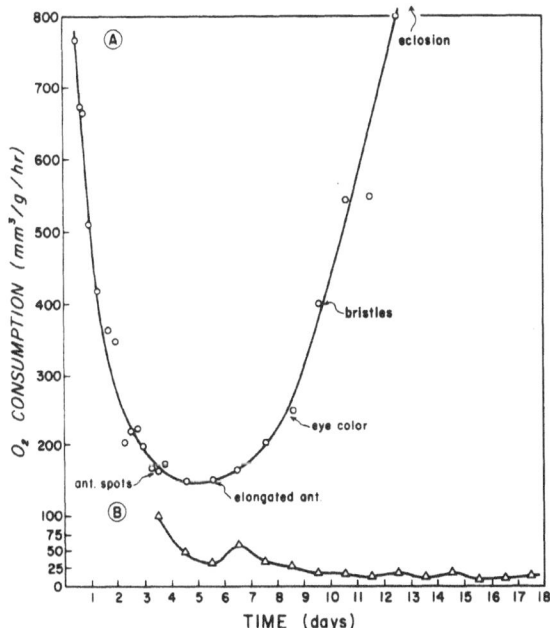

Fig.1   Comparison of the oxygen consumption rate in a nondiapausing
        individual from the time of puparium formation to adult
        eclosion with the rate in a group of 10 diapausing pupae of
        S. argyrostoma at 25°C (Denlinger et al., 1972).

ENVIRONMENTAL SIGNALS

Diapause Induction

    As in many other insects (Lees, 1955; de Wilde, 1962;
Danilevskii, 1965; Beck, 1968; Tauber and Tauber, 1976), short daylength provides the primary cue for diapause induction among flesh
flies living in temperate regions. The period sensitive to daylength can be very brief: exposure of pregnant females to short
daylength on the last two days of pregnancy and reinforcement with
two additional short days at the beginning of larval life is adequate
to induce a high incidence of diapause in pupae of Sarcophaga
crassipalpis (Fig. 2). In addition to S. crassipalpis (Denlinger,
1971), the extreme importance of maternal exposure to short day has
been recognized in S. argyrostoma and S. bullata (Denlinger, 1972a),
S. similis and Boettcherisca septentrionalis (Vinogradova, 1976),
and Sarcophaga peregrina (Ohtaki and Takahashi, 1972; Kurahashi
and Ohtaki, 1979).

    The photoperiodic sensitivity observed late in pregnancy re-
sults from direct perception of daylength by the embryos developing
in utero (Denlinger, 1971). Flesh flies are ovoviviparous: mature
eggs are ovulated, fertilized, and retained within the uterus of the
female for the duration of embryonic development (5 days at 25°C).
Embryos dissected from long-day mothers and cultured under short-day
conditions in an "artificial uterus" yield as high an incidence of
pupal diapause as embryos that develop in utero. The late embryos
apparently perceive their photoperiodic cues through the rather
heavily tanned black integument of the adult's abdomen. Over one
hundred eggs may be packed inside the uterus, thus further occluding
transmission of light. Yet, the photoperiodic response remains
operable at an external light intensity as low as 21 lux.

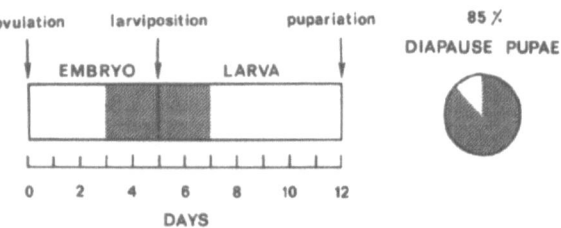

Fig.2   Stage in the development of S. crassipalpis that is sensitive
      to the photoperiodic cues used for programming pupal diapause
      at 25°C. Open area represents long photophase and the cross-
      hatched area represents short photophase (adapted from
      Denlinger, 1971).

If the period of embryonic sensitivity is missed, it is still possible to induce diapause, but induction then requires coupling low temperatures with short daylengths throughout a long period of larval life (Denlinger, 1972a). In a series of elegant experiments focusing on larval sensitivity, Saunders (1971; 1973; 1978b) developed a model that accounts for an increase in diapause incidence as temperature decreases. To enter diapause, fly larvae must be exposed to 13-14 short day cycles. The low temperature effect can then be explained as a means of retarding development, hence increasing the number of short day cycles to which larvae are exposed.

Although Saunders' model holds up well for exposure of larvae to diapause inducing conditions, it remains to be tested in situations which incorporate the important period of embryonic sensitivity. Experiments with S. bullata (Denlinger, 1972a) suggest that prolongation of larval life may be part of the syndrome of pupal diapause rather than the cause of diapause. As shown in Fig. 3, larvae reared at 25° stop feeding around day 5 regardless of whether they are reared under long or short day conditions as embryos and larvae. However, the duration of the wandering period is markedly different under different conditions. Long day (nondiapause destined) larvae pupariate within 1 or 2 days after leaving the food, but short day (diapause destined) larvae reared at the same temperature wander much longer and gradually pupariate over a two week period. Not all pupae produced under short day conditions enter diapause, but the proportion of diapause increases with increased time of wandering. Since the embryos are exposed to short days in this experimental design, the program for pupal diapause is already determined early in larval life (Fig. 2). This experiment thus indicates that the observed delay in pupariation cannot be construed as the cause of, but, rather, an effect of the diapause program. Similar delays of pre-diapause development have been noted among several insect species (Beck, 1968).

The precise period of daylight perceived as "short-day" is variable and highly dependent on latitude. For two populations of S. bullata collected in North America at 38° 30'N and 40° 15'N the critical photoperiod is 13.5 hrs (Fig. 4). A critical photoperiod of 12-13 hrs is documented for several species of flesh flies collected at 35-38° N or S (Vinogradova and Zinovjeva, 1972; Ohtaki and Takahashi, 1972; Roberts and Warren, 1975), and a critical photoperiod of 14-15 hrs is reported for several species originating from 45-50° N (Vinogradova and Zinovjeva, 1972; Ternovoy, 1978). Thus, as in many other insects (Danilevskii, 1965; Bradshaw, 1976), the critical photoperiod appears to increase at higher latitudes. Ohtaki and Takahashi (1972) also detected a shift in critical photoperiod related to temperature: higher rearing temperatures decrease the critical photoperiod. Temperature can thus modify the critical photoperiod and provide a degree of flexibility in the photoperiodic response.

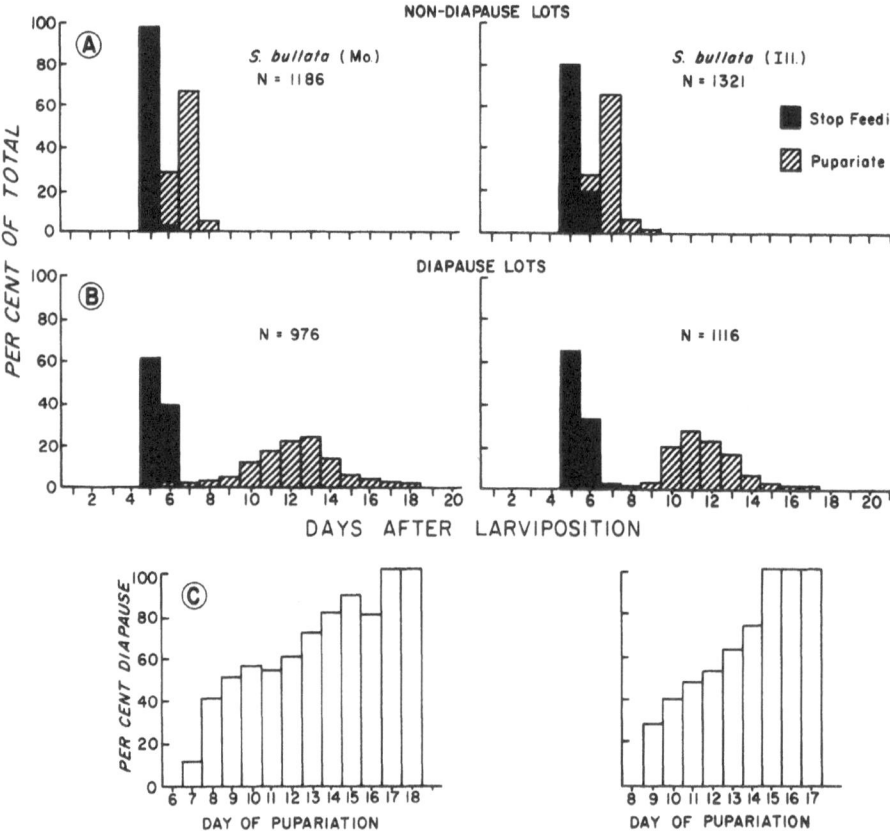

Fig.3 Delay of pupariation and correlation with incidence of pupal
       diapause in two strains of S. bullata. Nondiapausing lots
       originated from long-day mothers and diapausing lots origi-
       nated from short-day mothers. All larvae were reared under
       short days at 26°C (Denlinger, 1972a).

Seasonal changes of temperature and daylength are slight in
the tropics, yet the·potential for diapause has also been widely
observed among tropical flesh flies (Denlinger, 1974; 1978; 1979).
Although flies living within 10° of the equator fail to respond
to photoperiodic cues, diapause is readily induced by low, but
ecologically relevant, temperatures. The strongest diapause response is elicited when larvae are exposed to cool temperatures
throughout larval development. Transferring wandering 3rd instar
larvae or young pupae to lower temperatures is insufficient to
produce any developmental changes. Among tropical flies daytime
temperature is more important in determining diapause than night

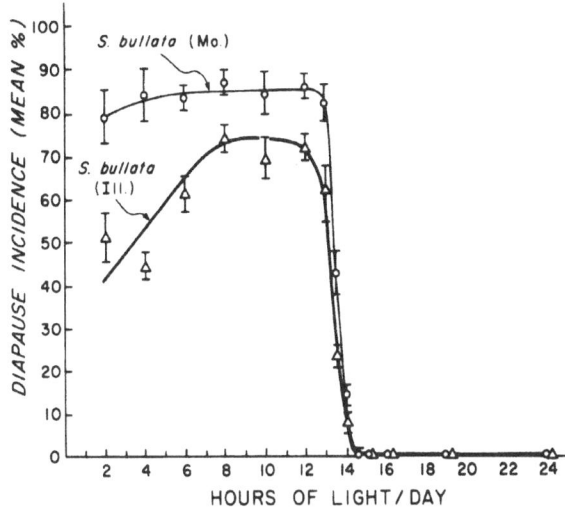

Fig.4   Critical photoperiod at 25°C in wild populations of S.
        bullata from Illinois (40° 15'N) and Missouri (38° 30'N)
        (Denlinger, 1972a).

temperature. (Reliance upon daytime temperature provides an accurate mechanism for identifying the coldest season in East Africa
(Denlinger, 1979).) This is in contrast to several temperate region insects which respond more strongly to cold night temperatures
(Beck, 1968; Chippendale et al., 1976; Greenfield and Karandinos,
1976).

     Moisture content of the larval food exerts a slight effect on
the incidence of diapause in both temperate and tropical flies
(Denlinger, 1972a; 1979). Adding water to a diet of liver increases
diapause while desiccation reduces diapause. However, the moisture
level can only modify the diapause response, and, without appropriate
cues of photoperiod or temperature, high moisture content is an
ineffective stimulant.

     Sex can also affect expression of diapause: males of S. bullata
from North America enter diapause more readily than females (Denlinger, 1972a). Although the sex ratio is nearly 1:1 among flies
representing 0 or 100% diapause, in a batch with a diapause incidence
less than 10%, 87% of the diapausing flies were males. At an incidence of diapause over 90%, 75% of the nondiapausing flies were
female. In contrast, the sex ratio of a tropical species,
Poecilometopa spilogaster, remained nearly 1:1 even at a low diapause incidence (Denlinger, 1979). We have not yet identified the
selective advantage of the skewed sex ratio in the temperate zone,

but apparently the same type of selective pressure is not present
in tropical regions.

The genetic basis for the fly's response to environmental cues
remains largely unexplored. Many years of laboratory inbreeding
and inadvertant selection have produced stocks of Pseudosarcophaga
affinis (House, 1967) and S. bullata (Denlinger, 1972a) with a low
incidence of diapause and a diapause of short duration. In experiments with P. spilogaster the incidence of diapause dropped from 80%
to 10% within 10 generations by actively selecting for nondiapause
(Denlinger, 1979). Kurahashi and Ohtaki (1977) crossed two strains
of S. peregrina, a temperate strain having a known photoperiodic
response and a tropical strain that failed to respond to photoperiod. A portion of the F_1 progeny entered diapause in response
to photoperiod, but the critical daylength was 2 hrs shorter than in
the diapausing parent. Although experiments completed thus far
convey little genetic information, we can be assured that many
features of diapause have a genetic basis. Flesh flies offer interesting untapped potential for genetic analysis of diapause since
the phenotype has been well described, reproductive rates are high,
generation time is short, chromosome number is small, and some
chromosome mapping (Whitten, 1969) has already been completed.

Diapause Termination

Photoperiod is ineffective in stimulating initiation of adult
development (Denlinger, 1972a; Ohtaki and Takahashi, 1972), and
indeed it would appear to be an ecologically inappropriate environmental cue for pupae that are buried 4-8 cm underground (unpublished
observations based on recovery depth of 38 pupae).

Termination of diapause is highly dependent upon temperature.
Although chilling is not a prerequisite for adult development as
it is in Hyalophora cecropia (Williams, 1956) and does not elicit
an immediate developmental response, chilling at 5-10°C does
significantly shorten diapause (Fraenkel and Hsiao, 1968a; Ohtaki
and Takahashi, 1972; Ternovoy, 1978). But, at constant temperatures ranging from 12 to 28°C, pupae can break diapause spontaneously
(Fraenkel and Hsiao, 1968a; Denlinger, 1972a; 1974; 1979; Ternovoy,
1978). The range of diapause duration is frequently very broad at
a constant temperature, but the mean values are quite characteristic
of each population. At 17°C, mean duration of diapause varied from
61 days in a laboratory strain of S. bullata to 227 days for S.
argyrostoma (Denlinger, 1972a). Diapause duration decreases at
higher temperatures: S. crassipalpis remained in diapause 118
days at 17°C, 70 days at 25°, and 57 days at 28°C (Denlinger, 1972a).
A switch to higher temperature further shortens diapause (Fraenkel
and Hsiao, 1968a; Denlinger, 1972a): diapause in S. crassipalpis
was reduced to 40 days by transferring pupae to 25° after a 20-day
exposure to 17°C.

Diapause duration in tropical flesh flies is likewise temperature dependent (Denlinger, 1974; 1979). While pupae of P. spilogaster remained in diapause 235 days at 12°C, the period was reduced to 73 days at 18°C. But, in marked contrast to temperate
species, short exposure (4 days) to 25°C is adequate to stimulate
adult development in P. spilogaster. The 4 days at 25°C may be
offered as a single block of time or as a series of smaller periods
at weekly intervals. The net effect is the same: when pupae have
accumulated the equivalent of 4 days at 25°C, adult development will
ensue. This type of temperature summation suggests a model in which
high temperature leads to accumulation of a stable product that
either triggers development or gradually erodes a developmental
inhibition.

The length of diapause does not appear to be influenced by
factors acting prior to the onset of diapause. Different photoperiods, larval rearing temperatures, and moisture levels in the
food fail to affect diapause duration. Pupae from a batch of flies
with a low incidence of diapause remain in diapause as long as pupae
from a batch representing a high incidence of diapause (Denlinger,
1972a). For each individual, diapause thus emerges as a threshold
character with a typical "all-or-none" response.

Phenology of Diapause

Field observations with flesh flies (Fig. 5) support the conclusions derived from laboratory experiments. In central Illinois
(40° 15'N) adults of S. bullata emerge in mid-May and trapping records indicate that adults can be collected until late October.
Laboratory-produced adults taken into the field as early as April

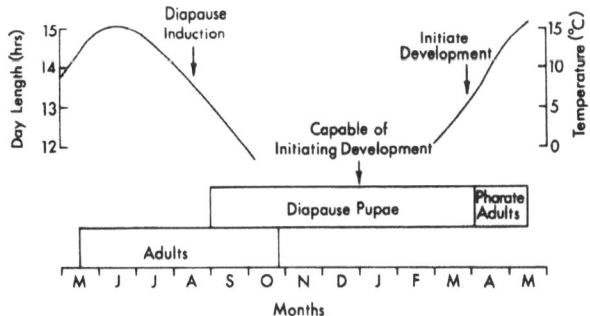

Fig.5   Seasonal distribution and phenology of development in a wild
        population of S. bullata from Illinois (adapted from
        Denlinger, 1972b).

(daylength shorter than 13 1/2 hrs) survive and their progeny yield a high incidence of diapause. Although this experiment is not ecologically relevant it indicates that the fly cannot differentiate between short, but increasing, daylength in the spring and short decreasing daylength in the autumn. During the long, warm days of June and July less than 2% entered diapause. In August daylength again drops below the laboratory-determined critical daylength (13 1/2 hrs). Twenty percent entered diapause in mid-August and nearly 100% in September. Diapause is thus initiated far earlier than the seasonal onset of cool temperature.

The significance of the "fixed" duration of diapause observed in the laboratory is apparent from the field studies. Pupae enter diapause when it is still warm, and the long latency period (133 days for S. bullata at 25°C) enables them to avoid the consequences of a premature response to warm temperature. By January 1, pupae that entered diapause in September were capable of initiating adult development immediately when transferred to 25°C (Fig. 5), but the cold temperature of the ensuing winter months prevented its onset. Outside, the first signs of adult development were detected in early April when the mean air temperature exceeded 5°C. The period of adult differentiation requires only 10 days at 25°C, but in the field adults did not emerge until mid-May. The median day of emergence was the same for pupae entering diapause in August and September (Denlinger, 1972b). Although the processes of diapause development may be completed earlier for the August pupae, suppression of development by the cold winter months provides a synchronization mechanism allowing uniform springtime emergence of the adults.

Field experiments also verified the preponderance of diapausing males at a low diapause incidence (Denlinger, 1972b). The small diapause segment of the August pupae (20%) was 79% male, whereas males represented only 45% of the September pupae (99% diapause).

From laboratory experiments with tropical flesh flies (Denlinger, 1974; 1979) we would predict that diapause near the equator is regulated by cool temperature. Since laboratory temperatures required to induce diapause are at the fringe of ecologically relevant temperatures in the tropics we would also predict a low expression of diapause and a less conspicuous role for diapause in the life histories of tropical flies. Indeed, these predictions were verified by rearing two species of flesh flies outside in Nairobi, Kenya (1° 18'S) (Denlinger, 1978). All individuals proceeded through the first experimental year without any sign of developmental interruption, and both Sarcophaga inzi and P. spilogaster successfully completed over seven generations. In the second experimental year temperatures during the coldest months, July and August, were slightly lower, and a few pupae entered diapause both months. Further evidence for a cool-temperature induced diapause can be garnered from

a five-year record of a Malaise trap operated in the Nairobi
National Park (Denlinger, 1980): the number of captured adults
drops sharply in September, suggesting that a portion of the
population may have been shunted toward pupal diapause during the
preceding cold months.

TRANSDUCTION OF ENVIRONMENTAL SIGNALS

An enormous void exists in our knowledge of diapause: how
are environmental signals received and transduced into information
that is later incorporated into the developmental program? In
temperate zone flesh flies, the photoperiodic cues are received
well in advance of pupal diapause (Fig. 2), thus implying an effective mechanism for storage and retrieval of environmental
information.

The location of the clock in flesh flies and the action spectrum of the photoreceptors remains undefined, but parameters of the
clock have been intensely researched by Saunders (1973; 1975; 1976;
1978a; 1978b). Resonance experiments (photoperiod is constant but
length of dark period is varied) and night-interruption experiments
(brief pulses of light during the dark period) using larvae of S.
argyrostoma provide evidence that the photoperiodic response involves a circadian clock. The clock shares many properties of the
external-coincidence model described by Pittendrigh (1966). Light
is involved not only in entrainment of a circadian oscillator but
it must be present during a particular light-sensitive phase of the
circadian cycle. The oscillation is entrained primarily by dusk at
photoperiods longer than 12 hrs, but at less than 12 hrs, the
oscillation is entrained to the entire photoperiod. The clock of
S. argyrostoma has an additional hour-glass component that requires
at least 6 hours of light in order to register a short day. Thus,
Saunders concludes that the clock is neither a simple hour-glass nor
a simple oscillator, but a subtle combination of both. It remains
to be seen whether similar results are obtained when the important
period of embryonic sensitivity is included in the experimental
design. If the same basic model remains valid, addition of the
sensitive embryonic period should reduce (from 14 days to 4 days)
the required number of inductive cycles.

When an adequate number of short days have been received (Fig.
2), the diapause program remains rather rigidly fixed throughout
larval life and a wide variety of physical manipulations including
exposure to long daylength, prolongation of larval life by wet treatment, desiccation, exsanguination, inflation of the body with air,
starvation (31 vs 128 mg pupae), cold shock, and heat shock (1 day
at 33°C) fail to reverse the decision to enter diapause (Denlinger,
1976). An exception is exposure to high temperature following
pupariation: both S. argyrostoma (Gibbs, 1975) and S. crassipalpis
(Denlinger, 1976) reverse their decisions to enter diapause when

shifted to high temperature around the time of pupariation. Shaking
the larvae also exerts a curious diapause-averting effect. When
mounted on an oscillating platform shaker (160 oscillations/min),
larvae of S. crassipalpis reared under diapause-inducing conditions
(23°C, 12:12 L:D, undisturbed controls = 95.6% diapause, N = 161)
entered diapause in significantly lower numbers (50.2% diapause, N
= 182). It is not yet clear whether transfer to high temperature
or shaking exert effects on the stored photoperiodic information or
act as direct stimuli on the neuroendocrine centers.

        Several chemicals are effective in preventing diapause (Den-
linger, 1976). Topical application of a high dose (10 µg) of
juvenile hormone analogue on newly ecdysed third instar larvae significantly reduced the incidence of pupal diapause, but later in the
third instar the hormone analogue had no effect in averting diapause.
The injection of 1.0 µg 20-hydroxy-ecdysone into wandering third
instar larvae was effective, and as little as 0.01 µg averted diapause if the injection was delayed until pupae were in the young
phaenerocephalic stage (Denlinger, 1976; Gibbs, 1976). Cholera
toxin, a stimulant of adenylate cyclase, prevented diapause when
injected into larvae shortly before pupariation (Denlinger, 1976),
and large doses of imidazole or pilocarpine exerted a similar effect.
Topical application of acetone or hexane at the stage of the cryp-
tocephalic or young phaenerocephalic pupa was also effective in
averting diapause (Denlinger et al., 1980). Many of these chemical
agents are most effective near the time of diapause inception, as
was observed with the high temperature effect. Thus, rather than
disrupting transduction of environmental cues they too may be
acting as direct stimulants of the neuroendocrine centers at a
particularly sensitive stage. Juvenile hormone, however, merits
further investigation as a candidate for interfering with transduction.

HORMONAL MILIEU AT DIAPAUSE INCEPTION

Absence of Molting Hormone

        The prolonged wandering period in diapause destined larvae
(Fig. 3) suggests that such larvae are endocrinologically distinct
from larvae destined for continuous development, but the basis for
that difference has not yet been explored. Events following pupariation have received the most attention. In Calliphora erythrocephala,
Shaaya and Karlson (1965) reported two peaks of molting hormone (MH)
activity: one peak around the time of puparium formation that is
involved in both pupariation and pupation, and a later peak coincid-
ing with the onset of adult development. Fraenkel and Hsiao (1968b)
predicted the absence of the second peak in diapausing Sarcophaga,
and indeed Ohtaki and Takahashi (1972) and Walker and Denlinger
(1980) verified the correctness of this idea (Fig. 6). In both

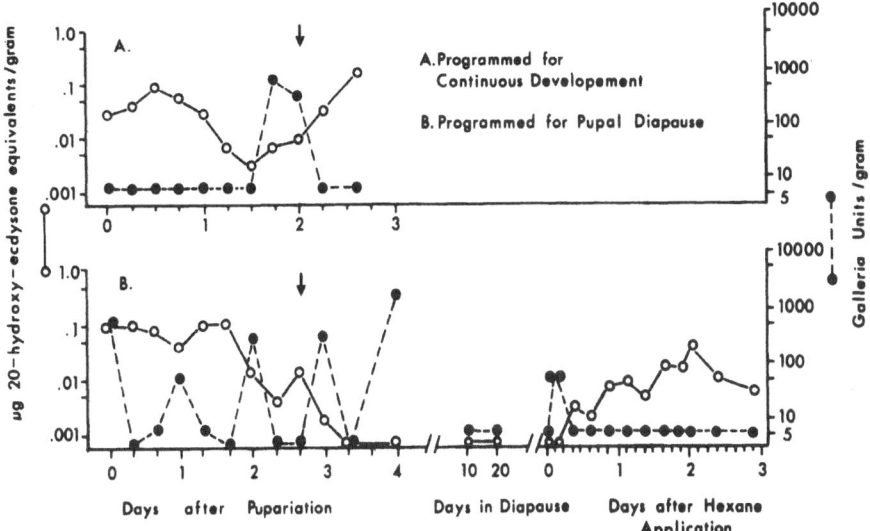

Fig.6   Juvenile hormone and ecdysone titers in <u>S</u>. <u>crassipalpis</u> pro-
         grammed for (A) continuous development without intervention of
         pupal diapause (25°C, 15:9 L:D) and (B) pupal diapause (20°C
         12:12 L:D). Time scales are adjusted to align comparable
         developmental stages; arrow indicates time of head eversion
         (Walker and Denlinger, 1980).

diapause and nondiapause destined flies MH activity begins to rise
several hours before pupariation (Ohtaki and Takahashi, 1972),
reaches a maximum of 0.1 µg 20-hydroxy-ecdysone equivalents/g body
weight, and then declines (Fig. 6). The MH titer remains low in
diapause destined pupae but rises again in pupae programmed for
continuous development. A very small amount of 20-hydroxy-ecdysone
injected into diapausing pupae at this early stage readily elicits
development (Denlinger, 1976; Gibbs, 1976).

     Evidence for inactivity of the prothoracic gland in diapause
is also available from experiments with imaginal discs and
glandular transplants. A number of investigators have demonstrated
that MH is required for evagination of imaginal discs (Fristrom et
al., 1973; Postlethwait and Schneiderman, 1970; Mandaron, 1973;
Ohmori and Ohtaki, 1973), and it has been shown that leg discs of
<u>S</u>. <u>crassipalpis</u> transplanted ectopically onto nondiapausing hosts
evaginate readily (Sivasubramanian, 1977). By contrast, discs
transplanted onto diapausing hosts (ecdysone-deficient) fail to
evaginate. Ring glands or brain-ring gland complexes dissected
near the time of pupariation readily induce pupariation in isolated larval abdomens, but transplants from diapausing pupae fail

to elicit the same response in S. bullata (Zdarek and Fraenkel,
1971). As in other insects, production and release of MH in flesh
flies is controlled by a prothoracicotropic neurosecretion from the
brain (Zdarek and Fraenkel, 1971). Thus, a diversity of experiments
point to a shut-down of the brain-prothoracic gland axis at the
inception of pupal diapause. In this regard, the results are in
complete accord with Williams' (1947; 1952) classic experiments
with diapausing pupae of Hyalophora cecropia.

Presence of Juvenile Hormone

The juvenile hormone (JH) titer is strikingly different after
pupariation in diapause and nondiapause destined flesh flies (Fig.
6). In nondiapause destined flies, the JH titer in whole body
homogenates is low (about 5 Galleria units: 1 G.U. = 3 x 10^{-7} μg
C18 JH), but consistently detectable during the first few days, and
then a single spike of JH activity (1000 Galleria units) precedes
the rise in MH associated with onset of adult development. In
contrast, JH activity is high at the time of pupariation in diapause destined flies and then drops sharply to undetectable levels.
Activity again rises 24 hrs after pupariation. Such 24 hour pulses
(with maximum activity exceeding 1000 G.U.) continue for at least
five days. Since each extraction was based on pooled samples that
had pupariated at different times throughout the day, the pulses
are in synchrony with time of pupariation rather than sidereal time.

Although JH appears to play an important role in larval diapause
of several insects (Chippendale, 1977), its presence in diapause
destined flies was unexpected and raises the interesting possibility
that pupal diapause involves more than failure of the brain to
stimulate the prothoracic gland. Several potential roles for JH in
diapause development are discussed later.

Cyclic AMP

Cholera toxin, a stimulant of adenylate cyclase (Finkelstein,
1973), provides a potent tool for generating cyclic AMP. When the
toxin is injected into flesh fly larvae destined for pupal diapause,
flies actually reverse their decision to enter diapause (Denlinger,
1976). This result suggests an important role for cyclic AMP in
the developmental decision.

To explore the basis for the toxin effect, brains and ring
glands from untreated diapause and nondiapause destined flies were
analyzed for cyclic AMP over a period of several days following
pupariation (Fig. 7). Cyclic AMP levels are much higher in flies
destined for continuous development. When brains and ring glands
are assayed separately, elevated levels of cyclic AMP are apparent
in both tissues, but the bulk of activity (59% of total activity or
87% of activity calculated on basis of mg protein content) can be

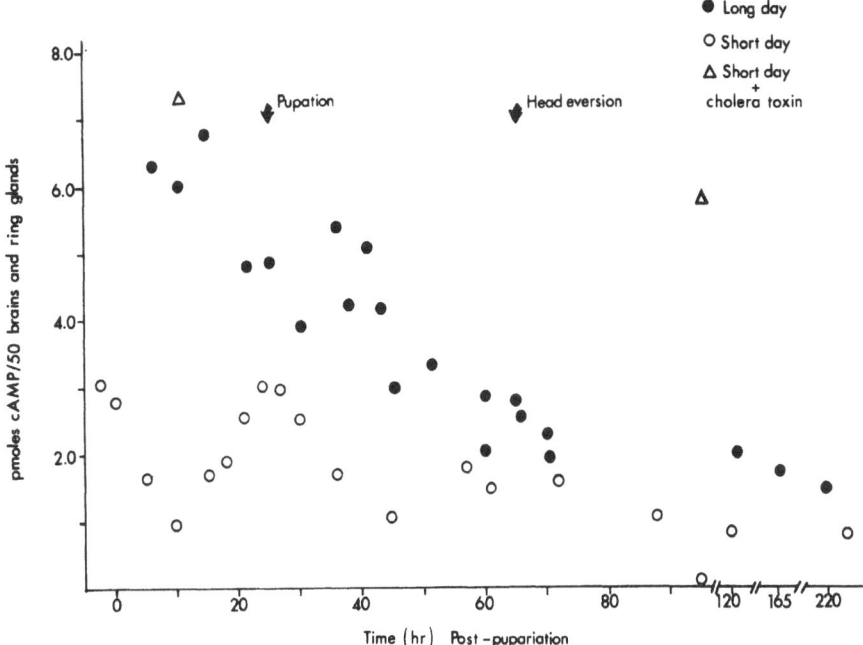

Fig.7   Levels of cyclic AMP in brains and ring glands dissected
        from S. crassipalpis at various intervals after puparium
        formation at 20°C.   Solid circles represent long-day (non-
        diapause) flies, open circles represent short-day (diapause-
        destined) flies, and open triangles represent short-day flies
        injected with cholera toxin on the day before pupariation
        (Gnagey and Denlinger, 1980).

attributed to the ring gland.   Injecting cholera toxin into diapause
destined larvae elevates cyclic AMP to nondiapause levels (Fig. 7),
and such flies fail to enter pupal diapause.   Like many other peptide
hormones (Robison et al., 1971; Greengard, 1978), prothoracicotropic hormone is likely to employ cyclic AMP as a second messenger
in stimulating the prothoracic gland to produce ecdysone.   High
levels of cyclic AMP have also been detected in stimulated prothoracic glands of Manduca sexta (Vedeckis et al., 1976).   Although
we cannot yet eliminate other potential roles for cyclic AMP in
promoting development, the low level that characterizes diapause
destined flies is consistent with the idea that the prothoracic gland
remains unstimulated in diapause.

Assays for cyclic GMP over the same time interval show no major
differences in cyclic GMP levels between diapause and nondiapause
destined flies.   Within 100 hrs after pupariation the level of

cyclic GMP varies between 1-2 pmoles/50 brains and ring glands
(Gnagey and Denlinger, 1980).

DIAPAUSE DEVELOPMENT

    With his term "diapause development" Andrewartha (1952) cap-
tured the idea that diapause is not a static state, but should
be regarded as a dynamic progression of events leading ultimately
to the resumption of active development. In flesh flies the dynamic
nature of diapause can be observed by examining patterns of oxygen
consumption and changes in sensitivity to hormones and other chemicals.

    Measurements of O_2 consumption of groups of 10 diapausing pupae
show very consistent mean rates of uptake (Fig. 1), but measurements
made on single individuals at 25°C demonstrate distinct infradian
cycles with periodicities of 3.7 days for S. argyrostoma, 5.1 days
for S. crassipalpis (Fig. 8), and 7.5 days for S. bullata (Denlinger
et al., 1972). At 18°C, the mean periodicity in S. crassipalpis
increases to 9.8 days, but the period length changes markedly
throughout diapause (Fig. 9). At the beginning of diapause the O_2

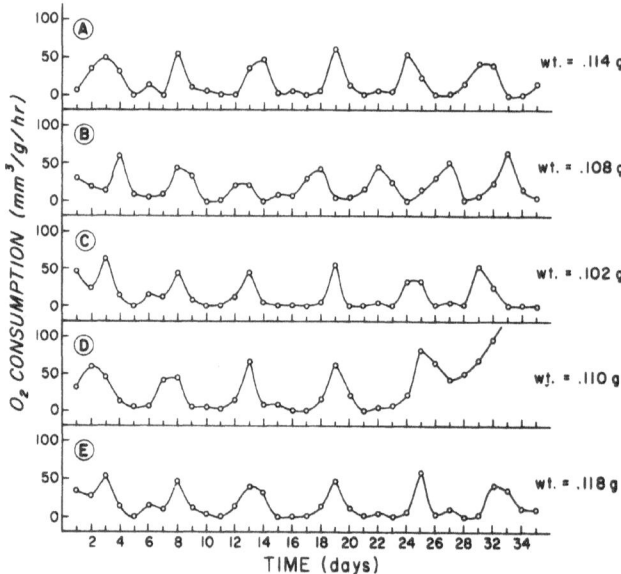

Fig.8 Infradian cycles of oxygen consumption in individual dia-
       pausing pupae of S. crassipalpis at 25°C (Denlinger et al.,
       1972).

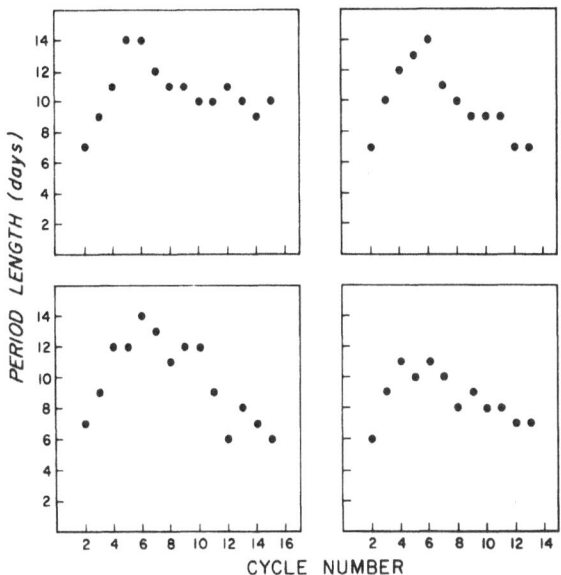

Fig.9   Change in periodicity of oxygen consumption cycles throughout the course of pupal diapause in four individuals of S.
crassipalpis at 18°C (Denlinger et al., 1972).

peaks are close together.   They gradually separate but as the
breaking of diapause is approached, the cycles again become shorter.
This pattern has been observed in all individuals examined and under-
scores the presence of a regular progression of events during the
course of diapause.   Oxygen consumption cycles, although somewhat
more erratic, have also been detected during diapause in tropical
flesh flies (Denlinger, 1979).   Consumption of O_2 does not appear
to be cyclic in diapausing Saturniid pupae (Schneiderman and
Williams, 1953; Buck and Keister, 1955), but similar infradian cycles
have recently been recorded in diapausing pupae of Pieris brassicae
(Crozier, 1979).

A number of organic solvents, including acetone and hexane, are
capable of stimulating development when applied topically to diapausing flesh fly pupae (Zdarek and Denlinger, 1975; Denlinger, 1979;
Denlinger et al., 1980).   As shown in Fig. 10, hexane is a potent
stimulant for diapausing pupae at any age, but the response to
acetone is age dependent.   The high sensitivity to acetone observed
at inception of diapause drops sharply two days later and remains
low during the remainder of diapause.

The response of diapausing pupae to 20-hydroxy-ecdysone also
suggests that sensitivity varies in relation to pupal age.   At

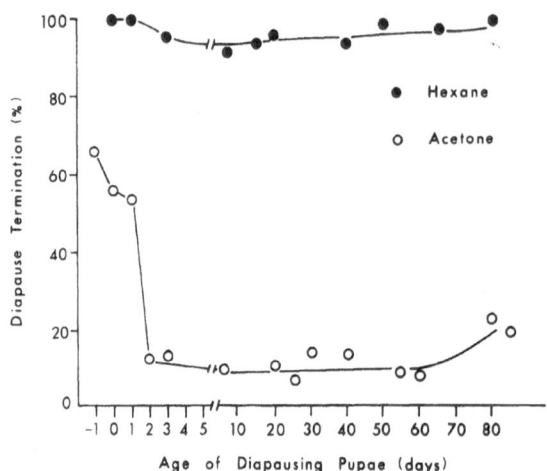

Fig.10   Response of diapausing pupae of S. crassipalpis at different ages to a 5 µl topical application of hexane or
acetone (adapted from Denlinger et al., 1980).

the inception of diapause a small amount of hormone (around 0.01
µg/pupa) is adequate to initiate development in S. argyrostoma
(Gibbs, 1976) and S. crassipalpis (Denlinger, 1976).   Two or
three weeks later 40-100 X more 20-hydroxy-ecdysone is required to
elicit the same response.   Although very young diapausing pupae
clearly differ from older pupae in their response to 20-hydroxy-
ecdysone, detailed studies of sensitivity changes as reported on
diapausing Lepidoptera pupae (Bodnaryk, 1977; Bradfield and
Denlinger, 1980) have not yet been completed for flesh flies.   These
results coupled with those from the acetone and high temperature
experiments suggest that the pupal neuroendocrine system is particularly sensitive to stimulation at the very onset of diapause, but
soon thereafter loses some of its capacity to respond.

A dose of ecdysone (Fraenkel and Hsiao, 1968b) or 20-hydroxy-
ecdysone (Zdarek and Denlinger, 1975; Gibbs, 1976) that is too
small to immediately terminate diapause can still shorten the duration of fly diapause (Fig. 11).   Thus, the response to ecdysteroids
is not "all or none" but a graded response that is dose dependent.
This type of response raises the possibility that MH could be released at a low level throughout diapause producing covert effects
(Ohtaki et al., 1968; Zdarek and Fraenkel, 1970) that ultimately
lead to diapause termination.   Although MH is not detected in diapausing fly pupae using the Musca bioassay (Fig. 6), the more sensitive radioimmune assay shows low levels of MH in diapausing pupae
of H. cecropia (McDaniel, 1979).   Dose response curves determined at

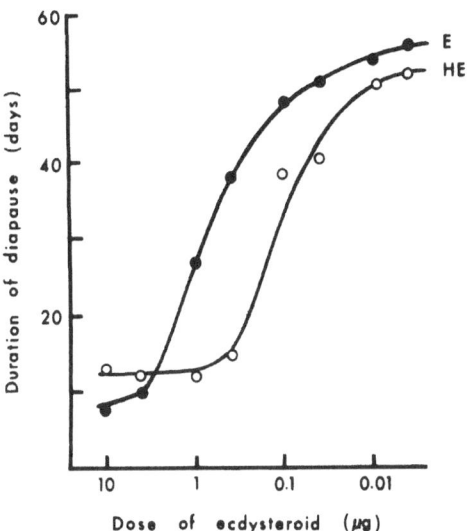

Fig.11   Duration of diapause after injection of graded doses of
         20-hydroxy-ecdysone (E) or 5,20-dihydroxy-ecdysone (HE)
         into 26-day-old diapausing pupae of S. crassipalpis (adapted
         from Zdarek and Denlinger, 1975).

different ages of diapause could readily test the merit of pro-
posing a potential role for ecdysone in diapause development: if
the idea is correct, older pupae should terminate diapause with
injections of much lower doses of 20-hydroxy-ecdysone.

     JH is another viable candidate for involvement in diapause
development. The pulses of JH activity observed early in diapause
(Fig. 6) may persist throughout diapause. Preliminary studies show
some JH activity to be present in mid-diapause (day 10 and 20, see
Fig. 6), but titer measurements using highly synchronous pupae are
needed to verify JH pulses throughout the diapause period. The
24 hr regularity of the observed JH pulses suggests an interesting
mechanism for measuring passage of time. An effect elicited by
JH could gradually accumulate until a threshold is reached and thus
play a critical role in determining the duration of diapause. If
such is the case, it should be possible to mimic passage of time
with exogenous JH. A JH analogue applied to diapausing pupae does
indeed shorten diapause (Zdarek and Denlinger, 1975), and application of a JH analogue to third instar larvae on the day before
pupariation can reduce the duration of pupal diapause from 80 to
40 days (Fig. 12). The presence of JH pulses and the effect of JH
in shortening diapause are features apparently not shared by all
insects having a pupal diapause (Bradfield and Denlinger, 1980).

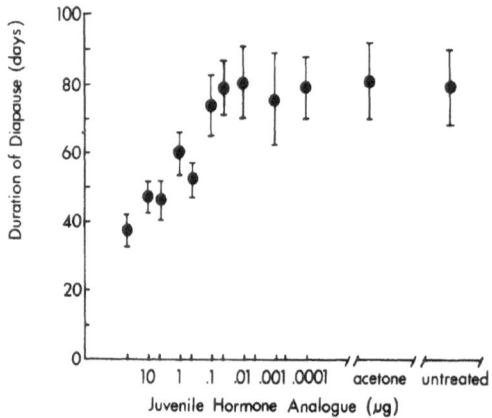

Fig.12 Duration of pupal diapause (mean ± S.D. days) in S.
        crassipalpis treated topically with graded doses of a
        juvenile hormone analogue (ZR515) on the day before
        puparium formation.

     If experimental evidence continues to support roles for juve-
nile hormone and ecdysone in diapause development of flies, such
evidence will not negate a major role for the brain. Neither the
corpora allata nor the prothoracic gland are likely to function
autonomously. Activity of both glands is regulated by the brain
and the hormonal products from the ring gland could be envisioned
to act on the brain by a feedback mechanism. Both JH and ecdysteroids have been documented to exert a trophic effect on neurosecretory cells of the brain (Agui and Hiruma, 1977; Hiruma et al.,
1978; Takeda, 1978).

INITIATION OF ADULT DEVELOPMENT

Ecdysteroids

     The effectiveness of ecdysone, 20-hydroxy-ecdysone, or 5,
20-dihydroxy-ecdysone in breaking diapause has been reported for
several species of flesh flies (Fraenkel and Hsiao, 1968b; Zdarek
and Denlinger, 1975; Gibbs, 1976; Denlinger, 1979). A dose response curve (Fig. 11) shows that a rather high dose is required to
terminate diapause immediately and lower doses result in a delayed
effect. The amount of hormone required to elicit an immediate
response produces developmental defects characteristic of hyper-
ecdysonism: misoriented bristles and underdeveloped antennae, eyes,
mouthparts, and genitalia (Zdarek and Denlinger, 1975; Gibbs, 1976).
Dividing the dose of ecdysteroid into several smaller doses injected on different days increases the effectiveness of the hormone
and reduces hyperhormonal abnormalities.

Measurement of the MH titer verifies a rise in MH at diapause
termination (Fig. 6). Using hexane as a tool for breaking diapause
(Denlinger et al., 1980), large groups of synchronous flies were
extracted at precise intervals after hexane stimulation. The MH
titer begins to rise within 9 hrs of stimulation and remains elevated for several days.

Flies were neck-ligated at daily intervals following hexane
treatment to assess the temporal requirements for the brain-ring
gland complex. As shown in Fig. 13, a three day exposure to MH
is required before 50% of the flies can successfully complete adult
development. Thus, the stimulant for adult development is not
merely a single pulse of ecdysone but the sustained presence of
hormone over a three day interval. This observation, which is
in agreement with observations in several Lepidoptera (Williams,
1952; Claret et al., 1977; Hsiao and Hsiao, 1977; Bradfield, 1980)
and Diptera (Borst et al., 1974), helps to explain the difficulty
of breaking diapause with a single injection of ecdysteroid.

Juvenile Hormone

In both nondiapausing and diapausing pupae the rise in MH
responsible for initiating adult development is directly preceded
by a brief pulse of JH (Fig. 6). JH apparently is not playing a
direct role in stimulating the prothoracic gland since JH, by itself, will not promptly terminate diapause (Fraenkel and Hsiao,
1968b; Zdarek and Denlinger, 1975; Denlinger, 1979). It is more
likely that JH is playing an indirect role, perhaps by priming the
tissue (Flanagan and Hagedorn, 1977) to respond to MH. Such a role

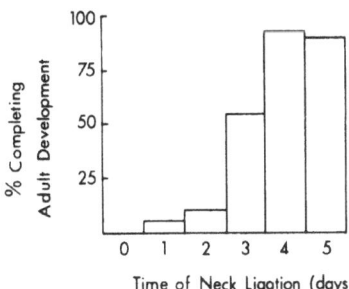

Fig.13   Percent of S. crassipalpis completing adult development at
         20°C when neck ligated at daily intervals following stimula-
         tion of development with a 5 μl topical application of
         hexane.

is consistent with the dramatic synergism observed when a JH analogue
is applied to diapausing flesh fly pupae simultaneously with 20-
hydroxy-ecdysone (Fig. 14). In the presence of JH analogue, as
little as 0.05 μg 20-hydroxy-ecdysone is sufficient to break diapause, while 5 μg is required to elicit the same response in the
absence of JH.

Cyclic GMP

Cyclic GMP and its derivatives promptly initiate adult development when injected into diapausing pupae (Fig. 15), whereas cyclic
AMP and its derivatives are inert (Denlinger and Wingard, 1978).
The effectiveness of cyclic GMP is further enhanced when injected
in concert with a low level of 20-hydroxy-ecdysone: the effect is
synergistic (Fig. 16). In contrast, dibutyryl cyclic AMP anta-
gonizes the effect of a 20-hydroxy-ecdysone injection. Opposing
actions for cyclic AMP and cyclic GMP have been noted in certain
vertebrate tissues (Goldberg et al., 1975), and Bodnaryk (1975)
found a similar opposing action for the two cyclic nucleotides in
pupal diapause termination of the armyworm Mamestra configurata.

The efficacy of cyclic GMP and its potency when combined with
20-hydroxy-ecdysone suggest that elevation of the level of this
cyclic nucleotide may be an important link in the chain of events
leading to initiation of adult development. However, brains and
ring glands assayed for cyclic GMP and cyclic AMP at short intervals
after hexane stimulation show no increase in either cyclic nucleotide
during the first 24 hrs after treatment (Fig. 17). Only after the
MH titer has already risen can a slight, but significant, cyclic AMP

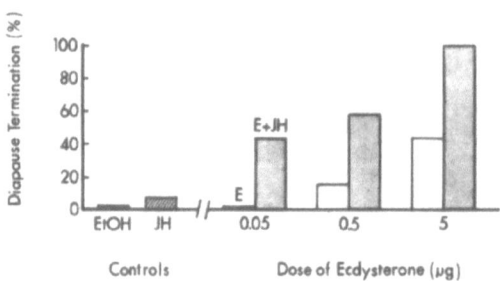

Fig.14   Diapause termination in pupae of S. inzi injected with
         various doses of 20-hydroxy-ecdysone (E) in 10% ethanol
         or treated with a combination of 20-hydroxy-ecdysone and a
         50 μg topical application of the juvenile hormone analogue
         ZR515 (Denlinger, 1979).

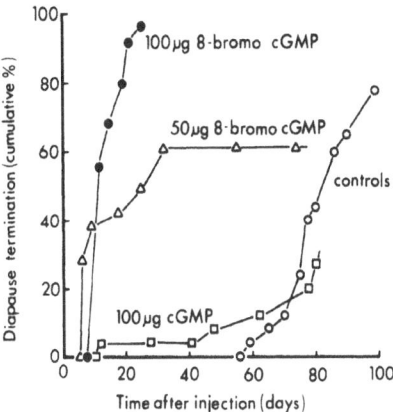

Fig.15   Rate of diapause termination at 19°C in pupae of S.
         crassipalpis injected with cyclic GMP, 8-bromo cyclic GMP,
         or distilled water (Denlinger and Wingard, 1978).

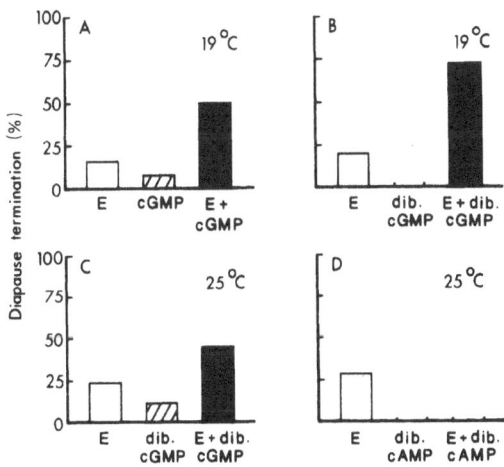

Fig.16   Interaction of 20-hydroxy-ecdysone and cyclic nucleotides
         in termination of pupal diapause in S. crassipalpis. Pupae
         were injected with 0.1 μg 20-hydroxy-ecdysone (E), a cyclic
         nucleotide, or a combination of the two. Cyclic nucleo-
         tide dose was 100 μg except in C where 50 μg was used.
         Nucleotides tested include cyclic GMP (cGMP) dibutyryl cyclic
         GMP (dib. cGMP), and dibutyryl cyclic AMP (dib. cAMP)
         (Denlinger and Wingard, 1978).

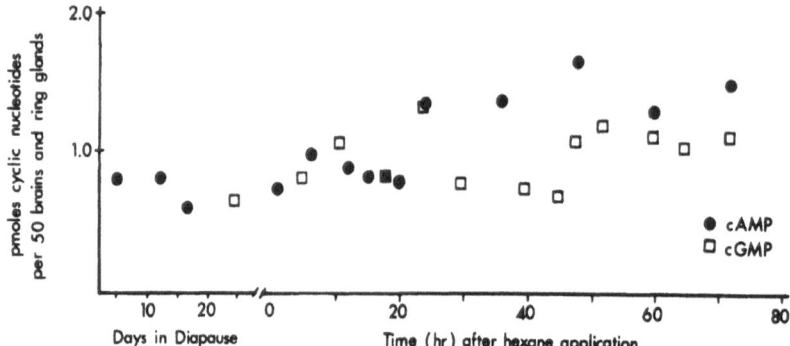

Fig.17   Levels of cyclic AMP and cyclic GMP in the brains and
         ring glands of S. crassipalpis during diapause and at various
         intervals following stimulation of adult development with
         hexane (Gnagey and Denlinger, 1980).

elevation be detected.   Thus a role in activating the neuroendocrine
system seems unlikely.   An interacting role with MH at the level of
the target tissue is another possibility, and, indeed, levels of
cyclic GMP in whole body homogenates rise steadily for at least
three days following hexane stimulation (Fig. 18).   Levels of
cyclic AMP in whole body homogenates are much lower and remain

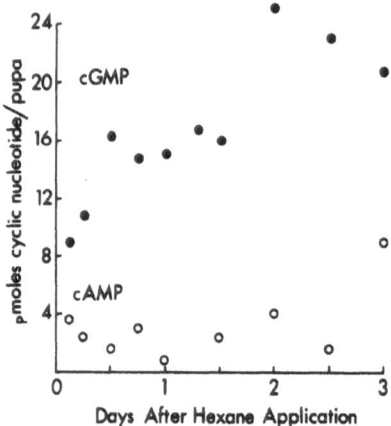

Fig.18   Whole body levels of cyclic AMP and cyclic GMP in S.
         crassipalpis at various intervals following stimulation of
         adult development with hexane (Gnagey and Denlinger,
         1980).

consistently low during the same interval. Preliminary results
thus suggest a potential interaction of ecdysteroids and cyclic
GMP in non-endocrine tissue.

Our failure to detect a rise in cyclic AMP in the brain-ring
gland complex prior to MH release presents another interesting
puzzle. High cyclic AMP in the ring gland is important shortly
after pupariation to avoid pupal diapause and promote immediate adult
development (Fig. 7), yet ring glands of diapausing pupae do not
show a similar rise of cyclic AMP prior to initiation of adult
development. Although the result is puzzling it is consistent with
the observed ineffectiveness of exogenous cyclic AMP and cholera
toxin in breaking diapause (Denlinger and Wingard, 1978). The
apparent discrepancy may indeed reflect two distinctly different
mechanisms for activating the prothoracic gland.

PROSPECTS

The experiments with flesh flies started so unassumingly by
Fraenkel have fostered the pursuit of a broad approach to diapause
that has the potential of culminating in a detailed understanding
of the ecology and physiology of pupal diapause in a discrete
taxonomic group. The large size of flesh flies, ease of laboratory
rearing, high reproductive rate, rapid generation time, and wide
distribution offer many advantages desirable in an experimental
animal. Already, work on the clock mechanism, the discovery of
cyclic JH pulses, results with cyclic nucleotides, and the peculiar
oxygen consumption cycles observed in flies point to the danger of
assuming that the Saturniid model for pupal diapause is completely
applicable to other taxa. Perhaps we should not expect pupal diapause in flies to be identical to moths. Diapause has evolved
independently many times, and the regulating mechanisms may likewise reflect an interesting diversity.

ACKNOWLEDGEMENTS

This work was supported in part by the Science and Education
Administration of the U.S. Department of Agriculture under Grant
No. 7800595 from the Competitive Research Grants Office. I thank
M. P. Murtaugh, V. C. Henrich, A. L. Gnagey, and J. Y. Bradfield
for critical reading of the manuscript.

REFERENCES

Agui, N. and Hiruma, K., 1977, In vitro activation of neurosecretory
    brain cells in Mamestra brassicae by β-ecdysone, Gen. Comp.
    Endocrinol. 33:467.
Andrewartha, H. G., 1952, Diapause in relation to the ecology of
    insects, Bio. Dev. 27:50.

Beck, S. D., 1968, "Insect Photoperiodism," Academic Press, New York.

Bodnaryk, R. P., 1975, Interaction of cyclic nucleotides and ecdysterone in breaking the pupal diapause of the Bertha armyworm, Mamestra configurata Wlk., Life Sci. 16:1411.

Bodnaryk, R. P., 1977, Stages of diapause development in the pupae of Mamestra configurata based on β-ecdysone sensitivity index, J. Insect Physiol. 23:537.

Borst, D. W., Bollenbacher, W. E., O'Connor, J. D., King, D. S., and Fristrom J. W., 1974, Ecdysone levels during metamorphosis of Drosophila melanogaster, Dev. Biol. 39:308.

Bradfield, J. Y., IV., 1980, Temporal requirements for the brain and prothoracic gland in adult development of the tobacco hornworm, in preparation.

Bradfield, J. Y., IV, and Denlinger, D. L., 1980, Diapause development in the tobacco hornworm: a role for ecdysone or juvenile hormone?, Gen. Comp. Endocrinol., in press

Bradshaw, W. E., 1976, Geography of photoperiodic response in diapausing mosquito, Nature 262:384.

Buck, J. and Keister, M., 1955, Cyclic CO_2 release in diapausing Agapema pupae, Biol. Bull. 109:144

Chippendale, G. M., 1977, Hormonal regulation of larval diapause, Ann. Rev. Entomol. 22:121.

Chippendale, G. M., Reddy, A. S., and Catt, C. L., 1976, Photoperiodic and thermoperiodic interaction in the regulation of the larval diapause of Diatraea grandiosella, J. Insect Physiol. 22:823.

Claret, J., Dray, F., and Porcheron, P., 1977, Critical period and ecdysone titers in the pupae of Pieris brassicae, Experientia 33:1389.

Coppel, H. C., House, H. L., and Maw, M. G., 1959, Studies on dipterous parasites of the spruce budworm, Choristoneura fumiferana (Clem.) (Lepidoptera:Tortricidae) VII. Agria affinis (Fall.) (Diptera:Sarcophagidae), Can. J. Zool. 37:817.

Crozier, A. J. G., 1979, Supradian and infradian cycles of oxygen uptake in diapausing pupae of Pieris brassicae, J. Insect Physiol. 25:575.

Danilevskii, A. S., 1965, "Photoperiodism and seasonal development of insects," Oliver and Boyd, Edinburgh.

Denlinger, D. L., 1971, Embryonic determination of pupal diapause in the flesh fly Sarcophaga crassipalpis, J. Insect Physiol. 17:1815.

Denlinger, D. L., 1972a, Induction and termination of pupal diapause in Sarcophaga (Diptera:Sarcophagidae), Biol. Bull. 142:11.

Denlinger, D. L., 1972b, Seasonal phenology of diapause in the flesh fly Sarcophaga bullata, Ann. Entomol. Soc. Am. 65:410.

Denlinger, D. L., 1974, Diapause potential in tropical flesh flies, Nature 252:223.

Denlinger, D. L., 1976, Preventing insect diapause with hormones
    and cholera toxin, Life Sci. 19:1485.
Denlinger, D. L., 1978, The developmental response of flesh flies
    (Diptera:Sarcophagidae) to tropical seasons: variation in
    generation time and diapause in East Africa, Oecologia 35:105.
Denlinger, D. L., 1979, Pupal diapause in tropical flesh flies:
    environmental and endocrine regulation, metabolic rate and
    genetic selection, Biol. Bull. 156:31.
Denlinger, D. L., 1980, Seasonal and annual variation of insect
    abundance in the Nairobi National Park, Kenya, Biotropica,
    in press.
Denlinger, D. L. and Wingard, P., 1978, Cyclic GMP breaks pupal
    diapause in the flesh fly Sarcophaga crassipalpis, J. Insect
    Physiol. 24:715.
Denlinger, D. L., Campbell, J. J., and Bradfield, J. Y., 1980,
    Stimulatory effect of organic solvents on initiating develop-
    ment in diapausing pupae of the flesh fly, Sarcophaga crassi-
    palpis, and the tobacco hornworm, Manduca sexta, Physiol. Entom.
    5:7.
Denlinger, D. L., Willis, J. H., and Fraenkel, G., 1972, Rates and
    cycles of oxygen consumption during pupal diapause in Sarcophaga
    flesh flies, J. Insect Physiol. 18:871.
de Wilde, J., 1962, Photoperiodism in insects and mites, Ann. Rev.
    Entomol. 7:1.
Finkelstein, R. A., 1973, Cholera, Crit. Rev. Microbiol. 2:553.
Flanagan, T. R. and Hagedorn, H. H., 1977, Vitellogenin synthesis
    in the mosquito: the role of juvenile hormone in the develop-
    ment of responsiveness to ecdysone, Physiol. Entom. 2:173.
Fraenkel, G. and Hsiao, C., 1968a, Manifestations of a pupal
    diapause in two species of flies, Sarcophaga argyrostoma and
    S. bullata, J. Insect Physiol. 14:689.
Fraenkel, G. and Hsiao, C., 1968b, Morphological and endocrinologi-
    cal aspects of pupal diapause in a flesh fly, Sarcophaga
    argyrostoma (Robineau-Desvoidy), J. Insect Physiol. 14:707.
Fristrom, J. W., Logan, W. R., and Murphy, C., 1973, The synthetic
    and minimal culture requirements for evagination of imaginal
    discs of Drosphila melanogaster in vitro, Dev. Biol. 33:441.
Gibbs, D., 1975, Reversal of pupal diapause in Sarcophaga
    argyrostoma by temperature shift after puparium formation,
    J. Insect Physiol. 21:1179.
Gibbs, D., 1976, The initiation of adult development in Sarcophaga
    argyrostoma by β-ecdysone, J. Insect Physiol. 22:1195.
Gnagey, A. L. and Denlinger, D. L., 1980, Brain and ring gland
    cyclic AMP and cyclic GMP levels during pupal diapause induction
    and termination in flesh flies, in preparation.
Goldberg, N. D., Haddox, M. K., Nicol, S. E., Glass, D. B., Sanford,
    C. H., Kuehl, F. A., and Estensen, R., 1975, Biological regu-
    lation through opposing influences of cyclic GMP and cyclic AMP:
    the Yin Yang hypothesis, Adv. Cyclic Nuc. Res. 5:307.

Greenfield, M. D. and Karandinos, M. G., 1976, Oviposition rhythm
    of Synanthedon pictipes under a 16:8 L:D photoperiod and
    various thermoperiods, Environ. Entomol. 5:712.
Greengard, P., 1978, "Cyclic Nucleotides, Phosphorylated Proteins,
    and Neuronal Function," Raven Press, New York.
Hiruma, K., Yagi, S., and Agui, N., 1978, Action of juvenile hor-
    mone on cerebral neurosecretory cells of Mamestra brassicae
    in vivo and in vitro, Appl. Entomol. Zool. 13:149.
House, H. L., 1967, The decreasing occurrence of diapause in the
    fly Pseudosarcophaga affinis through laboratory-reared genera-
    tions, Can. J. Zool. 45:149.
Hsiao, T. H. and Hsiao, C.,1977, Simultaneous determination of
    molting and juvenile hormone titers of the greater wax moth,
    J. Insect Physiol. 23:89.
Kurahashi, H. and Ohtaki, T., 1977, Crossing between nondiapausing
    and diapausing races of Sarcophaga peregrina, Experientia 33:
    186.
Kurahashi, H. and Ohtaki, T., 1979, Induction of pupal diapause
    and photoperiodic sensitivity during early development of
    Sarcophaga peregrina larvae, Jap. J. Med. Sci. Biol. 32:77.
Lees, A. D., 1955, "The Physiology of Diapause in Arthropods,"
    Cambridge University Press, Cambridge.
Mandaron, D., 1973, Effects of α-ecdysone, β-ecdysone and inoko-
    sterone on the in vitro evagination of Drosophila leg discs
    and the subsequent differentiation of imaginal integumentary
    structures, Dev. Biol. 31:101.
McDaniel, C. N., 1979, Haemolymph ecdysone concentrations in
    Hyalophora cecropia pupae, dauer pupae and adults, J. Insect
    Physiol. 25:143.
Ohmori, K. and Ohtaki, T., 1973, Effects of ecdysone analogues on
    development and metabolic activity of wing disks of the
    fleshfly, Sarcophaga peregrina, in vitro, J. Insect Physiol.
    19:1199.
Ohtaki, T., Milkman, R. D., and Williams, C. M., 1968, Dynamics
    of ecdysone secretion and action in the fleshfly Sarcophaga
    peregrina, Biol. Bull. 135:322.
Ohtaki, T. and Takahashi, M., 1972, Induction and termination of
    pupal diapause in relation to the change of ecdysone titer in
    the fleshfly, Sarcophaga peregrina, Jap. J. Med. Sci. Biol. 25:
    369.
Pittendrigh, C. S., 1966, The circadian oscillation in Drosophila
    pseudoobscura pupae: a model for the photoperiodic clock, Z.
    Pflanzenphysiol. 54:275.
Postlethwait, J. H. and Schneiderman, H. A., 1970, Induction of
    metamorphosis by ecdysone analogues: Drosophila imaginal disks
    cultured in vivo, Biol. Bull. 138:47.
Roberts, B. and Warren, M. A., 1975, Diapause in the Australian
    flesh fly Tricholioproctia impatiens (Diptera:Sarcophagidae),
    Aust. J. Zool. 23:563.

Robison, G. A., Butcher, R. W., and Sutherland, E. W., 1971, "Cylic AMP," Academic Press, New York.

Roubaud, E., 1922, Etudes sur le sommeil d'hiver préimaginal des Muscides, Bull. Biol. Fr. Belg. 56:455.

Saunders, D. S., 1971, The temperature-compensated photoperiodic clock 'programming' development and pupal diapause in the flesh-fly, Sarcophaga argyrostoma, J. Insect Physiol. 17:801.

Saunders, D. S., 1973, The photoperiodic clock in the flesh-fly, Sarcophaga argyrostoma, J. Insect Physiol. 19:1941.

Saunders, D. S., 1975, 'Skeleton' photoperiods and the control of diapause and development in the flesh-fly, Sarcophaga argyrostoma, J. Comp. Physiol. 97:97.

Saunders, D. S., 1976, The circadian eclosion rhythm in Sarcophaga argyrostoma: some comparisons with the photoperiodic "clock," J. Comp. Physiol. 110:111.

Saunders, D. S., 1978a, An experimental and theoretical analysis of photoperiodic induction in the flesh-fly, Sarcophaga argyrostoma, J. Comp. Physiol. 124:75.

Saunders, D. S., 1978b, Internal and external coincidence and the apparent diversity of photoperiodic clocks in the insects, J. Comp. Physiol. 127:197.

Schneiderman, H. A. and Williams, C. M., 1953, The physiology of insect diapause. VII. The respiratory metabolism of the Cecropia silkworm during diapause and development, Biol. Bull. 105:320.

Shaaya, E. and Karlson, P., 1965, Der Ecdysontiter wahrend der Insektenentwicklung. II. Die-postembryonale Entwicklung der Schmeissfliege Calliphora erythrocephala Meig., J. Insect Physiol. 11:65.

Sivasubramanian, P., 1977, Evagination of imaginal discs in the fleshfly Sarcophaga crassipalpis: hormonal control in vivo, Wilhelm Roux's Arch. 183:101.

Stone, A., Sabrosky, C. W., Wirth, W. W., Foote, R. H., and Coulson, J. R., 1965, "A Catalog of the Diptera of America North of Mexico," U. S. Government Printing Office, Washington, D. C.

Takeda, N., 1978, Hormonal control of prepupal diapause in Monema flavescens (Lepidoptera), Gen. Comp. Endocrinol. 34:123.

Tauber, M. J. and Tauber, C. A., 1976, Insect seasonality: diapause maintenance, termination, and post-diapause development, Ann. Rev. Entomol. 21:81.

Ternovoy, V. I., 1978, A study of the diapause in Wohlfahrtia magnifica (Diptera, Sarcophagidae), Entomol. Rev. 57:328.

Vedeckis, M. V., Bollenbacher, W. E., and Gilbert, L. I., 1976, Insect prothoracic glands: a role for cyclic AMP in the stimulation of α-ecdysone secretion, Molec. Cell. Endocrinol. 5:81.

Vinogradova, E. B., 1976, Embryonic photoperiodic sensitivity in two
    species of fleshflies, Parasarcophaga similis and Boettcherisca
    septentrionalis, J. Insect Physiol. 22:819.
Vinogradova, E. B. and Zinovjeva, K. B., 1972, The control of
    seasonal development in parasites of blow-flies. 1. Ecological
    regulation of pupal diapause in sarcophagids (Diptera, Sarco-
    phagidae), in: "Host-Parasite Relationships in Insects,"
    Nauka, Leningrad. (in Russian)
Walker, G. P. and Denlinger, D. L., 1980, Juvenile hormone and moult-
    ing hormone titers in diapause and nondiapause destined flesh
    flies, J. Insect Physiol., in press.
Whitten, J. M., 1969, Coordinated development in the footpad of the
    fly Sarcophaga bullata during metamorphosis: changing puffing
    patterns of the giant cell chromosomes, Chromosoma 26:215.
Williams, C. M., 1947, Physiology of insect diapause. II. Inter-
    action between the pupal brain and the prothoracic glands in
    the metamorphosis of the giant silkworm, Platysamia cecropia,
    Biol. Bull. 93:89.
Williams, C. M., 1952, Physiology of insect diapause. IV. The
    brain and prothoracic glands as an endocrine system in the
    Cecropia silkworm, Biol. Bull. 103:120.
Williams, C. M., 1956, Physiology of insect diapause. X. An
    endocrine mechanism for the influence of temperature on the
    diapausing pupa of the Cecropia silkworm, Biol. Bull. 110:201.
Zdarek, J. and Denlinger, D. L., 1975, Action of ecdysoids, juvenoids,
    and non-hormonal agents on termination of pupal diapause in
    the flesh fly, J. Insect Physiol. 21:1193.
Zdarek, J. and Fraenkel, G., 1970, Overt and covert effects of
    endogenous and exogenous ecdysone in puparium formation of
    flies, Proc. Nat'l. Acad. Sci. USA 67:331.
Zdarek, J. and Fraenkel, G., 1971, Neurosecretory control of ecdysone
    release during puparium formation of flies, Gen. Comp. Endocrin.
    17:483.

BLOOD-BORNE FACTORS CONTROLLING PUPARIATION IN FLIES

Jan Zdarek
Institute of Entomology
Czechoslovak Academy of Sciences
Prague, Czechoslovakia

DISCOVERY OF THE PUPARIATION FACTORS

It was 35 years after Gottfried Fraenkel ligated his first
blowfly larva (Fraenkel, 1935) that I got the opportunity to work
in his laboratory. By this time, in 1970, the ligation experiment had become generally known as the Calliphora-test, and had
served as the major bioassay for the isolation and identification
of the molting hormone, ecdysone (Butenandt and Karlson, 1954;
Huber and Hoppe, 1965). When I joined the laboratory, the validity
of the test was being challenged by a group at Harvard (Ohtaki
et al., 1968), and one of my first tasks was to reinvestigate certain of the test conditions (Fraenkel and Zdarek, 1970).

To do this, I had to make thousands of ligations, because, as
anyone familiar with the Calliphora test knows, only a minority of
larvae ligated behind the CNS yield preparations suitable for the
test, i.e., with the front region contracted and tanned as a puparium
and the hind region remaining larval. As for the rest, many do not
tan at all, some tan only in the hind part, and some tan both anteriorly and posteriorly. These useless specimens were always thrown
away, but on one such occasion, we noticed a conspicuous asynchrony
in the darkening rate of the front and hind parts of specimens which
were tanning on both sides of the ligature. It was this observation
which initiated the investigations leading to the discovery of the
pupariation factors.

Specimens with tanned hind parts can be purposely produced by
ligating well after the critical period for the action of ecdysone:
that is, at a time when the whole body is already committed to

pupariation. Certain of these larvae will fail to tan anteriorly, a condition which is attributed to respiratory deficiencies caused by injury to the tracheal system during ligation (Ratnasiri and Fraenkel, 1974a;b).

However, a majority of the postcritically ligated larvae will tan in both parts. In such cases, the anterior part invariably starts to tan at about the same time as nonligated controls, independent of the period between ligation and tanning, whereas the posterior part does not tan until 1-2 hours later.

What causes the delay?

In a series of simple experiments we eliminated the nervous system as a causative agent. Disconnection of the hind part from the CNS led to tanning without the normal barrel-shaped puparial contraction, but did not delay tanning. When a ligature was tightly applied and immediately removed, the shape of the puparium remained larval due to denervation of the posterior section, but both sections tanned simultaneously. Thus, the delay in tanning behind the ligature must have been due to the absence in the hind part of a blood-borne factor, produced, and/or stimulated to be released by the front part.

Indeed, injection of only a small amount (1 µl of hemolymph from a pupariating larva accelerated tanning in the hind part to the extent that it began earlier than in the front part. Similar accelerating effects were obtained from injection of homogenates of tissues known to produce, store, or release neurosecretory materials (e.g., pars intercerebralis of the brain, ring gland, corpora cardiaca, or crayfish eye stalks) (Zdarek and Fraenkel, 1969). Moreover, non-ligated larvae, when injected with the materials mentioned above not more than approximately 4 hours before the expected onset of pupariation, pupariated much sooner than controls. These precociously produced puparia were almost normal, showing the longitudinal contraction and smoothening of the surface before tanning. All of this led to the conclusion that there existed one or more blood-borne "pupariation factors", presumably of neurosecretory origin, which participate in the control of formation and tanning of the puparium in cyclorrhaphous Diptera.

In the initial paper on the subject, only one factor was considered. We called it "a neurohormone accelerating formation and tanning of the fly puparium" (Zdarek and Fraenkel, 1969). In the following paper two separate factors, accelerating tanning and retraction, were recognized and named X_t and X_r, respectively (Fraenkel et al., 1972). After further purification, these two activities were named the puparium tanning factor (PTF) and the anterior (segments) retraction factor (ARF), according to the

effects the assayed materials produced in the tests then used
(Sivasubramanian et al., 1974). More recently, when the effects of
active materials (hemolymph, CNS extracts) were reinvestigated by
continuously recording muscular activity, more effects were recognized, and these have been preliminarily attributed to two other
neurotropic factors, namely the pupariation immobilization factor
(PIF) and the pupariation stimulation factor (PSF). This article,
then, is a short review of our current knowledge of the occurrence,
identity, mode of action and biological function of the pupariation
factors.

THE PUPARIUM TANNING FACTOR (PTF)

This is the first discovered factor (Zdarek and Fraenkel, 1969),
and thus far, the most intensively investigated. Since the information regarding this material has been in print for several years,
I shall give here only a brief recapitulation of the most relevant
facts. PTF activity is most easily tested in a fleshfly (e.g.,
Sarcophaga bullata) ligated 3-4 hours before pupariation, when the
cuticle of the hind spiracular plate just begins to tan. Materials
to be tested are injected into the hind part, and PTF activity is
evaluated by recording the length of the pretanning period in the
anterior (A) and posterior (P) regions. In controls, the ratio P/A
is greater than 1, usually about 1.5, because the hind part tans
with a delay. Accelerated tanning in the injected hindpart signifies
PTF activity in the injected materials. With the most active
materials P/A has value of about 0.5.

PTF activity was shown in tissue with neurosecretory activity
such as brain and corpora cardiaca, and was found to be present in
large amount in the hemolymph of fly larvae during the first hours
of tanning (after pupariation).

It has been purified to some extent from both the hemolymph and
CNS of Sarcophaga bullata, and in view of the similarity in properties of the material from both sources, PTF is thought to originate
in the CNS. Its molecular weight is well over 300,000 daltons, and
it has the properties of a protein, being precipitated by TCA and
ammonium sulfate, and destroyed by proteases. It is relatively
stable to heating and can be lyophilized without damage
(Sivasubramanian et al., 1974).

PTF appears in the hemolymph of a ligated hindpart after the
beginning of tanning when the ligation has been made after the
critical period of ecdysone release. It also appears if the animal
is ligated before that time and subsequently injected in the hind
part with ecdysone. In the latter case, the CNS is, of course,
absent, and in order to maintain the presumption of a neurosecretory origin it has been postulated that the factor is stored

in peripheral nerve terminals and released into the hemolymph under
the influence of ecdysone (Sivasubramanian et al., 1974). Tanning
per se is not a prerequisite for the appearance of PTF in the hemolymph, since puparia injected with an inhibitor of dopa decarboxylase
("αMDH") do not tan but do accumulate blood PTF.

It has recently been shown that certain cyclic nucleotides mimic
the effect of PTF, and that accelerated tanning can proceed in the
absence of PTF when certain phenolic intermediates in the overall
reaction: tyrosine→N-acetyl dopamine are injected (Fraenkel et al.,
1977). This evidence, and that provided through the use of inhibitors of transcription and translation have led to the suggestion
that PTF operates at the level of the gene in exerting some unknown
control over the reaction: tyrosine→DOPA (Seligman et al., 1977).
It is obvious that further investigation of this activity must be
pursued before the role of PTF at the molecular level is established.

THE FACTORS AFFECTING PUPARIATION BEHAVIOR

Formation of the puparium, a unique phenomenon in the insect
world, is the result of a stereotyped behavior accompanied by
irreversible deplasticization and hardening of the larval cuticle.
The process starts with a growing immobilization of a crawling larva.
Then, the three anterior segments of the immobile larva bearing
the cephalopharyngeal apparatus ("the mouth hooks") are retracted
so that the cornicles of the anterior spiracles assume a position
at the front end of the puparium. Finally, in the last 20-minute
period, the major morphological transformations take place. The
larva contracts to the regular barrel-shaped puparium and, at the
same time, the surface becomes smooth due to "shrinkage" of the
cuticle. After this, the new shape gets fixed by sclerotization
and tanning of the cuticle (Zdarek and Fraenkel, 1972).

As already mentioned, the hemolymph of pupariating larvae and
larval CNS extracts contain factors which precociously induce
pupariating behavior. A wandering larva injected with active
material 4 or less hours before the expected onset of puparial
contraction almost immediately stops crawling, retracts the
anterior segments, and subsequently contracts the whole body to a
more or less perfect puparial shape. The factor responsible for
acceleration of this retraction behavior was called the ARF by
Sivasubramanian et al. (1974).

Certain important characteristics of the process of pupariation cannot be seen and appreciated by mere observation. A very
efficient technique for recording the changes concerned with
muscular activity and with transformation of the cuticle is a

barographic method which continuously measures the internal pressure changes during pupariation (Slama, 1976; Zdared et al., 1979). The barogram reveals that changes in internal pressure are basically of two types: highly specific reversible pulsations of various kinds, reflecting different patterns of muscular contraction, and an irreversible increase of the pressure baseline caused by shrinkage and loss of elasticity of the cuticle, associated with ultrastructural reorganization and sclerotization, respectively (Fig. 1A).

Recently, our use of barograms in the study of pupariation behavior precociously induced by hemolymph injection has revealed great differences from normal behavior, some of the patterns of neuromuscular activity being reduced or missing and others extended or otherwise altered. Fig. 1 illustrates the differences in records of pupariating larvae accelerated by hemolymph from larvae beginning to contract (B), from puparia already contracted and beginning to tan (orange puparia) (C), and by homogenates of the CNS (D). In the first case: (B), the immobilization (i) and retraction (r) stages are greatly prolonged, but the contraction (c) stage is normal. The blood from orange puparia (record C) hastens and shortens both preliminary stages (immobilization [i] and retraction [r]) and the contraction stage (c) is greatly extended and shows abnormal neuromuscular patterns. A characteristic feature of pupariation accelerated by the CNS extract (D) is a reversal of behavioral patterns; contraction precedes retraction and immobilization.

The barograph has also provided new insights into the effects of hemolymph from animals at various stages in the pupariation process on neural activity, as tested in the young wandering larva (5-24 hours before pupariation). The effects can generally be classified as immobilizing or stimulating. The former are manifested by sluggishness and cessation of locomotion for periods raning from a few minutes to one hour (Fig. 2B). The immobilization is sometimes accompanied by a temporary retraction. This activity has been attributed to a pupariation immobilizing factor (PIF) which is found in the hemolymph of orange puparia and, on the basis of its chemical properties, may be identical with ARF (Fraenkel, unpublished). Stimulating effects, preliminarily attributed to a pupariation stimulating factor (PSF), have been observed when blood from just contracting larvae, or homogenates of CNS are tested on younger wandering larvae (Fig. 2C).

Thus, it appears that compounds other than the anterior retraction factor (ARF) are present in hemolymph at various times during pupariation.

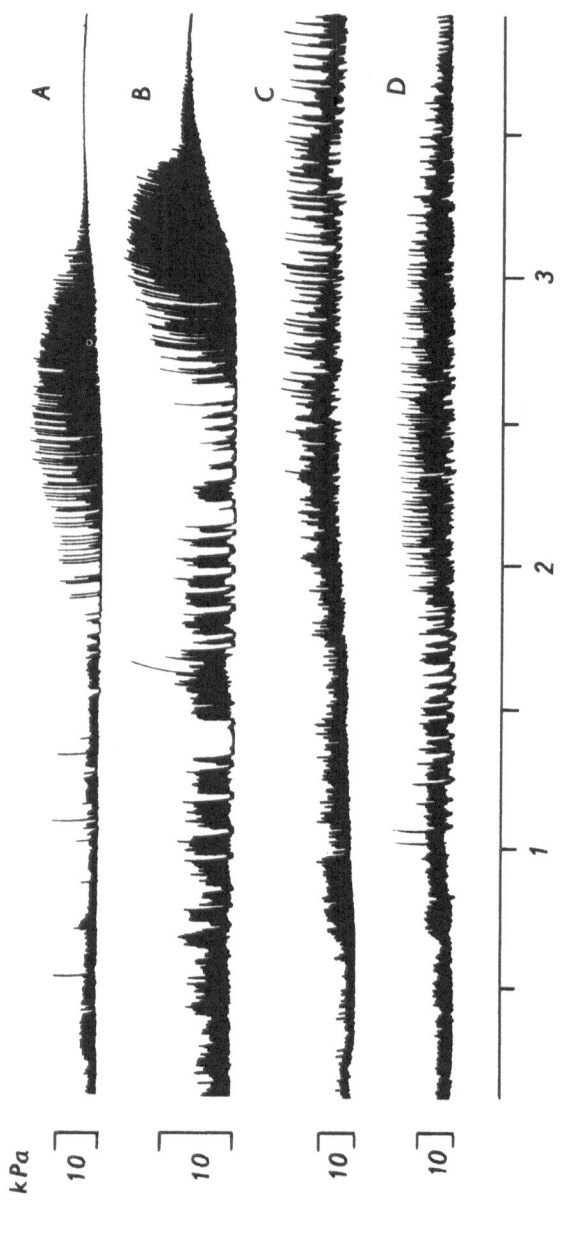

Fig.1   Barograms of the changes in internal pressure during pupariation of larvae of <u>Sarcophaga</u> <u>bullata</u> injected with different materials approximately 3 hours before expected pupariation. (A) – hemolymph of larvae of the same age (controls); (B)– hemolymph of just contracting puparia; (C) – hemolymph of 2 hour old puparia; (D) – CNS homogenate (5 μl of hemolymph was injected in (A)–(C), and 2CNS equivalents in 5 μl of the Ringer solution in (D)).

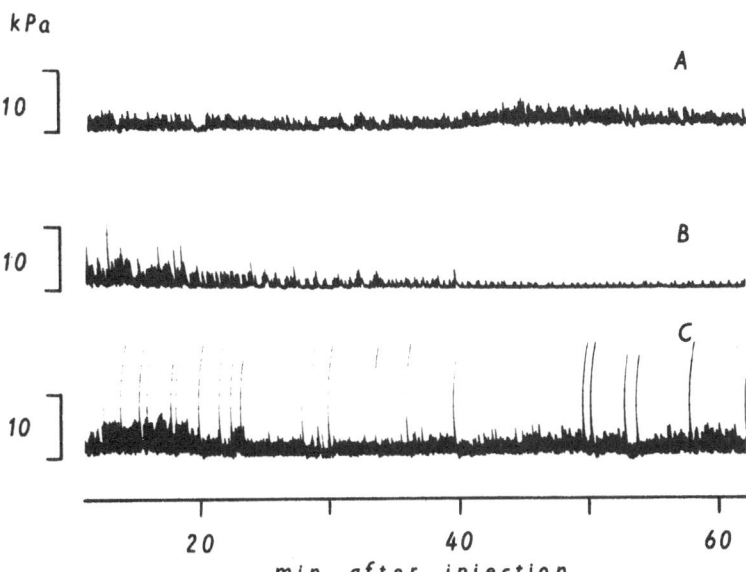

Fig2. Barograms of the changes in internal pressure in wandering
      larvae of <u>Sarcophaga</u> <u>bullata</u> after injections of; (A.) 5 µl
      hemolymph of larvae of the same age (control), (B.) 5 µl
      hemolymph from orange puparia containing the PIF, and (C.)
      5 µl hemolymph of contracting puparia containing PSF.

WHAT IS THE ROLE OF THE PUPARIATION FACTORS?

As stated previously, formation of the puparium is a complex process involving specific patterns of muscular activity and precisely timed ultrastructural rearrangements in the cuticle. The muscular and cuticular systems work in concert and any disharmony may result in lethal morphological deviations (Zdarek and Fraenkel, 1972). Thus, an efficient and accurate control system is of primary importance. On the basis of available evidence we assume that the blood-borne pupariation factors control the onset of pupariation and participate in the coordination of its performance.

The beginning of pupariation is also timed with respect to environmental conditions. In unfavorable conditions, such as, for instance, excessive moisture, pupariation is inhibited through inhibition of molting hormone release (Ohtaki, 1966). However, the molting hormone does not give a signal for immediate pupariation. The puparium is never formed earlier than about 8 hours after critical hormonal levels are reached, even if this is accomplished by injection of a massive dose of ecdysteroid. This means that there probably exists a more or less constant latent period, during which tissues concerned with pupariation (nerves, muscles, integument) become ready for the process. The pupariation factors, then, may be part of the homeostatic system ensuring a coordinated release of pre-programmed behavioral elements on the one hand, and biochemical reactions underlying sclerotization and tanning, on the other.

REFERENCES

Butenandt, A., and Karlson, P., 1954, Über die Isolierung eines Metamorphose-hormons der Insekten in kristallisierter Form. Z. Naturforsch. 9b:389-391.

Fraenkel, G., 1935, A hormone causing pupation in the blowfly, Calliphora erythrocephala, Proc. Roy. Soc. Ser. B 118:1-12.

Fraenkel, G., and Zdarek, J., 1970, The evaluation of the "Calliphora test" as an assay for ecdysone, Biol. Bull. Woods Hole 139:138-150.

Fraenkel, G., Zdarek, J., and Sivasubramanian, P., 1972, Hormonal factors in the CNS and hemolymph of pupariating fly larvae which accelerate puparium formation and tanning, Biol. Bull. Woods Hole 143:127-139.

Fraenkel, G., Blechl, A., Blechl, J., Herman, P., and Seligman, M., 1977, 3',5'cyclic AMP and hormonal control of puparium formation in the fleshfly, Sarcophaga bullata, Proc. Natl. Acad. Sci. USA 74:2182-2186.

Huber, R., and Hoppe, W., 1965, Zur Chemie des Ekdysons. VII. Die Kristall und Molekül-analyze des Insektenverpuppungshormons

Ecdyson mit der automatisierten Falt-Molekülmethode. Chem.
Ber. 98:2403-2424.

Ohtaki, T., 1966, On the delayed pupation of the fleshfly, Sarcophaga
peregrina Robineau-Desvoidy, Jpn. J. Med. Sci. Biol. 19:97-104.

Ohtaki, T., Milkman, R. P., and Williams, C. M., 1968, Dynamics of
ecdysone secretion and action in the fleshfly, Sarcophaga
peregrina, Biol. Bull. Woods Hole 135:322-334.

Ratnasiri, N. P., and Fraenkel, G., 1974a, Anterior inhibition of
pupariation in ligated larvae of Sarcophaga bullata and other
fly species: Incidence and expression, Ann. Entomol. Soc. Am.
67:195-203.

Ratnasiri, N. P., and Fraenkel, G., 1974b, The physiological basis
of anterior inhibition of puparium formation in ligated fly
larvae, J. Insect Physiol. 20:105-119.

Seligman, M., Blechl, A., Blechl, J., Herman, P., and Fraenkel, G.,
1977, Role of ecdysone, pupariation factors, and cyclic AMP
in formation and tanning of the puparium of the fleshfly, Sarcophaga bullata, Proc. Natl. Acad. Sci. USA 74:4697-4701.

Sivasubramanian, P., Friedman, S., and Fraenkel, G., 1974, Nature
and role of proteinaceous hormonal factors acting during
puparium formation of flies, Biol. Bull. Woods Hole 146:163-185.

Sláma, K., 1976, Insect hemolymph pressure and its determination,
Acta Ent. Bohemoslov. 73:65-75.

Zdarek, J., and Fraenkel, G., 1969, Correlated effects of ecdysone
and neurosecretion in puparium formation (pupariation) in
flies, Proc. Natl. Acad. Sci. USA 64:565-572.

Zdarek, J., and Fraenkel, G., 1972, The mechanism of puparium formation in flies, J. Exp. Zool. 179:315-324.

Zdarek, J., Sláma, K., and Fraenkel, G., 1979, Changes in internal
pressure during puparium formation in flies, J. Exp. Zool.
207:187-196.

HORMONAL CONTROL OF PUPAL ECDYSIS IN THE TOBACCO HORNWORM, MANDUCA SEXTA[1]

James W. Truman and Paul H. Taghert

Department of Zoology
University of Washington
Seattle, Washington 98195

Molting in insects is a complex process that begins with apolysis, extends through the production of a new cuticle, and ends with ecdysis and the expansion and hardening of the new skin. This process is controlled by an interplay between at least 5 hormones (see Riddiford and Truman, 1978, for a review), three of which are involved in the initiation and direction of the molt. The prothoracicotropic hormone (PTTH) from the brain exerts a tropic influence on the prothoracic glands, driving them to release α-ecdysone. This steroid is then converted by peripheral tissues to the active form, 20-hydroxyecdysone, which causes apolysis and the beginning of secretion of a new cuticle by the epidermis. The type of cuticle that is secreted is regulated by the titer of the third hormone, juvenile hormone. The final phases of the molt are controlled by 2 additional hormones: the eclosion hormone triggers the behavior involved in the shedding of the old cuticle, and bursicon stimulates the postecdysial tanning of the new cuticle.

Activities corresponding to PTTH, ecdysone, juvenile hormone, and bursicon have been found in various orders of insects and these hormones are likely to be involved in the molting processes of all insects. In contrast, the eclosion hormone has been implicated only in the control of adult eclosion of certain Lepidoptera (Truman, 1973a; Reynolds et al., 1979). Although fragmentary evidence in crickets (Carlson, 1977) and locusts (Miller and Mills,

---

[1]This paper is dedicated to Professor Gottfried Fraenkel who has been a pioneer in the field of insect endocrinology and whose work continues to set standards for experimental endocrinology.

1976) suggests that hormonal triggering of ecdysis may be more
widespread, the work on adult eclosion in moths still stands as the
only unequivocal demonstration of this phenomenon. This paper
examines the question of whether the involvement of eclosion hormone
in ecdysis is confined only to the eclosion of the imago in the
Lepidoptera.

In pharate adult moths, the time of eclosion is dictated by a
circadian clock that is located in the brain. This clock is photo-
sensitive in saturniid moths (Truman, 1972b) and temperature sensitive in the case of the sphinx moth, Manduca sexta, which emerges
from a subterranean pupation chamber (Lockshin et al., 1975). At
the end of adult development and at a species-specific time of day,
the clock stimulates the release of stored eclosion hormone from the
neurohaemal corpora cardiaca (CC) (Truman, 1978). This hormone has
a number of effects on the pharate moth, one of which is the release
of the behaviors used for the escape from the pupal cuticle and
cocoon (Truman, 1971).

When larval ecdyses were compared to adult eclosion they appeared to differ in at least three respects: (i) the control over
the timing of ecdysis; (ii) the role of the brain in this control;
and (iii) the amount of extractable eclosion hormone in the head
(Truman, 1972a).

A comparison of the time of ecdysis among various instars in
Manduca revealed that each instar ecdysed at a characteristic time
of day. However, with each successive instar, the time of ecdysis
shifted to later in the day and the distribution became progressively
broader. It was also noted that PTTH release, used to initiate each
molt, was gated and always occurred at approximately the same time
of day. From this information, it was concluded that ecdysis itself was not gated but occurred immediately following the completion
of the development of the new stage. The generally observed
synchrony in larval ecdysis was due to the larvae initiating the
molt at the same time, i.e., at the gated release of PTTH, 1.5 to 2
days before. As the larvae grew, they required more time to make the
larger cuticle for each succeeding instar. Thus, the ecdysis times
for later instars were progressively delayed and the distributions
broadened. Pupal ecdysis in Manduca showed relatively poor synchrony,
which appeared to be due to an extension of the principles that regulated the larval ecdyses.

A second way by which larval ecdyses differed from adult eclosion was in the importance of the brain in the timing of the behavior. As illustrated by an example using Antheraea pernyi (Fig. 1),
the removal of the brain from developing adults seriously disrupted
the timing of eclosion (Truman and Riddiford, 1970). By contrast,
in the case of the ecdysis of 5th instar Manduca, the removal of

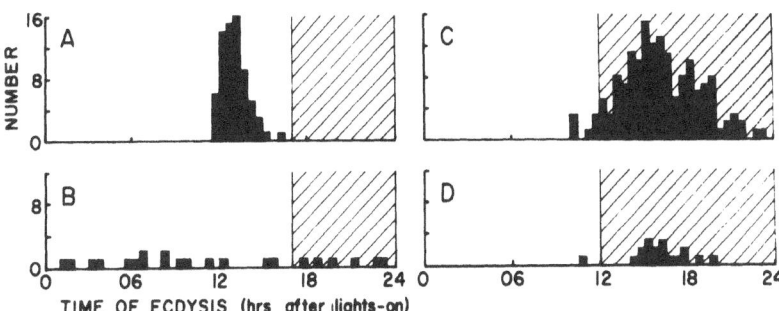

Fig.1   Distribution of adult eclosions in A.)   intact and B.)
         debrained <u>A</u>. <u>pernyi</u>. Distribution of ecdyses to the fifth
         instar in C.)   intact and D.)   decapitated <u>Manduca</u> <u>sexta</u>.
         Cross-hatched areas represent dark.

the entire head by neck ligation shortly after PTTH secretion (a-
bout 36 hours before ecdysis) had no effect on the timing of the
ecdysis attempts (Truman, 1972a). Thus, it appeared that once the
larval molt was initiated, the brain had no control over the timing
of the subsequent ecdysis.

    Pharate adults also differed from earlier stages in the amount
of eclosion hormone present in the head. In the pharate adult of
<u>Manduca</u> <u>sexta</u>, the brain and CC have substantial stores of eclosion
hormone. By contrast the same structures from prepupae had only
traces of hormone activity, as measured by their ability to provoke
eclosion of pharate adult <u>Antheraea</u> <u>pernyi</u> (Truman, 1973a). This
comparative lack of eclosion hormone in prepupae, along with the
ungated nature of larval and pupal ecdysis and the apparent unim-
portance of the brain for these ecdyses argued for substantial
differences between the control of adult eclosion and the earlier
ecdyses.

Effects of Eclosion Hormone on Pupal Ecdysis in Manduca sexta

    The low levels of eclosion hormone in prepupae did not ex-
clude the possibility that this hormone played a role in the control of pupal ecdysis behavior. This possibility was directly
tested by challenging prepupal <u>Manduca</u> <u>sexta</u> with extracts that
contained eclosion hormone (Truman et al., 1980). A time-table for
normal pupal development was established, and prepupae were staged
according to the extent of pre-ecdysial tanning of the pupal cuticle
and the degree of larval cuticle digestion. The stage selected
for the injection experiments was the "anterior shrinkage" (AS)
stage characterized by the appearance of distinct transverse folds
in the larval cuticle along the margin of the first abdominal segment. Prepupae entered stage "AS" about 3 1/2 hours before ecdysis.

"AS" stage prepupae were injected with various doses of an extract prepared from pharate adult CC. This extract is a rich source
of eclosion hormone (Truman, 1978) and caused precocious ecdysis of
the test animals--the ecdyses beginning from 47 to 130 minutes after
injection. The latency between injection and ecdysis was a function of the dose of hormone that was administered, with the largest
doses producing the shortest latencies. Fig. 2 presents a dose
response curve for the response of prepupae to various amounts of
pharate adult CC extract. Positive responses were scored as the
reciprocal of the latency (in minutes) multiplied by 10^4. Negative
responses were given a score of 51 (see Truman et al., 1980, for
explanation). Recorded in this manner, the scores increased linearly
with dosage over the range of 0.02 to 0.25 pharate adult CC equivalents (paCC equiv.) and showed a half maximal response at approximately 0.07 paCC equiv. It is of interest to note that the prepupae
were about 10 fold more sensitive to eclosion hormone than were
pharate adult A. pernyi (Truman, 1973a), so that the low levels of
hormone reported from Manduca sexta prepupae might be physiologically
relevant to the animals.

Injection of "AS" stage prepupae with highly purified eclosion
hormone resulted in essentially the same latencies as were seen
with the calculated equivalent dosage of the crude CC extract,
demonstrating that the response of prepupae to pharate adult CC
extract was due to the eclosion hormone in that extract.

Blood Titers of Eclosion Hormone Activity in Prepupae

Blood samples were taken from prepupae and the animals observed

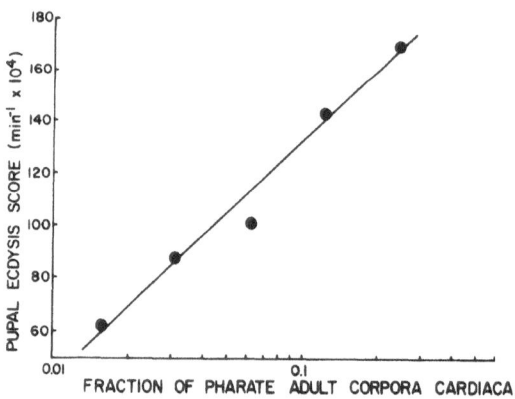

Fig.2   Relationship of the latency score of the pupal ecdysis re-
         sponse to the amount of active extract that was injected
         (From Truman et al., 1980).

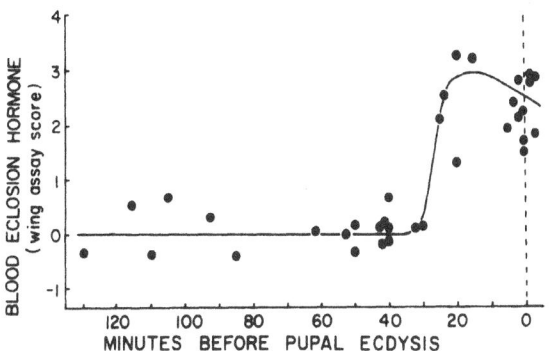

Fig.3   Eclosion hormone activity in the blood of <u>Manduca</u> prepupae
         prior to the onset of ecdysis behavior. Each point repre-
         sents the average of 2 assays on the blood from a single
         insect. Scores greater than 1 represent eclosion hormone
         activity (from Truman et al., 1980).

to determine the time of ecdysis. Eclosion hormone activity in the
blood samples was measured using the pharate adult isolated wing
bioassay (Reynolds, 1977; Reynolds and Truman, 1980). As seen in
Fig. 3, eclosion hormone activity appeared in the blood approximately 30 minutes before the onset of ecdysial behavior. The latency
between the appearance of hormone in the blood and the subsequent
onset of ecdysis behavior was much shorter than that seen for adult
eclosion in <u>Manduca</u>, in which the latency was approximately 2.5
hours (Reynolds et al., 1979).

Distribution of Eclosion Hormone Activity in the Prepupal Nervous System

The data presented to this point are consistent with the
conclusion that the eclosion hormone triggers pupal ecdysis. But
they still need to be reconciled with the finding that, at least
in the case of larval ecdysis, the head seems to play no direct
role in the timing of the behavior (Truman, 1972a). An examination
of the distribution of eclosion hormone in the nervous system of
prepupae (Taghert et al., 1980) presents a resolution of this
problem.

As seen in Table 1, eclosion hormone activity in the head of
the prepupa is confined to the brain. Interestingly, no activity is
detectable in the CC-CA complex, the region that is the richest
source of eclosion hormone in the pharate adult (Truman, 1973a;
1978). Likewise, there is no detectable activity in the subeso-
phageal ganglion. However, activity is found in the ventral nerve
cord, in both the thoracic and abdominal regions. The ventral nerve

Table 1. Eclosion hormone activity in the nervous system of <u>Manduca</u> before and after pupal ecdysis (Data from Taghert et al., 1980).

Organ	Before ecdysis			After ecdysis		
	#	Score* $(\bar{X} + \text{s.e.})$	Hormone amount (paCC equiv.)	#	Score* $(\bar{X} + \text{s.e.})$	Hormone amount (paCC equiv.)
Brain	13	$123 + 7$	0.09	10	$131 + 11$	0.10
CC–CA complex	10	$51 + 0$	<0.01		–	–
Subesophageal ganglion	9	$51 + 0$	<0.01		–	–
Thoracic ganglia	11	$85 + 11$	0.03		–	–
Abdominal ganglia	10	$133 + 5$	0.11	10	$85 + 11$	0.03

*Determined by the pupal ecdysis assay.

cord contains about 0.14 paCC equivalents, an amount greater than
that found in the brain. If the activity were uniformly distributed through the chain of ganglia, it would represent an average
of about 0.01 paCC equiv. per ganglion.

Thus, in prepupal Manduca there are 2 potential release sites
for the hormone activity, one in the brain and the other in the
chain of ventral ganglia. An indication of which is used during
pupal ecdysis was obtained by comparing the titer of stored hormone
in these 2 regions before and immediately after ecdysis. As seen
in Table 1, the activity in the brain did not change during ecdysis.
By contrast, the abdominal ganglia showed a 75% depletion of activity
during the same period. This sharp reduction indicated that the
hormone activity in the blood of ecdysing pupae most likely came
from the ventral nerve cord rather than from the brain.

This conclusion was tested by examining larvae that had their
brains removed shortly after entry into the wandering stage
(Taghert et al., 1980). They subsequently underwent pupal development and then ecdysed. At the time of ecdysis these debrained
animals showed blood titers of eclosion hormone activity similar
to those seen in control Manduca. These data, and the apparent
lack of depletion of stored material in the brain during the time
of ecdysis, indicate that the brain does not contribute to the blood-
borne hormone that triggers pupal ecdysis behavior.

Chemical Characteristics of the Prepupal Eclosion Hormone

The facts that adult eclosion hormone can trigger pupal
ecdysis and that hormone from the prepupa is active in the adult
eclosion hormone assay do not prove that the activities in
the 2 stages are due to the same molecule. Consequently, Taghert
et al. (1980) determined some of the chemical characteristics of
the prepupal activity from both the brain and the abdominal nerve
cord. Fig. 4 summarizes the data for the activity from the abdominal
nerve cord. Each fraction was tested for its ability to increase
the plasticity of isolated wings from pharate adult Manduca (Reynolds, 1977; Reynolds and Truman, 1980), to trigger pre-eclosion
or eclosion behaviors in isolated abdomens from Hyalophora cecropia
(Truman, 1978; Mumby et al., in preparation), and to evoke pupal
ecdysis behavior (Truman et al., 1980). As seen in Fig. 4A, when
pupal abdominal nerve cords were subjected to chromatography over
Sephadex G-50, only one peak of activity was detected. The peak
containing fractions showed activity in all 3 assays and had an
apparent molecular weight of 8500 daltons, the same as that determined for the eclosion hormone from the pharate adult (Reynolds
and Truman, 1980). The active peak from the Sephadex G-50 column
was then subjected to electrofocusing (Fig. 4B). Again, only one
peak of activity was observed, the peak fractions being active in

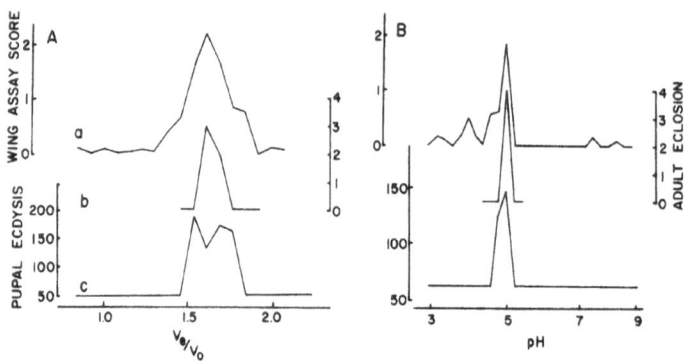

Fig.4   Distribution of biological activity in abdominal nervous
        system extracts that were subjected to A.) chromatography
        through Sephadex G-50 and B.) isoelectrofocusing. Fractions
        were tested for their ability to a) increase plasticity of
        Manduca wings, b) promote eclosion behavior, and c) trigger
        pupal ecdysis (data from Taghert et al., 1980).

all 3 assays and centering around a pH of about 5.0. When treated
in the same manner, adult eclosion hormone has shown the same pI
(Reynolds and Truman, 1980). Thus, using the criteria of apparent
molecular weight and charge, the material from the prepupal abdominal CNS has the same characteristics as does the adult hormone.
The experiments were repeated using prepupal brain activity and
the same properties were found.

      Aside from their similar chemical characteristics, all of
the hormones from the 2 stages show similar spectra of relative
activity in the 3 bioassays that were employed--two of them based
on pharate adult tissues and the other on a prepupal response.
It is most likely, therefore, that the various sources of eclosion
hormone activity are based on stores of the same or extremely
similar peptides.

The Extent of Eclosion Hormone Involvement in the Ecdyses of Insects

      Fig. 5 contrasts the role of eclosion hormone in the control
of pupal ecdysis and adult eclosion. In both cases, the hormone
triggers behavioral programs that enable the insect to shed its
old cuticle. However, the behavioral and physiological effects
of the hormone extend beyond the simple release of these ecdysial
programs. At the time of adult eclosion, it coordinates the
"turning-on" of many adult behavior patterns and reflexes (Truman,

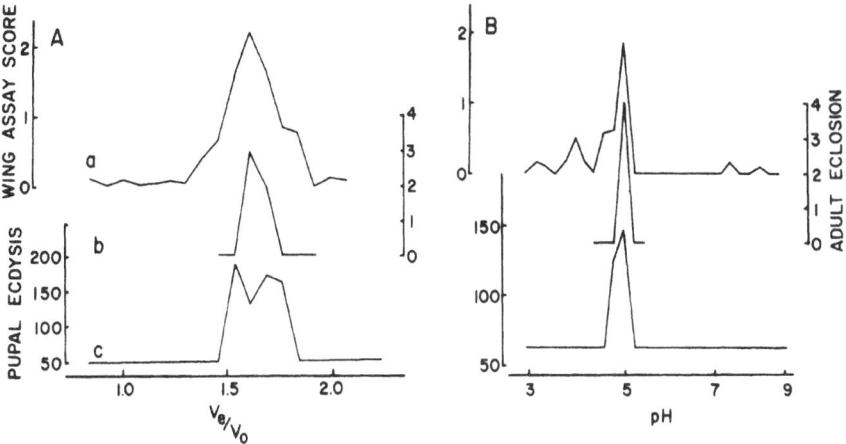

Fig.5 Comparison of the endocrine control of adult eclosion and pupal ecdysis.

1976). Similarly, at pupal ecdysis, certain pupal specific reflexes do not appear until the animal has been exposed to the hormone (Levine and Truman, unpublished). In the adult one of the
neural programs that is activated is the neuroendocrine reflex for
bursicon release (Truman, 1973b). At pupal ecdysis bursicon release
also depends on the prior release of eclosion hormone (Reynolds
et al., unpublished), indicating that this relationship may be a general feature of ecdysis. Two additional actions have been ascribed
to this hormone during adult eclosion. It induces an increase in
plasticity of the wing cuticle to facilitate inflation after
eclosion (Reynolds, 1977), and in the silkmoths it triggers the
postecdysial degeneration of the intersegmental muscles (Schwartz
and Truman, unpublished). Possible involvement of the eclosion
hormone in similar phenomena during pupal ecdysis has not been explored.

An intriguing difference between the 2 ecdyses is the site of
eclosion hormone release. The hormone that triggers adult eclosion
is released from the brain whereas that which triggers pupal ecdysis
comes from the ventral ganglia. These two ecdyses also differ
in that adult eclosion is controlled by an interaction between
the developmental state of the animal and a circadian clock whereas
pupal ecdysis is apparently triggered only by developmental factors.
These facts would imply that the brain-centered system is used for
ecdyses associated with a daily rhythm whereas the ventral ganglion
system is used for developmentally triggered ecdyses. Whether this
generalization is valid awaits the study of more insects.

Besides being used for adult eclosion and pupal ecdysis, there are preliminary data which implicate the hormone in the regulation of larval ecdysis (Copenhaver and Truman, unpublished). In addition, hormone activity has been detected in embryos of Hyalophora cecropia (Truman, unpublished) suggesting that it may be involved in regulation of some events during embryonic development or hatching.

The study of the distribution of eclosion hormone in insects other than Lepidoptera is very incomplete. An initial survey using adult eclosion of Antheraea pernyi as an assay for eclosion hormone failed to show activity in the brain CC-CA complexes of the cockroach Leucophaea maderae or the bug Pyrrhocoris apterus (Truman, 1973a). We have recently reinitiated a search for eclosion hormone activity in non-lepidopterans using the Manduca pupal ecdysis assay which is 10-fold more sensitive than the one based on A. pernyi. Also, we are now cognizant of the fact that hormone may be stored in the ventral ganglia instead of (or in addition to) the brain. At this time we have found hormonal activity in nervous systems from orthopterans and from hemipterans. We are still a long way from establishing the full extent of the use of this hormone but its presence in 2 hemimetabolous insects in addition to the (holometabolous) Lepidoptera suggests that the ecdysis promoting role of this hormone may be widespread throughout the Class Insecta.

ACKNOWLEDGEMENTS

The unpublished work was supported by grants from NSF (BMS 75-02272) and NIH (R01 NS 13079). PHT was supported by a NSF Predoctoral Fellowship. JWT was supported by a NIH Career Development Award (1 K04 NS 00193).

REFERENCES

Carlson, J. F., 1977, The imaginal ecdysis of the cricket (Telegryllus oceanicus)-I. Organization of motor programs and roles of central and sensory control, J. Comp. Physiol. 115:299-317.

Lockshin, R. A., Rosett, M., and Srokose, K., 1975, Control of ecdysis by heat in Manduca sexta, J. Insect Physiol. 21:1799-1802.

Miller, P. L. and Mills, P. S., 1976, Some aspects of the development of breathing in the locust, in: "Perspectives in Experimental Biology," Vol. I. "Zoology," P. Spencer-Davies, ed., Pergamon Press, Oxford, pp. 199-208.

Reynolds, S. E., 1977, Control of cuticle extensibility in the wings of adult Manduca at the time of eclosion: effects of eclosion hormone and bursicon, J. Exp. Biol. 70:27-39.

Reynolds, S. E., Taghert, P. H., and Truman, J. W., 1979, Eclosion hormone and bursicon titres and the onset of hormonal responsiveness during the last day of adult development in Manduca sexta (L.), J. Exp. Biol. 78:77-86.

Reynolds, S. E. and Truman, J. W., 1980, Eclosion hormone, in:
    "Insect Neurohormones,: T. A. Miller, ed., Springer-Verlag,
    New York, in press.
Riddiford, L. M. and Truman, J. W., 1978, Biochemistry of insect
    hormones and insect growth regulators, in: "Insect Biochemis-
    try," M. Rockstein, ed., Academic Press, New York, pp. 307-357.
Taghert, P. H., Truman, J. W., and Reynolds, S. E., 1980, Physiol-
    ogy of pupal ecdysis in the tobacco hornworm, Manduca sexta.
    II. Chemistry, distribution, and release of eclosion hormone
    at pupal ecdysis, J. Exp. Biol., in press.
Truman, J. W., 1971, Physiology of insect ecdysis. I. The
    eclosion behaviour of saturniid moths and its hormonal re-
    lease, J. Exp. Biol. 54:805-814.
Truman, J. W., 1972a, Physiology of insect rhythms. I. Circadian
    organization of the endocrine events underlying the moulting
    cycle of larval tobacco hornworms, J. Exp. Biol. 57:805-820.
Truman, J. W., 1972b, Physiology of insect rhythms. II. The silk-
    moth brain as the location of the biological clock controlling
    eclosion, J. Comp. Physiol. 81:99-114.
Truman, J. W., 1973a, Physiology of insect ecdysis. II. The assay
    and occurrence of the eclosion hormone in the Chinese Oak
    Silkmoth, Antheraea pernyi, Biol. Bull. 144:200-211.
Truman, J. W., 1973b, Physiology of insect ecdysis. III. The
    relationship between the hormonal control of eclosion and of
    tanning in the tobacco hornworm, Manduca sexta, J. Exp. Biol.
    58:821-829.
Truman, J. W., 1976, Development and hormonal release of adult
    behavior patterns in silkmoths, J. Comp. Physiol. 107:39-48.
Truman, J. W., 1978, Rhythmic control over endocrine activity
    in insects, in: "Comparative Endocrinology," P. J. Gaillard
    and H. H. Boer, eds., Elsevier/North Holland, Amsterdam,
    pp. 123-148.
Truman, J. W. and Riddiford, L. M., 1970, Neuroendocrine control
    of ecdysis in silkmoths, Science 167:1624-1626.
Truman, J. W., Taghert, P. H., and Reynolds, S. E., 1980, Physiol-
    ogy of pupal ecdysis in the tobacco hornworm, Manduca sexta.
    I. Evidence for control by eclosion hormone, J. Exp. Biol.,
    in press.

SYMPOSIUM ON INSECT NUTRITION

INTRODUCTION

J. G. Rodriguez

University of Kentucky
Lexington, Kentucky

It is indeed an honor to have had a role in organizing this Nutrition Symposium to commemorate Dr. Gottfried Fraenkel's contributions to the field of insect nutrition. Surely this man has possessed a green thumb in cultivating his many fields of interest throughout his career. A glimpse into his research accomplishments causes the eye to blink in wonderment at his intellectual capacity to explore, his tenacity to prove, and his intense dedication to the objective at hand.

Within the scope of this presentation, his accomplishments in the field of insect nutrition can be dealt with only in most general of terms. In this context, then, it can be said that G. Fraenkel sorted out for the first time the basic food requirements in terms of dietetic constituents, proteins, carbohydrates, sterols, vitamins, fatty acids in such insects as Tribolium confusum, Tenebrio molitor, Sitodrepa panicea, Lasioderma serricorne, Oryzaephilus surinamensis, Ptinus tectus, Dermestes vulpinus, Ephestia kuehniella, E. cautella, E. elutella, Tineola biselliella, Plodia interpunctella, Phormia regina, Musca domestica, Sarcophaga bullata, and Xenopsylla cheopis. The result of this research generally was that basic nutritional requirements were very similar. His studies on Vitamin B requirements showed that they are not specific. He was probably the first to show requirements for unsaturated fatty acids and the first to prove that metabolic water is a significant source of water in certain stored product insects.

A highlight of his career was his discovery of a growth factor required for the meal worm, Tenebrio molitor; this developed to be carnitine and although the presence of carnitine in a large variety of biological materials was known in the mid-forties, insights about

185

its biological importance were not realized until the pioneering work
of G. Fraenkel and his collaborators opened the door.

I am probably biased on this point, but perhaps the biggest
contribution that G. Fraenkel made to science was his contribution
in the role of secondary plant substances in the host selection of
phytophagous insects. The focus on chemistry of host specificity
stemmed from the recognition that foliage contains all known required
food constituents for plant feeding insects and yet does not serve
as a suitable diet for most of these insects. His classic paper,
"The Raison d'Etre of Secondary Plant Substances" in 1959 (Science
129:1466), stated this eloquently and from this "position paper",
there emanated about 20 more specific papers from his laboratory.
Later, in his 1969 paper, "Evaluations of Our Thoughts on Secondary
Plant Substances" (Ent. Exp. and Appl. 12:473), he generalized his
conclusions with the thought that " . . . host selection is guided
by the presence or absence of secondary plant substances, and that
qualitatively or quantitatively nutrients can play only a very
minor role, if one at all, in this context."

In this nutrition symposium, we have imposed the previously
mentioned limitations of time and space. The task we have taken
is to present a wide spectrum of research activity within the
general field of nutrition, representative of research areas whose
geneses are traceable also to G. Fraenkel's work.

In his opening presentation on essential fatty acids for
mosquitoes, R. H. Dadd also discusses some of the background per-
taining to essential fatty acids for insects generally, recalling
the work of Fraenkel and Blewett three decades ago in defining
requirements of oils and fatty acids for Ephestia. Dadd
exquisitely demonstrated that arachidonic acid induced the
emergence of strong flying adults of six species of mosquitoes.
Without it the flight index was either very low or nil. He suggests an interesting parallel: to the extent that 3 adult female
mosquitoes can be considered carnivores and cats cannot utilize
linoleic acid, both require arachidonic or similar acid.

An area of nutrition that has been most worthy of intense
investigation has been that of artificially rearing parasitic
insects, especially Hymenoptera, as the implications in pest
management are obvious. S. N. Thompson has accomplished much
during the last decade in his investigations in the nutritional
physiology of parasitic wasps, especially in the development of
in vitro culture of the ichneumonid, Exeristes roborator. In his
present paper he deals comprehensively with dietary carbohydrate
and lipid nutrition and metabolism of metazoan parasites. He
discusses the complexities of carbohydrate-lipid interactions and
compares metazoan utilization of nutrients.

Some of the most exquisite work in nutritional genetics has been done by B. W. Geer and associates. He has methodically added to the data bank in nutritional physiology and biochemical genetics using <u>Drosophila</u> <u>melanogaster</u> as the model system. Geer and associates now describe their work with the regulation of the pentose shunt enzymes. The genetic tools available in their system are extensive, and these have been most advantageous as they continue their studies in the genetic determinants associated with dietary modulation of the enzymes of the pentose phosphate cycle.

The protein, keratin, is extremely resistant to proteinases and only relatively few insects are capable of digesting it. As was found by Fraenkel and Blewett three decades ago for clothes moth larvae, J. E. Baker found that black carpet beetle larvae, likewise, do not grow on clean woolen cloth. He showed that the addition of vitamins, minerals and sterol increased the consumption index and approximate digestibility coefficient, enabling black carpet beetle larvae to grow. In the present paper, Baker describes the meticulous undertaking of isolating and characterizing the properties of the digestive enzymes in the black carpet beetle larva midgut.

Host plant resistance (HPR) is considered a strategy and a major component of insect pest management. Workers in this field concern themselves with plant-insect interactions and the secondary chemicals referred to by G. Fraenkel. John C. Reese examines the general question of insect dietetics and the complexities of plant-insect interactions. He puts into perspective the recent advances made in this field, discussing such aspects as pheromone production by phytophagous insects, sequestration of plant products as defensive chemicals, feeding deterrents in sorghum and cotton, and the interactions between allelochemics and nutrients.

It is precisely in the question of interaction between allelochemics and nutrients and HPR generally, that measurements of food consumption and utilization can become such crucial points in the investigation, for these quantitative measurements will provide data on qualitative effects. The paper by Marcos Kogan and Jose R. P. Parra provides vital comparisons of the various techniques that have been used by various investigators, for example, in determining the effect of allelochemics on levels of HPR.

ESSENTIAL FATTY ACIDS FOR MOSQUITOES, OTHER INSECTS AND

VERTEBRATES

R. H. Dadd

Department of Entomological Sciences
University of California
Berkeley, California 94720

INTRODUCTION

As with so much else in insect nutrition, the first demonstration that unsaturated fatty acid was essential in the diet of an insect came from the work of the man we here honor, Gottfried Fraenkel. In the late 1940's he and his associate, M. Blewett, showed that a polyunsaturated fatty acid, either linoleic or linolenic acid, was needed by certain flour moths of the genus Ephestia (Fraenkel and Blewett, 1946). A similar requirement for one or other of these fatty acids has since been demonstrated for many insect species (Dadd, 1977).

Recently, arachidonic acid, but neither linoleic nor linolenic acids, was found essential for a mosquito, Culex pipiens (Dadd and Kleinjan, 1979a). Arachidonic acid is the fatty acid of central physiological importance to warm-blooded vertebrates, which can utilize it directly, if present in the diet, or derive it metabolically from other dietary essential fatty acids, of which linoleic acid is the most widespread (Alfin-Slater and Aftergood, 1971). In contrast, arachidonic acid has not been found an adequate substitute for linoleic acid for those few insects other than mosquitoes examined in this respect (Dadd, 1973). The mosquito requirement thus appears unique among insects hitherto studied, and, on the face of it, more akin to the essential fatty acid needs of vertebrates than of other insects. It is timely, therefore, to review and speculate on the essential fatty acid needs of mosquitoes in the context of our current understanding of essential fatty acid nutrition and metabolism in both insects and vertebrates.

ESSENTIAL FATTY ACIDS FOR INSECTS

Three decades ago Fraenkel and Blewett found that Anagasta
(Ephestia) kuehniella reared on artificial diets grew poorly and
failed to emerge properly from the pupa unless the diet contained
certain oils. Without oil, fully developed pharate moths were
usually present within the pupa, and the essential lesion in pupal
failure appeared to be an inability to free the new adult cuticle
from the old pupal integument. This condition was prevented by
wheatgerm oil or its saponifiable fraction, but not by various
unsaturated vegetable or animal fats and oils, nor by highly
unsaturated fish oils, though the latter optimized larval growth
rate as effectively as wheatgerm oil. Eventually it was shown
that the characteristic failure during pupal adult eclosion resulted from a lack of linoleic acid, which wheatgerm oil provides
in abundance. This failure was averted also by another common
polyunsaturated acid of seed oils, linolenic acid, which was rather
more potent than linoleic acid. Arachidonic and docosahexaenoic
acids, characteristic polyunsaturates of animal and fish oils known
to prevent vertebrate fatty acid deficiency symptoms, did not prevent the striking failure of adult emergence, though as effective
as linoleic acid for optimizing the larval growth rate. By contrast,
the mono-unsaturated oleic acid was ineffective for both larval
growth optimization and normal adult emergence. Beneficial effects
of linoleic acid were enhanced with alpha-tocopherol or ascorbic
acid in the diet. Though alpha-tocopherol appeared to have a small
intrinsic growth-promoting effect, their major action was ascribed
to antioxidant protection of the very labile unsaturated fatty acids,
since a similar boost to the effect of linoleic acid was obtained
by other antioxidants such as propyl gallate (Fraenkel and Blewett,
1946; 1947).

In the years following these discoveries with Ephestia many
other insects were found also to require polyunsaturated fatty acid,
sometimes for good larval growth, but more particularly for normal
adult emergence or proper adult maturation and reproduction. Table
1 lists by order the numbers of insect species that have been investigated with respect to a need for polyunsaturated fatty acid and
the number of these that were found to require it. The data of
Table 1 are drawn mainly from my previous summary tables (Dadd,
1977) augmented by a few more recent publications (Turunen, 1974;
1976; Hou and Hsiao, 1978; Sivapalan and Gnanapragasam, 1979).
For more detail on effects of fatty acid deficiency in these
various insects than is given in the following synopsis, the
comprehensive bibliography in Dadd (1977) should be consulted.

In the absence of dietary unsaturated fatty acid most of the
Lepidoptera tested showed the classic pupal/adult emergence lesions
first described by Fraenkel, though not all showed retarded larval

Table 1.   Numbers of species tested for unsaturated fatty acid re-
           quirement in various orders of insect. Species indif-
           ferent to fatty acid but known to depend on intracellular
           symbiotes (e.g., all Aphids) are excluded.

Order	No. of species investigated	No. shown to require polyunsaturated fatty acid
Lepidoptera	18	15
Orthoptera	5	5
Coleoptera	9	3
Hymenoptera	2	2
Diptera excluding mosquitoes	8	0
TOTAL	42	25

growth. Of Orthoptera studied, acridid grasshoppers deprived of
fatty acid had a faulty adult emergence reminiscent of the lepidopteran syndrome and sometimes exhibited retarded larval growth.
Cockroaches deprived of unsaturated fatty acid produced abnormal
oothecae, from which a second generation of nymphs might emerge,
but pale-colored, weak, and short-lived. A cricket, and those
beetles that needed essential fatty acid, showed signs of deficiency mainly in terms of delayed growth and reduced fecundity,
sometimes not obvious until more than one generation of deprivation had been imposed. Both of two parasitoid Hymenoptera recently
studied nutritionally required linolenic acid for proper adult
emergence. Among Homoptera, many aphids have been reared for
generations in the absence of dietary fatty acids, but, since all
these insects have intracellular symbiotes whose metabolic contributions to the host insect are largely unknown, one cannot
conclude that aphids dispense with polyunsaturated fatty acids.
Finally, of several Diptera other than mosquitoes tested, none
showed any indication of need for polyunsaturated fatty acid, though
Agria houseii grew slightly faster with small amounts of various
common fatty acids, particularly oleic, in the diet (House and
Barlow, 1960), and larval growth of Drosophila melanogaster and
Aedes aegypti was improved by dietary lecithin (Sang, 1956; Golberg
and DeMeillon, 1948).

      Where fatty acid was indubitably found essential (i.e., where
deficiency resulted in deformity or failure at the pupal stage or

or adult molt) the requirement could be satisfied by either linoleic
or linolenic acids, whereas the common saturated or monounsaturated
fatty acids (palmitic, stearic, oleic) were always ineffective. In
the context of extensive information now accumulated on the fatty
acid biosynthetic capabilities of insects (Gilbert, 1967; Fast,
1970; Downer, 1978) it can readily be understood why only the di-
and tri-unsaturated linoleic and linolenic acids should be
essential, for in all species studied, the common C16 and C18
saturated and monoenoic acids can be synthesized from acetate or
derived from preexisting fatty acids, whereas no insect has been
shown unequivocally to synthesize linoleic or linolenic acids de novo
or to derive them by desaturation of other fatty acids. This being
so, if an insect species did not require dietary polyunsaturated
fatty acid, it would follow that it must have no physiological need
for polyunsaturates.

     Since the ultimate physiological function(s) of essential fatty
acids are not known even in those insects for which a dietary need
has been established, it may not appear surprising that some insects
might neither synthesize polyunsaturates nor need them in the diet.
However, the data of Table 1 show that a majority of insects investigated with respect to essential fatty acid have needed it,
suggesting that the need, clearly widespread and found with increasing frequency as the years pass, might indeed prove general if
rigorous testing over more than one generation of deprivation were
carried out with species that were apparently indifferent to polyunsaturated fatty acid in previous studies. In those species which
are known to require unsaturated fatty acid, the requirement becomes manifest mainly during late development, at the pupal/adult
molt or as reproductive deficiency. It may, then, be suspected that
a requirement could often remain cryptic in the usual short-term,
single-generation growth experiment; sufficient unsaturated fatty
acids for normal larval development could well be carried over from
the egg, especially if the major physiological requirement is related to metamorphosis and adult functions, as seems generally the
case and is especially strongly indicated in those instances where
metamorphic failure has been averted by ingestion of essential fatty
acid only shortly before pupation (Grau and Terriere, 1971; Rock
et al., 1965; Dadd and Kleinjan, 1979a).

     Thus, the general picture which early developed was that many
insects required a dietary polyunsaturated fatty acid for some
crucial though undefined function, and that the dietary requirement could be satisfied by linoleic or linolenic acids, unsynthesiz-
able by all insects critically examined. Up to this point, the
situation resembled that for vertebrates; but whereas the essential
fatty acid deficiency of vertebrates could be alleviated equally
well or better with dietary arachidonic acid and related longchain
polyunsaturates, this seemed not to be so for insects. As noted

above, the initial studies with Ephestia found that neither arachidonic nor docosahexaenoic acids were able to induce normal pupal/adult ecdysis. In Table 2 are listed the effects on adult emergence and larval growth for all insect species that have been tested with long-chain (C20 and C22) polyunsaturates in place of linoleic or linolenic acids. Although improved larval growth was observed in several cases, in none was the specific adult failure obviated. In this respect, therefore, the action of fatty acids in insects appeared to differ fundamentally from that in vertebrates, in which dietary linoleic or linolenic acids are converted to longer chained, more unsaturated members, these latter being the entities of ultimate physiological importance as components of specialized lipid membrane structures and as precursors of the ubiquitous hormone-like prostaglandins. Analyses of vertebrate fatty tissues reveal substantial amounts of longchain polyunsaturates, particularly in phospholipids, and a characteristic symptom of deficiency in these animals is a reduction in the proportions of particular fatty acids such as arachidonic acid (see next section). Fatty acid analyses of tissue lipids are now available for perhaps 100 insect species (Fast, 1964; 1970), but only rarely have any longchain polyunsaturates been recorded, which would be expected if they were of no physiological significance, unless derived adventitiously from the diet. However, knowing now that arachidonic or certain related fatty acids are essential for mosquitoes, it is perhaps surprising that none were detected by the many fatty acid analyses carried out on whole mosquito tissues, though unidentified long-chain fatty acids were recorded in one study and several C20 and C22 polyunsaturates were recently identified in mosquito tissue culture cell lines (see Dadd and Kleinjan, 1979a for review of this literature). Most probably the gas chromatographic methods used were insensitive to trace levels of fatty acids, and with this possibility in mind one may well wonder how much weight to attach to the apparent absence of arachidonic and other long-chain polyunsaturates in the great majority of the analyses made on other insects.

Before the recent discovery of a need for unusual fatty acids in mosquitoes, another complexity in insect fatty acid requirements had emerged over the years. This concerns the equivalency or otherwise of linoleic and linolenic acids. Fraenkel's trail-blazing studies already had suggested that while either linoleic or linolenic acids could individually avert pupal/adult failure in Ephestia, linolenic acid seemed more potent on a dosage basis. Similar observations were subsequently recorded for other Lepidoptera and, as summarized in Table 3, several species were eventually found to require linolenic acid specifically to avert pupal/adult failure. Linoleic acid was totally unable to avert molting failure in these species, but sometimes was necessary in addition to linolenic acid for optimal larval growth. These findings suggest that the essential fatty acid requirement of some insects is not a simple,

Table 2. Species with which C20 or C22 polyunsaturates were tested as substitutes for linoleic or linolenic acids.

Species	Fatty Acid tested	Induced normal adults	Improved larval growth
LEPIDOPTERA			
Anagasta (Ephestia) kuehniella [a]	arachidonic	no	yes
Anagasta (Ephestia) kuehniella [a]	docosahexaenoic	no	yes
Galleria mellonella [b]	arachidonic	no	detrimental
Trichoplusia ni [c]	arachidonic	no	not recorded
Trichoplusia ni [c]	eicosapentaenoic	no	yes
Trichoplusia ni [c]	docosahexaenoic	no	yes
Carpocapsa pomonella [d]	arachidonic	no	not recorded
Argyrotaenia velutinanaella [e]	arachidonic	no	detrimental
ORTHOPTERA			
Schistocerca gregaria [f]	arachidonic	no	yes
Locusta migratoria [f]	arachidonic	no	yes

[a] Fraenkel and Blewett, 1946
[b] Dadd, 1964
[c] Chippendale et al., 1964
[d] Rock, 1967
[e] Rock et al., 1965
[f] Dadd, 1961

Table 3. Numbers of lepidopteran species with specified fatty
          acid requirement for normal adult emergence.

No fatty acid required	Either linoleic or linolenic	Linolenic Specifically	Both linolenic and linoleic
3	6 + 3[a]	5	1

[a]Linolenic more potent on dosage basis than linoleic.
References: Same as cited in text for Table 1.

unifunctional matter physiologically, and this is now further com-
plicated by the yet different needs of mosquitoes. Before reviewing
the mosquito requirement, to place it in context it will be helpful
to consider the essential fatty acid requirement of vertebrates and
the distinctive features of those fatty acids which can alleviate
the vertebrate deficiency syndrome to a greater or lesser extent.

VERTEBRATE ESSENTIAL FATTY ACID REQUIREMENTS

     This summary of five decades of work on vertebrate essential
fatty acid nutrition and metabolism draws mainly from recent reviews by Mead (1970), Guarnieri and Johnson (1970), Alfin-Slater
and Aftergood (1971), Holman (1977), Sprecher (1977), and Tinoco
et al. (1979). The main effects of a lack of essential fatty acids
in warm-blooded vertebrates are listed in Table 4. Gross sympto-
mology in severe deficiency is indicated by items 1-4, which general
types of disorder have been observed in several laboratory species
of mammals and birds though manifesting variously in detail. Accompanying these gross dysfunctions, a characteristic biochemical
change occurs. Relative proportions of tissue fatty acids, particularly those of phospholipids, undergo modification. As indicated by items 5 and 6 of the table, arachidonic acid declines,
accompanied by changes in various other fatty acids, notably an
increase in 5,8,11-eicosatrienoic acid resulting from chain lengthening and desaturation of oleic and other ω9 acids when linoleic and
related ω6 fatty acids, the preferred substrates of the fatty acid
desaturases concerned, are reduced in amount. The resulting alteration in the fatty acid triene/tetraene ratio provides a good
index of essential fatty acid deficiency. Items 7 and 8 refer to
a group of micro-physiological lesions thought to result from
changes in the physical properties of subcellular lipid membranes
concomitant with their altered fatty acid constitution: lipid
fluidity changes, lipid-bound enzymes show altered activities, and
the membranes become leaky. Probably most gross pathologies can
be traced ultimately to such deterioration in cell permeability and
enzyme activity.

Table 4. Symptoms and effects of essential fatty acid deficiency
          in mammals.

Gross pathologies:

    1.   Reduced growth.
    2.   Epidermal lesions (e.g., scaly tail of rat).
    3.   Infertility.
    4.   Increased water intake due to increased loss through skin.

Biochemical changes:

    5.   Reduced arachidonic acid, especially in phospholipid.
    6.   Increases in fatty acids derived from oleic acid.

Subcellular lesions:

    7.   Increased membrane permeability of cell organelles.
    8.   Altered activities for various membrane-bound enzymes.

Other effects:

    9.   Prostaglandin precursors diminished.

    Item 9 notes that all essential fatty acids are also potential
precursors of physiologically active prostaglandins, ubiquitous hormone-like substances involved in the regulation of many and diverse
physiological functions in vertebrate tissues of all kinds. Because
of the close identity between the essentiality and prostaglandino-
genic capability of fatty acids, it has been suggested that the need
for essential fatty acids is fundamentally just a need for prostaglandin precursors. However, attempts to alleviate fatty acid deficiency with dietary or injected prostaglandins have so far failed,
and no consistent changes in prostaglandin levels have been observed
in fatty acid deficient animals, though some recent evidence indicates that gross dermal lesions of deficiency may be ameliorated
by topical application of prostaglandin (Lands et al., 1977). It
seems most likely, therefore, that although essential fatty acids
are necessary precursors of prostaglandins, they also directly
subserve other necessary physiological functions, most probably as
essential components of specific membrane lipids.

    Of the dozens of unsaturated fatty acids that have been tested
with various warm-blooded vertebrates very many can at least partially
alleviate some of the deficiency symptoms listed in Table 4. The
essential features of those that banish all symptoms are indicated
in Fig. 1. They possess 2 or more cis double bonds in divinyl

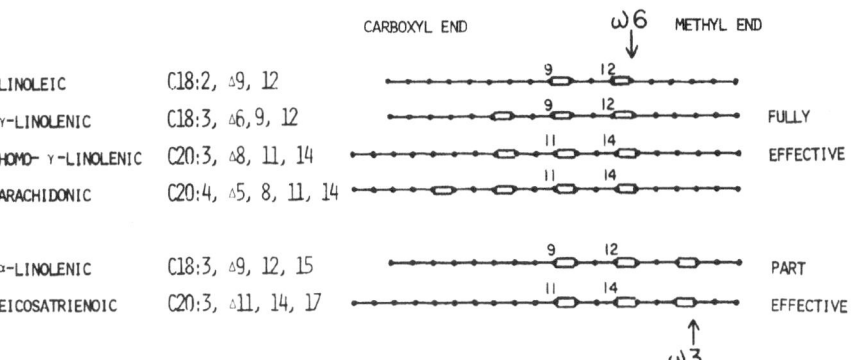

1)   18 OR MORE CARBONS

2)   2 OR MORE CIS DOUBLE BONDS IN DIVINYL METHANE RHYTHM

3)   Δ9, 12 STRUCTURE FOR C18, Δ11, 14 FOR C20, OR Δ13, 16 FOR C22 ACIDS

4)   FULL ACTIVITY ONLY WITH ω6 ACIDS; PARTIAL ACTIVITY WITH ω3 FAMILY

Fig.1   Diagrams of chain structure for some fully effective and
partly effective essential fatty acids for warm-blooded
vertebrates. Four structural features of the major fully
effective ω6 fatty acids are listed.

methane rhythm with the double bond nearest the methyl end of the
chain ending on carbon 6 from the methyl termination. Because of
this last feature, all such fatty acids are said to comprise the
ω6 family; the longer ones, such as arachidonic acid, can be derived
from the shorter ones, ultimately from linoleic acid, by 2-carbon
chain elongation and the insertion of extra double bonds towards
the carboxyl end of the chain. A noteworthy exception to this
general situation is found in cats, which, lacking suitable desaturases, cannot convert linoleic acid to higher members of the ω6
family and so require a dietary source of arachidonic acid (Rivers
et al., 1975).

In the diagram I include for comparison ordinary α-linolenic
acid and a C20 trienoic acid that was semi-active in our tests with
mosquitoes. It will be seen that these have their last double bond
terminating on the 3rd carbon from the methyl end of the fatty acid
chain, and such fatty acids are grouped as the ω3 family. Analogously with the ω6 series, the higher members of the family are
derivable from linolenic acid by chain elongations and desaturations
at the carboxyl end. Though members of this ω3 family of

polyunsaturated fatty acids alleviate the warm-blooded vertebrate
deficiency syndrome to a great extent, the lower members -- e.g.,
linolenic or 11,14,17-trienoic acids -- are not completely satis-
factory cures for all symptoms. On the other hand, the ω3 family,
but not the ω6, are the most important essential fatty acids for
many fish. Whether the ω3 fatty acids are independently and additionally essential for warm-blooded vertebrates is currently a
matter of contention (Tinoco et al., 1979); recent evidence suggests
that for the normal composition of brain lipids linolenic as well as
linoleic acid may be needed in the diet (Galli et al., 1976).

MOSQUITO ESSENTIAL FATTY ACIDS

     For several years past we have used aseptic, fully-defined basal
dietary media for nutritional studies with Culex pipiens. A complete
listing of the components of the basal diet used in work discussed
here is given in Dadd and Kleinjan (1978). It contains the usual
array of nutrients required by the generality of insects, plus a few
unusual but essential components such as the nucleotides and asparagine (Dadd and Kleinjan, 1977; Dadd, 1978; 1979). For purposes of
this discussion the important thing to note is that the only lipid
in it is cholesterol. With this basal diet, newly hatched larvae
can be grown to adults with 90% or better survival, taking about 9
days to pupation at a temperature of 29 ± 1°C, a performance comparable with that achieved with many crude culture regimes. However,
when adults reared on such synthetic diets emerge from the pupa,
they are weak, cannot fly, and generally are trapped at the surface
of the slightly viscid medium.

     Strongly flying adults were first obtained using basal medium
supplemented with about 30 mg% of mammalian beta-lipoprotein. The
flight-inducing factor was then shown to occur in the chloroform/
methanol lipid extract from whole lipoprotein. As phospholipids
contribute about half the lipid of lipoprotein, a variety of
lecithins and cephalins were next tested. From this an interesting
pattern emerged. High Flight Indices, defined below, were obtained
with all animal lecithins and cephalins tested, but not with the
vegetable lecithins nor with the synthetic dipalmitoyl lecithins
and cephalins (Dadd and Kleinjan, 1978). What distinguished between these flight-active and inactive phospholipids? Inspection
of the literature revealed that the animal phospholipids are
characterized by substantial amounts of arachidonic acid and other
very long chain polyunsaturated fatty acids; the synthetic phospholipids had only palmitic acid, and the vegetable lecithins had
linoleic and linolenic acids as their longest and most unsaturated
fatty acids. We therefore tested the effect of adding pure arachidonic acid as a supplement to basal dietary medium. But before
considering this crucial experiment a few comments on methodology
are in order.

Procedures for the preparation of dietary media, the surface sterilization of eggs, and the inoculation of aseptic, newly-hatched larvae into autoclaved media can be found in previous publications (Dadd et al., 1973; Dadd and Kleinjan, 1976). For each dietary treatment of a growth experiment, 20 newly-hatched larvae were inoculated in pairs into each of 10 culture tubes containing 5 ml of autoclaved medium. Tubes were kept thereafter in an incubator at 27-28°C, and developmental stages attained by the larvae were recorded periodically, every day once pupation commenced until adult emergence or death at an earlier stage. For all experimental treatments considered here, differences in survival and developmental rate up to adult emergence were trivial, and usually 80-100% of starting first instar larvae became adult. Hence interest focused on the ability of newly emerged adults to fly, and to record this, all adults, on the day they emerged, were designated into one of the following categories: 1) Trapped -- if found collapsed at the slightly sticky surface of the medium; 2) Standers -- if standing free on the surface of the medium but unable to fly when jarred; 3) Hoppers -- if able to jump from the surface, often with wings beating, but then immediately falling back; 4) Fliers -- if able to fly well away from the surface without immediately falling back. To condense individual mosquito flight records into one index for each treatment we calculate a Flight Index as follows. Adults designated as trapped, standers, hoppers and fliers are assigned flight values of zero, 0.25, 0.5 and 1, respectively, and the sum of such values for a treatment is expressed as a percentage of the total number of emerged adults for that treatment. The Flight Indices so obtained range between 0, if all adults were trapped, to 100%, if all were fliers. Reference to Table 6 of Dadd and Kleinjan (1978) will clarify how these Flight Indices are derived from raw flight records.

As mentioned above, the original basal diet contained only one lipid component, cholesterol, and in the earliest experiments with fatty acids they were simply added to the medium, like cholesterol, after dispersion into a stock solution via a small amount of ethanol. Subsequently it became convenient to prepare stock suspensions of fatty acids intimately mixed into liposome-like preparations with an antioxidant, ascorbyl palmitate, and synthetic lecithin; this greatly extends the stability of the readily oxidized polyunsaturated fatty acids both during storage and after incorporation into experimental diets (Dadd and Kleinjan, 1979b). Synthetic dipalmitoyl lecithin slightly increases the rate of larval development (Dadd and Kleinjan, 1978), but neither it nor ascorbyl palmitate induce any flight activity in emergent adults. Thus the basal diet for most of the experiments here discussed contained synthetic lecithin and ascorbyl palmitate at 1 mg and 0.4 mg respectively per 100 mg of medium.

The crucial experiment which first demonstrated that arachidonic acid induced the emergence of strong, flying adults of C. pipiens

included treatments in which arachidonic acid was incorporated in
the diet from the start, others in which it was added aseptically
after larvae had grown to the final instar without it, and, for
comparison, other treatments with linoleic acid incorporated from
the start, with and without subsequent arachidonic acid supplementation (Dadd and Kleinjan, 1979a). Linoleic acid alone was
ineffective, whereas all treatments that incorporated arachidonic
acid, whether present from the start or added only during the last
larval instar, gave very high Flight Indices. The dosage plots of
Fig. 2 show that an optimal effect on flight induction may be obtained with 0.05 mg per 100 ml of dietary medium when the fatty acid
is protected by ascorbyl palmitate, and some effect is detectable
with as little as 0.003 mg arachidonic acid per 100 ml.

Results of an investigation into the flight-promoting capabilities of a variety of saturated and unsaturated fatty acids,
currently in press (Dadd, 1980), are summarized in Table 5. Fatty
acids studied are listed in 3 groups: Inactive -- those giving
Flight Indices no higher than in the absence of fatty acid; Semi-
active -- giving Flight Indices consistently higher than in corresponding treatments lacking fatty acid, due to appreciable numbers
of standing or hopping adults, but with only an occasional true
flier; Active -- giving high Flight Indices with a majority of adults
designated fliers.

All saturated and monoenoic acids studied were inactive. The
two dienoic acids, both ω6, were considered semi-active, as were
the two ω3 trienoic acids. Most interestingly, the ω6 isomers of
these trienoic acids both proved to be flight active. Besides
arachidonic acid (ω6 tetraenoic), the pentaenoic and hexaenoic acids,
both ω3, were highly active.

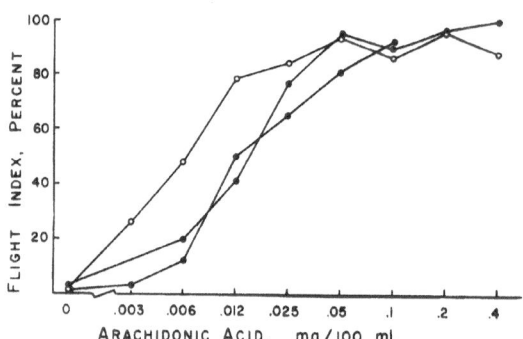

Fig. 2   Flight indices for C. pipiens reared in media with serial half
         dilutions of arachidonic acid. Open circles (Dadd and Kleinjan,
         1979a): arachidonic acid (A.A.) fresh from storage, ascorbyl
         palmitate (A.P.) added to diet. Closed circles: stock suspen-
         sions of A.A./A.P./lecithin, as in Dadd and Kleinjan, 1979b.

Table 5. Fatty acids tested for flight inducing activity with
          Culex pipiens. Inactive: adults mostly trapped, with
          occasional standers and hoppers but not fliers. Semi-
          active: adults commonly stand or hop, and occasionally
          fly weakly. Active: adults predominantly fliers.

	Flight Index range	Rating
none	0 - 4	inactive
C16:0 (palmitic)	0 - 3	inactive
C18:0 (stearic)	0 - 7	inactive
C20:0 (arachidic)	0 - 3	inactive
C22:0 (behenic)	0 - 1	inactive
Δ9-C18:1 (oleic)	0 - 4	inactive
Δ6-C18:1 (petroselenic)	0 - 3	inactive
Δ11-C20:1	0 - 1	inactive
Δ13-C22:1 (erucic)	0 - 1	inactive
Δ9,12-C18:2 (linoleic)	0 -14	semi-active
Δ11,14-C20:2	4 - 9	semi-active
Δ9,12,15-C18:3 (α-linolenic)	6 -23	semi-active
Δ11,14,17-C20:3	0 -39	semi-active
Δ6,9,12-C18:3 (γ-linolenic)	33-79	active
Δ8,11,14-C20:3 (homo-γ-linolenic)	23-68	active
Δ5,8,11,14-C20:4 (arachidonic)	69-100	active
Δ5,8,11,14,17-C20:5 (eicosapentaenoic)	70-83	active
Δ4,7,10,13,16,19-C22:6 (docosahexaenoic)	71-78	active

All saturated and monoenoic acids studied were inactive. The
two dienoic acids, both ω6, were considered semi-active, as were
the two ω3 trienoic acids. Most interestingly, the ω6 isomers of
these trienoic acids both proved to be flight active. Besides
arachidonic acid (ω6 tetraenoic), the pentaenoic and hexaenoic acids,
both ω3, were highly active.

In searching for a feature shared in common by the 5 flight-
active fatty acids it was noted that they all contain a group of
3 double bonds ending on carbon 6 from the methyl termination.
However, they are not all in the ω6 family of fatty acids, for an
additional double bond at the methyl end, making them ω3, in the
pentaenoic and hexaenoic acids, does not negate activity, and
neither do additional double bonds towards the carboxyl end as in
the C22:6 and C20:5 acids and in arachidonic acid itself. The
group of semi-active fatty acids is also characterized by a common
structure, comprising the two double bonds, nearest the methyl

termination, of the active structure of 3 double bonds; this semi-active structure is not deactivated by the presence of an additional double bond towards the methyl end, so that semi-active fatty acids can also be both ω3 and ω6.

If the foregoing structure/activity relationships hold generally, one would anticipate that other fatty acids with three cis double bonds at positions ω6, ω9, and ω12 from the methyl termination would make C. pipiens fly, and that if only two of these double bonds were present at positions ω6 and ω9, newly emerged adults would frequently stand or hop but rarely fly. Three recently acquired C22 polyunsaturated fatty acids allowed us to test this prediction, and the results of experiments comparing them with the previously tested docosahexaenoic acid are set forth in Table 6. The C22:4 acid was fully flight-effective; the C22:3 acid was clearly semi-active; and the C22:2 acid was also incipiently active. If their chain structures are now compared with those of all fatty acids previously found active or semi-active, as shown diagrammatically in Fig. 3, it can be seen that the C22 tetraenoic acid has precisely the same structure from the methyl end as arachidonic acid; the C22 trienoic acid is analogous to the semi-active C20 and C18 trienoic acids; and the C22 dienoic acid terminates analogously to the semi-active C20 and C18 dienoic acids.

Table 6.   Induction of flight in C. pipiens by various C22 poly-unsaturated fatty acids at a concentration of 0.4 mg per 100 ml of dietary medium. T, S, H and F are abbreviations respectively for trapped adults, standers, hoppers and fliers.

Fatty acid	Flight Index for 3 experiments			Adult categories 3 experiments combined			
	A	B	C	T	S	H	F
None	1	4	4	47	5	1	0
Δ13,16-C22:2	8	13	10	33	10	5	0
Δ13,16,19-C22:3	26	12	24	27	18	8	3
Δ7,10,13,16-C22:4	84	93	90	5	1	1	48
Δ4,7,10,13,16,19-C22:6	94	85	87	2	6	1	51
Arachidonic	79	95	86	6	2	0	48

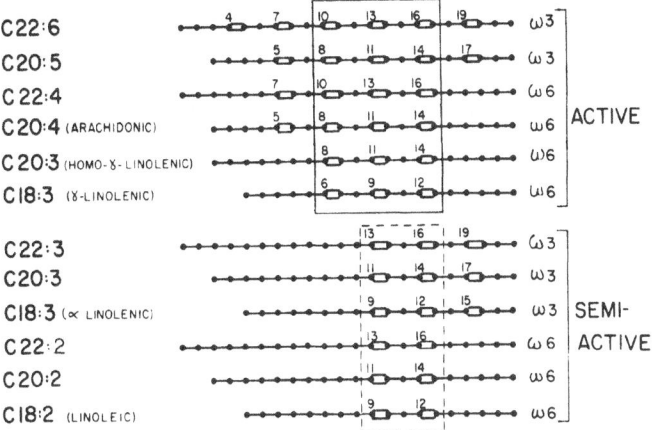

Fig.3 Diagrams of the chain structure of active and semi-active
       fatty acids for Culex pipiens.

From these studies it appears that the crucial structural
feature of a fatty acid which can fulfill the essential requirement
of C. pipiens is a group of 3 cis double bonds in divinyl methane
rhythm terminating on carbon 6 from the methyl end of the chain.
The possession of only the 2 double bonds of this grouping nearest
the methyl end appears to support incipient flight activity, on
the evidence of an increased number of standing and hopping adults,
and such fatty acids we consider semi-active. Additional double
bonds on either side of these groupings in no way diminish activity.
On the contrary, they seem to enhance it, for among active fatty
acids, the hexaenes, pentaenes and tetraenes generally gave higher
Flight Indices than the trienes; and among semi-active fatty acids
the trienes appear more effective than the dienes (Dadd, 1980, and
Table 6 of this paper).

We are curious to know what might be the flight-activity status
of fatty acids containing the two double bonds neareast the carboxyl
end of the active group of 3 double bonds, with or without additional
double bonds in the carboxyl direction, but no suitable examples have
been obtainable. Such fatty acids would be members of the ω9 family
whose root member is oleic acid; all are devoid of essential fatty
acid activity for vertebrates, and, so far as oleic acid is concerned, also for insects. Perhaps of most interest would be the
triene, Δ5,8,11-C20:3, the fatty acid which becomes reciprocally
prominent as arachidonic acid diminishes with the onset of severe
deficiency in mammals. Of interest in this connection is the frequent observation that insects deprived of linoleic and linolenic
acids tend to synthesize or accumulate increased proportions of oleic
acid, as if in some way compensating for the lack of polyunsaturates.

Up to this point my discussion of mosquito essential fatty acids
has been based entirely on work done with C. pipiens, and one may
wonder whether a similar situation prevails in other species. It
seems likely that it does. Table 7 presents Flight Indices for 5
other species of mosquito for which comparisons of rearings on
synthetic diet with and without arachidonic acid have been made.
Except for Culex tarsalis (Dadd and Kleinjan, unpublished observations) and Culiseta incidens (Dadd et al., in press) the results
given exemplify only a single experiments carried out with each
species; nevertheless, in all cases except for Aedes aegypti it is
apparent that adults failed to emerge as fliers without fatty acid
in the diet but were able to fly when arachidonic acid was incorporated. Though considerable numbers of A. aegypti became flying

Table 7. Developmental parameters and flight indices for 6 species
of mosquito reared on synthetic diet with and without
arachidonic acid at concentrations of 0.4 mg per 100 ml
of dietary medium.

Species	Arachidonic acid?	Avg # of molts per starter	% Adults	Avg days to pupation	Flight Index
Culex pipiens	no	4.95	95	10.1	3
	yes	4.81	95	9.8	95
Aedes sierrensis	no	4.86	93	18.5	0
	yes	5.00	100	18.1	59
Aedes aegypti	no	4.90	95	11.8	32
	yes	5.00	100	12.2	77
Culiseta incidens	no	4.71	71	14.3	0
	yes	4.73	87	12.2	43
Culex tarsalis	no	2.41	32	16.2	0
	yes	3.11	62	13.3	71
Culiseta inornata	no	2.80	0	25.0	-
	yes	3.76	41	18.7	61

adults without arachidonic acid, far greater numbers emerged as
fliers when it was present, and this is the same pattern observed
in more extensive studies of this species when reared on synthetic
diet with and without animal lecithin, the activity of which is most
likely due to the arachidonic acid which it contains (Sneller, 1978,
Ph.D. Thesis); of particular interest in these studies is the
finding that females reared without animal lecithin had markedly
reduced longevity and deposited fewer eggs than those provided
during larval growth with animal lecithin.

These tests with various species show only that they have an
essential fatty acid requirement which can be satisfied by arachidonic acid, but say nothing about the possible utilization of other
fatty acids. However, for C. tarsalis enough data is already available to show that its pattern of utilization of fatty acids will
closely approximate that of C. pipiens; for example, linoleic and
linolenic acids failed to induce flight, whereas gamma-linolenic,
homo-gamma-linolenic and docosahexaenoic acids were flight-active.

In Table 7 various measures of larval development are given.
Earlier in this section it was stated that effects of arachidonic
acid on development up to adult emergence were trivial in C. pipiens,
and this is exemplified by the typical set of data given for this
species. A similar situation prevails with Culiseta incidens,
Aedes sierrensis, and Aedes aegypti, but it is otherwise for Culex
tarsalis and Culiseta inornata. As indicated by lower average
numbers of molts, by lower percentages to reach adulthood, and by
longer periods to pupation for larvae reared on diets lacking
arachidonic acid, it is clear that essential fatty acids are involved in other aspects of development than just the ability of
adults to fly. This is most pronounced for C. inornata, few larvae
of which survived to pupate without arachidonic acid, and none to
become adult (Dadd et al., in press). With C. tarsalis, for which
we have extensive data, a distinct retardation in development can
usually be detected by the 2nd or 3rd larval instar in arachidonic
acid-deficient diets. Thus, although the pattern of development
for C. pipiens and several other species indicates that the major
demand for essential fatty acid comes at metamorphosis, also shown
by the ability of late larval supplementation with arachidonic acid
to induce adult flight, it is also probable that a physiological
requirement for some essential fatty acid occurs throughout development. On this view, differences between species with respect to the
stage of development when the requirement becomes manifest probably
reflect species differences in the amount of reserve essential fatty
acids laid down in the egg, with A. aegypti being an extreme example
of such reserves being sufficient to assure the viability of many
adults even under complete larval deprivation.

GENERAL DISCUSSION AND SPECULATION

From a structure/function point of view the essential fatty acids for Culex pipiens, and doubtless many other mosquitoes, have much in common with those of vertebrates. In Fig. 4 are shown the two series of fatty acids which are essential for warm-blooded vertebrates and many fish, respectively, with an indication of the route whereby the various members of the ω6 and ω3 families are derived. An out-standing difference between the utilization patterns of mosquitoes and vertebrates is that linoleic and linolenic acids, the parent members of the warm-blooded and fish series of essential fatty acids, are unable to satisfy the mosquito requirement. This suggests that C. pipiens cannot carry out the chain elongations and/or desatura-tions that vertebrates manage. However, there is a vertebrate precedent for this, since cats, lacking the appropriate desaturase, cannot utilize linoleic acid but must have a dietary source of higher members of the ω6 family such as arachidonic acid (Rivers et al., 1975), a restriction on their fatty acid versatility that has clearly evolved as a correlate of carnivory. To the extent that mosquitoes can be considered carnivorous -- with blood-sucking female adults and filter-feeding larvae which may be considered to micro-predate on protists -- there is an interesting parallel here.

ω6 SERIES FOR WARM-BLOODED VERTEBRATES:

$\Delta 9,12-18:2 \rightarrow \Delta 6,9,12-18:3 \rightarrow \Delta 8,11,14-20:3 \rightarrow \Delta 5,8,11,14-20:4$
    linoleic   γ-linolenic*   homo-γ-linolenic*   arachidonic*

ω3 SERIES FOR SOME FISH:

$\Delta 9,12,15-18:3 \rightarrow \Delta 6,9,12,15-18:4 \rightarrow \Delta 8,11,14,17-20:4$
    α-linolenic
                                                    $\Delta 5,8,11,14,17-20:5$
                                                    eicosapentaenoic*

$\Delta 4,7,10,14,16,19-22:6 \leftarrow \Delta 7,10,13,16,19-22:5$
docosahexaenoic*   docosapentaenoic

Fig.4   Pathways for the sequential derivation of essential fatty acids of the ω6 and ω3 families from their respective parent fatty acids, linoleic and linolenic acids. Asterisks mark fatty acids found to be flight-active for C. pipiens.

Turning to similarities between vertebrates and insects other than mosquitoes, in both cases either linoleic or linolenic acid is utilizable. However, although it is established that vertebrates metabolize these parent fatty acids to physiologically active higher members of the series which, if administered as the sole dietary fatty acid, can completely fulfill all essential fatty acid functions, this appears, on current evidence, not to be the case with insects. Accepting this, it would then follow that insects in general diverge markedly from vertebrates in their ultimate physiological fatty acid requirement, for in insects this would be linoleic or linolenic acids as such, rather than derived higher polyunsaturates. In this context, the mosquito requirement for higher polyunsaturates specifically, with an inability to utilize linoleic/linolenic acid, would be envisaged as a special evolutionary divergence from the generality of insects that came to parallel in many respects the pattern of vertebrate requirements. One can imagine that a change in phospholipid membrane physiology to a dependence on arachidonic acid in addition to linoleic/linolenic acid might first have evolved, with a subsequent loss of the original insectan linoleic/linolenic dependence. This is no doubt a tenable view, but before embracing it wholeheartedly to the exclusion of an alternative interpretation offered below that would better unify insect and vertebrate requirements, the evidence suggesting a fundamental difference between the ultimate physiological fatty acid requirements of insects and vertebrates should be critically examined.

This evidence is basically of three sorts. First, nearly all insect fatty acid analyses have failed to detect polyunsaturates higher than linoleic and linolenic acids. Earlier in this paper, and at greater length elsewhere (Dadd and Kleinjan, 1979a), allusion was made to uncertainties about this evidence that arise if long-chain polyunsaturates were present in tissues only in very low concentrations below the sensitivities of the analytical methods employed. Our work shows the effective dietary requirement of C. pipiens for arachidonic acid to be very low; this indicates that tissue concentrations in healthy mosquitoes could also be very low, and this could explain why its presence in mosquito tissue analyzed by gas chromatography was generally overlooked, though recently detected in mosquito tissue culture cell lines, and by a bioassay I discuss below. Were longchain polyunsaturates physiologically important to other insects but present in only trace amounts, and perhaps only transiently metabolized from linoleic acid in connection with particular physiological events, then they too might well remain undetected in routine analyses for the major bulk fatty acids.

Secondly, essentially negative results were obtained in those few cases where longchain polyunsaturates were tested as substitutes for linoleic or linolenic acids in insects known to require

these later. This is really the weightiest evidence against the
possibility of a physiological need for the higher polyunsaturates
in the majority of insects. However, in all cases examined, arachidonic acid (and in 2 Lepidoptera, docosahexaenoic acid also)
was incorporated into synthetic diets at the start of growth experiments that ran for 2 to 4 weeks; since most of these insects were
Lepidoptera (see Table 2), which substantial evidence suggests require fatty acid mainly at metamorphosis, the experimental polyunsaturates, especially prone to oxidation the greater the unsaturation, may well have been degraded from the diet before the time of
critical need. Against this it could be argued that in the same
experiments linoleic and linolenic acids were still in sufficiently
active concentrations to produce positive results; but it may be
that linoleic and linolenic acids are less readily degraded, being
less unsaturated, and that they might possibly be taken up by the
tissues during early larval growth and stored for later conversion
to polyunsaturates in a way that preformed polyunsaturates are not.

Thirdly, there is circumstantial support by default. That is
to say, it cannot be argued that certain physiological functions
of insects require the formation from dietary fatty acids of higher
fatty acids as, for example, the ubiquitous occurrence of prostaglandins in vertebrates requires the prior formation of polyunsaturated eicosaenoic acids. Virtually nothing is known of the
metabolism of linoleic and linolenic acids in insects, nor is there
any information on their cellular physiological functions beyond
the frequent observation that they preferentially accumulate in
phospholipids. The one exception to this is the recent delineation
of a physiological role for prostaglandin in crickets (Destephano
and Brady, 1977), which would seem to entail the prior formation of
arachidonic acid from their abundant linoleic acid (Meikle and Mc-
Farlane, 1965).

Because of the foregoing pregnabilities in the evidence pointing
to a basic difference in fatty acid physiology between most insects
and vertebrates, and especially when one considers the awkward
corollary that one group of insects, mosquitoes, must quite separately have evolved an essential fatty acid physiology more akin to that
of vertebrates than to the general insectan pattern from which it
derived, it is worth considering an alternative, more unifying,
hypothesis. This would propose, basically, that insects, like
vertebrates, had an ultimate physiological requirement for long-
chain polyunsaturates of either or both the ω6 and ω3 families for
specific lipid membrane functions. Further, in view of the very
small amounts of essential fatty acids required by mosquitoes (Dadd
and Kleinjan, 1979a), the phenomena of semi-active and fully active
fatty acids in both insects and vertebrates, and the recent intru-
sion of prostaglandins into insect physiology (Destephano and Brady,
1975; Loher, 1979), it would entertain the possibility that insects,

like vertebrates, might additionally require specific polyunsaturates
as precursors for hormone-like substances such as prostaglandins.
Within such a general physiological need for long-chain polyunsaturates, differences of emphasis with respect to the fatty acid
family of prime importance, ω6 or ω3, may be discerned among insects
as among vertebrates; thus many Lepidoptera require linolenic acid
specifically in the diet rather than linoleic acid, a situation
analogous to that for fish among vertebrates. Finally, the inability
of mosquitoes to utilize dietary linoleic and linolenic acids as
substitutes for arachidonic acid would, by analogy with the feline
case, find its explanation simply as an evolutionary loss of the
enzymes to convert parent fatty acids to the physiologically necessary higher polyunsaturates in a taxon that had evolved a dietetic
facies, carnivory, based on food containing abundant, preformed
long-chain polyunsaturates.

This hypothesis predicts certain testable outcomes which, if
positive, would suggest it has real substance. If the hypothesis
is correct, arachidonic acid or similar higher polyunsaturates should
be demonstrable in the generality of insects, contrary to most
chemical analyses hitherto carried out; such demonstrations would
be most persuasive for herbivorous species unable to obtain long-
chain polyunsaturates adventitiously from their food. Currently
we are using a bioassay which examines whether lipid extracts from
various insects can, when added to fatty acid-deficient basal diet,
induce flight in C. pipiens; if flight is induced, the presumption
is that the lipid extract supplied one or other of the flight-active
polyunsaturates. The outcome for several species we have so far
been able to examine is summarized in Table 8, which shows that most
contained fatty acids that were fully flight-active for C. pipiens.
In several cases, such as the roaches, the polyunsaturates might
have been absorbed adventitiously from the stock rearing food, but
this is unlikely for species such as Galleria mellonella and
Spodoptera exigua which were stock-reared on food devoid of animal
matter. Thus this survey already indicates the presence of higher
polyunsaturates in a number of species where it is unlikely to have
come directly from the food and therefore would presumably have been
metabolized from the linoleic and linolenic acids present in and
absorbed from the stock rearing diets.

With regard to the possibility that a function of essential
fatty acids in insects as in vertebrates might devolve around the
provision of precursors for hormone-like entities, we have also
started to look into the matter of prostaglandins. Do they have a
significance for insects beyond their demonstrated function in
crickets? In particular, we are examining whether they can replace or spare essential fatty acid for mosquitoes. Outright
substitution of various prostaglandins for arachidonic acid in the
diet of C. pipiens has been entirely negative in result, as indeed

Table 8. Ability of lipid extract from various species of insect
to induce flight activity in <u>Culex</u> <u>pipiens</u>.

ORTHOPTERA:

    <u>Periplaneta</u> <u>americana</u>   positive
    <u>Leucophaea</u> <u>maderae</u>   positive
    <u>Teleogryllus</u> <u>commodus</u>   positive
    <u>Melanoplus</u> <u>sanguinipes</u>   negative

LEPIDOPTERA:

    <u>Galleria</u> <u>mellonela</u>   positive
    <u>Spodoptera</u> <u>exigua</u>   positive

DIPTERA:

    <u>Culex</u> <u>pipiens</u>   positive
    <u>Musca</u> <u>domestica</u>   borderline positive
    <u>Rhagoletis</u> <u>completa</u>   negative

HYMENOPTERA:

    <u>Apis</u> <u>mellifera</u>   negative

HOMOPTERA:

    <u>Myzus</u> <u>persicae</u>   toxic

similar types of experiments have been for vertebrates. However,
on the assumption that arachidonic acid subserves multiple functions, only one of which would be that of prostaglandin precursor,
and further hypothesizing that semi-active fatty acids might fulfill other functions though not precursors for active prostaglandins,
we are now testing prostaglandins in combination with semi-active
fatty acids. None of these combinations have yet approached full
flight activity, but some have given indications of increased
activity when compared with semi-active fatty acids alone. The
results to date must therefore be judged indeterminate, but worth
pursuing.

The biggest weakness of my hypothesis is, of course, the failure of previous investigators to achieve amelioration of the principal fatty acid deficiency symptom when long-chain polyunsaturates
were substituted for linoleic or linolenic acids in the diets of
insects requiring these later. Thus, the crucial test of the hypothesis will be a critical re-examination of one of these cases

using antioxidant-protected fatty acid preparations, supplementation of diet with fresh fatty acid shortly before metamorphosis, and
testing of both ω6 and ω3 polyunsaturates. Until this step is taken,
an understanding of essential fatty acid nutrition and function in
insects and a clarification of relationships to the requirements of
vertebrates will remain largely speculative. Though instances of
requirement have proliferated, in insects their physiological import
has not much advanced from where Fraenkel left it, except for the
aesthetically unsatisfactory complication that in the realm of
essential fatty acids mosquitoes seem more like feline vertebrates
than insects.

ACKNOWLEDGEMENTS

I thank J. E. Kleinjan for her excellent and unstinting assistance with the mosquito studies, which were supported in part by
Water Resource and Mosquito Control Funds of the State of California.

REFERENCES

Alfin-Slater, R. B. and Aftergood, L., 1971, Physiological functions
    of essential fatty acids, Progr. Biochem. Pharmacol. 6:214-244.
Chippendale, G. M., Beck, S. D., and Strong, F. M., 1964, Methyl
    linolenate as an essential nutrient for the Cabbage Looper,
    Trichoplusia ni (Hübner), Nature 204:710-711.
Dadd, R. H., 1961, The nutritional requirements of locusts - V.
    Observations on essential fatty acids, chlorophyll, nutri-
    tional salt mixtures, and protein or amino acid components
    of synthetic diets, J. Insect Physiol. 6:126-145.
Dadd, R. H., 1964, A study of carbohydrate and lipid nutrition in
    the wax moth, Galleria mellonella (L.), using partially
    synthetic diets, J. Insect Physiol. 10:161-178.
Dadd, R. H., 1973, Insect nutrition: current developments and
    metabolic implications, Ann. Rev. Ent. 18:381-420.
Dadd, R. H., 1977, Qualitative requirements and utilization of
    nutrients: Insects, in: "Handbook series in nutrition and
    food," Vol. 1, pp. 305-346, M. Rechcigl, ed., CRC Press,
    Cleveland.
Dadd, R. H., 1978, Amino acid requirements of the mosquito, Culex
    pipiens: asparagine essential, J. Insect Physiol. 24:25-30.
Dadd, R. H., 1979, Nucleotide, nucleoside and base nutritional
    requirements of the mosquito Culex pipiens, J. Insect Physiol.
    25:353-359.
Dadd, R. H., 1980, Essential fatty acids for the mosquito Culex
    pipiens, J. Nutr., in press.

Dadd, R. H., Friend, W. G., and Kleinjan, J. E., 1980, Arachidonic acid requirement for two species of Culiseta reared on synthetic diet, Canad. J. Zool., in press.

Dadd, R. H., Gomez, I., and Namba, M., 1973, Requirement for ribonucleic acid in a semisynthetic larval diet for the mosquito Culex pipiens, J. Med. Ent. 10:47-52.

Dadd, R. H. and Kleinjan, J. E., 1976, Chemically defined dietary media for larvae of the mosquito Culex pipiens (Diptera: Culicidae): effects of colloid texturizers, J. Med. Ent. 13:285-291.

Dadd, R. H. and Kleinjan, J. E., 1977, Dietary nucleotide requirements of the mosquito Culex pipiens, J. Insect Physiol. 23:333-341.

Dadd, R. H. and Kleinjan, J. E., 1978, An essential nutrient for the mosquito Culex pipiens associated with certain animal-derived phospholipids, Ann. Ent. Soc. Amer. 71:794-800.

Dadd, R. H. and Kleinjan, J. E., 1979a, Essential fatty acid for the mosquito Culex pipiens: arachidonic acid, J. Insect Physiol. 25:495-502.

Dadd, R. H. and Kleinjan, J. E., 1979b, Vitamin E, ascorbyl palmitate and propyl gallate protect arachidonic acid in synthetic diets for mosquitoes, Ent. Exp. Appl. 26:222-226.

Destephano, D. B. and Brady, V. E., 1977, Prostaglandin and prostaglandin synthetase in the cricket, Acheta domesticus, J. Insect Physiol. 23:905-911.

Downer, R. G. H., 1978, Functional role of lipids in insects, in: "Biochemistry of Insects", M. Rockstein, ed., Academic Press,, London.

Fast, P. G., 1964, Insect lipids: a review, Memoirs Ent. Soc. Canad. No. 37.

Fast, P. G., 1970, Insect lipids, Progr. Chem. Fats Lipids 11: 181-242.

Fraenkel, G. and Blewett, M., 1946, Linoleic acid, vitamin E and other fat-soluble substances in the nutrition of certain insects (Ephestia kuehniella, E. elutella, E. cautella and Plodia interpunctella (Lep.)), J. Exp. Biol. 22:172-190.

Fraenkel, G. and Blewett, M., 1947, Linoleic acid and arachidonic acid in the metabolism of the insects Ephestia kuehniella and Tenebrio molitor, Biochem. J. 41:475-478.

Galli, C., Spagnuolo, C., Agradi, E., and Paoletti, R., 1976, Comparative effects of olive oil and other edible fats on brain structural lipids during development, Lipids 1:237-243.

Gilbert, L. I., 1967, Lipid metabolism and function in insects, Adv. Insect Physiol. 4:69-211.

Golberg, L. and DeMeillon, B., 1948, The nutrition of the larvae of Aedes aegypti Linnaeus. 3. Lipid requirements, Biochem. J. 43:372-379.

Grau, P. A. and Terriere, L. C., 1971, Fatty acid profile of the cabbage looper Trichoplusia ni and the effects of diet and rearing condition, J. Insect Physiol. 17:1637-1649.

Guarnieri, M. and Johnson, R. M., 1970, The essential fatty acids, Adv. Lipid Res. 8:115-174.

Holman, R. T., 1977, The deficiency of essential fatty acids, in: "Polyunsaturated Fatty Acids," Chapter 9, pp. 163-182, W.-H. Kunau and R.T. Holman, eds., American Oil Chemists Society, Champaign.

Hou, R. F. N. and Hsiao, J. -H., 1978, Studies on some nutritional requirements of the Diamond Back Moth, Plutella xylostella L., Proc. Natl. Sci. Council ROC 2:385-390.

House, H. L. and Barlow, J. S., 1960, Effects of oleic and other fatty acids on the growth rate of Agria affinis (Fall.) (Diptera: Sarcophagidae), J. Nutr. 72:409-414.

Lands, W. E. M., Martin, E. H., and Crawford, C. G., 1977, Functions of polyunsaturated fatty acids: biosynthesis of prostaglandins, in: "Polyunsaturated Fatty Acids," pp. 193-228, W.-H. Kunau and R.T. Holman, eds., American Oil Chemists Society, Champaign.

Loher, W., 1979, The influence of prostaglandin E2 on oviposition in Teleogryllus commodus, Entomol. Exp. Appl. 25:107-108.

Mead, J. F., 1970, The metabolism of the polyunsaturated fatty acids, Progr. Chem. Fats Lipids 161-192.

Meikle, J. E. S. and McFarlane, J. E., 1965, The role of lipid in the nutrition of the house cricket, Acheta domesticus L. (Orthoptera: Gryllidae), Canad. J. Zool. 43:87-98.

Rivers, J. P. W., Sinclair, A. J., and Crawford, M. A., 1975, Inability of the cat to desaturate essential fatty acids, Nature 258:171-173.

Rock, G. D., 1967, Aseptic rearing of the codling moth on synthetic diets: ascorbic acid and fatty acid requirements, J. Econ. Ent. 60:1002-1005.

Rock, G. D., Patton, R. L., and Glass, E. H., 1965, Studies on the fatty acid requirements of Argyrotaenia velutinana (Walker), J. Insect Physiol. 11:91-101.

Sang, J. H., 1956, The quantitative nutritional requirements of Drosophila melanogaster, J. Exp. Biol. 33:45-72.

Sivapalan, P. and Gnanapragasam, N. C., 1979, The influence of linoleic and linolenic acid on adult moth emergence of Homona coffearia from meridic diets in vitro, J. Insect Physiol. 25:393-398.

Sneller, V. -P., 1978, Development of Brugia pahangi in Aedes aegypti: Effects of mosquito nutrition on filarial development, Ph.D. Thesis, University of California, Berkeley.

Sprecher, H., 1977, Biosynthesis of polyunsaturated fatty acids and its regulation, in: "Polyunsaturated Fatty Acids," pp. 1-18, W.-H. Kunau and R.T. Holman, eds., American Oil Chemists Society, Campaign.

Tinoco, J., Babcock, R., Hincenbergs, I., Medwadowski, B., Miljanich,
    P., and Williams, M. A., 1979, Linolenic acid deficiency,
    Lipids 14:166-173.
Turunen, S., 1974, Polyunsaturated fatty acids in the nutrition of
    Pieris brassicae (Lepidoptera), Ann. Zool. Fennici 11:300-303.
Turunen, S., 1976, Vitamin E: effect on lipid synthesis and accumu-
    lation of linolenate in Pieris brassicae, Ann. Zool., Fennici
    13:148-152.

EFFECTS OF DIETARY CARBOHYDRATE AND LIPID ON NUTRITION AND METABOLISM OF METAZOAN PARASITES WITH SPECIAL REFERENCE TO PARASITIC HYMENOPTERA

S. N. Thompson

Division of Biological Control
University of California
Riverside, California

INTRODUCTION

Comprehensive investigation on the nutrition and biochemistry of parasitic metozoans is largely restricted to a few species. Severe limitations are therefore placed on the conclusive assessment of their nutritional and biochemical nature as characteristic of a parasitic way of life. Furthermore, in those species studied, the possible adaptive significance of their nutrition and metabolism to parasitism has been elusive due to the difficulty in establishing a relationship to probable reproductive fitness as well as a lack of comparative knowledge concerning related free-living forms. Our present knowledge, nevertheless, indicates that parasitic animals have evolved a variety of life strategies and become physiologically and biochemically adapted to their hosts in novel ways. Lipid and carbohydrate nutrients play unusual and important roles in parasites as nutrients essential for growth and development and as factors affecting the success of the parasite-host relationship.

Nutritional studies have been carried out on the specific metabolic effects of dietary lipids and carbohydrates in a variety of metazoan parasites. Two basic approaches in such investigations have been pursued. The first or in vivo approach involves the manipulation of host diet and concurrent observation of resulting changes in the parasite. Establishment of specific nutritional relationships from such studies, however, is difficult without intimate knowledge of the intermediate effects of dietary manipulation on the host organism. The complexity of the in vivo approach has dictated the development of in vitro methods of investigation involving suitable culture techniques. This approach not only allows complete control over nutrient quality and quantity but enables the study of basal parasite metabolism in the absence of the regulating effects of host metabolism as well as in axenic and xenic culture with a variety of microorganisms.

In evaluating the above two approaches to nutritional investigation it is important to stress that the parasitic stages of organisms are closely integrated with the physiology and biochemistry of their hosts. Indeed, it is the biochemical and metabolic interactions which become the essence of a parasite-host relationship. The host may play an active metabolic role in influencing the development of the parasite in addition to its role as a nutrient source. Hence, it is not possible to gain a complete understanding of the parasite's nutritional and metabolic nature by examining its behavior apart from the host. The in vitro approach then provides an invaluable tool for evaluating the basic nutritional requirements of a parasite, but may give little insight into the nature of parasitism (Read, 1966).

Varying degrees of success have been achieved at rearing parasites under artificial conditions. Numerous techniques are now available for rearing and/or maintaining specific life stages of a variety of parasitic helminths as reviewed by Silverman (1965), but less success has been attained at rearing such parasites through their complete life cycles. This difficulty undoubtedly stems from the complex and varying nutritional environment of the various life stages and the difficulties in duplicating these conditions outside the host. The limited availability of culture procedures for rearing parasitic helminths has made the evaluation of nutritional requirements of these parasites by the in vitro method difficult. The validity of such investigations for determining nutritional requirements and resulting developmental problems has been discussed by Weinstein (1966), Rothstein and Nicholas (1969), and Silverman and Hansen (1971).

Investigation has proceeded in a similar manner with protelean insect parasites, specifically parasitic Hymenoptera and Diptera. However, more success has been achieved at rearing these parasites through their complete life cycles under artificial conditions, as reviewed by Mellini (1975) and House (1977). This success is in part the result of the simpler parasitic strategy of this group as compared with others. Protelean parasities are those in which only the larval stages are parasitic. Furthermore, many are not host species or tissue specific and are, therefore, perhaps less metabolically integrated with the host. Despite the development of artificial culture techniques for a number of species, nutritional investigation has only been made with three, the dipteran Agria housei and the hymenopterans, Itoplectis conquisitor and Exeristes roborator. Selected species from the above groups on which carbohydrate and lipid nutritional investigations have been made are listed in Table 1. Due to the highly unspecialized nature of the host "associations" of A. housei and the extensive consideration of its nutrition in reviews on general insect nutrition, including those of House (1974) and Dadd (1977), it will not be discussed here. Furthermore, the nutrition of adult parasitic Hymenoptera which are predaceous in nature (Bartlett, 1964; House, 1977) will not be considered.

CARBOHYDRATE

The utilization and nutritive value of dietary sugars have been demonstrated in the protelean species I. conquisitor (Yazgan, 1972) and E. roborator (Thompson, 1976a). In both cases, glucose was not an absolute requirement, but when the nutritional or caloric value of the other nutrients was low, glucose was necessary to maintain normal development. In the parasite E. roborator reared in vitro, the deletion of glucose had no effect on larval development when amino acids were fed at the 6% level with or without dietary lipid (Table 2). At lower amino acid levels, however, specifically 1 and 3%, carbohydrate was necessary to maintain survival equal to that at the higher level. At the 1% amino acid level no development took place without glucose. Since survival at this amino acid level could be totally restored by inclusion of carbohydrate, it is evident that the latter can spare amino acids as an energy source. Larval development of I. conquisitor reared in vitro was not affected by glucose deletion at low amino acid levels, but pupal survival was markedly reduced.

The effect of dietary glucose on the larval growth rate of E. roborator calculated as

$$GR = \frac{G}{TA}$$

where GR = growth rate, G = fresh weight gain of the parasite during the feeding period, T = development time, and A = mean fresh weight of the parasite during the feeding period, as described by Waldbauer (1968), is shown in Fig. 1. The growth rate increased in a typical sigmoid fashion with the maximum increase occuring between 2 and 4% dietary glucose levels. Unfortunately, sufficient data were not obtained for calculating growth rates for I. conquisitor. Additional in vitro dietary studies with E. roborator also demonstrated the nutritional value of sucrose, and fructose (Fig. 2).

Similarly, the utilization of a variety of mono- and disaccharides has been made in many species of parsitic helminths (Table 1). Many, such as the cestode, Hymenolepis diminuta, have an absolute requirement for carbohydrate to maintain growth.

Table 1. Carbohydrate and lipid nutritional studies in selected metazoan parasite species.

Taxonomic Group	Species	Stage	Investigation[a]	Nutrient	Reference
Acanthocephala	Acanthocephalus ranae	adult	in vitro	lipid	Hammond (1968)
	Moniliformis dubius	adult	in vitro	carbohydrate	Graff (1964)
		Cystacanth to adult	in vivo	carbohydrate	Dunagen (1962)
Cestoda	Hymenolepis diminuta	Cysticercoid to adult	in vivo	carbohydrate	Crompton and Nesheim (1973)
		Cysticercoid to adult	in vivo	carbohydrate	Dunkley and Metterick (1969)
		adult	in vivo	carbohydrate	Read, Schiller and Phifer (1958)
		adult	in vitro	carbohydrate	Laurie (1957)
		sexually immature adult	in vitro	lipid	Bailey and Fairbairn (1968)
		adult	in vitro	lipid	Lumsden and Harrington (1966)
Trematoda	Schistosoma mansoni	adult	in vitro	carbohydrate	Bruce et al. (1974)
		adult	in vitro	lipid	Meyer, Meyer and Bueding (1970)
		larva	in vitro	carbohydrate	Bueding (1952)
Nematoda	Trichinella spiralis	juvenile and adult	in vitro & in vivo	carbohydrate	Castro and Roy (1974)
		juvenile and adult	in vitro	carbohydrate	Ferguson and Castro (1973)
	Ascaris lumbricoides	adult	in vitro	carbohydrate	Rathbone and Rees (1954)
Diptera	Agria housei	larva	in vitro	carbohydrate	House (1956)
		larva	in vitro	lipid	House and Barlow (1960)
Hymenoptera	Itoplectis conquisitor	larva and pupal	in vitro	carbohydrate and lipid	Yazgan (1972)
	Exeristes comstockii	free-living adult	"in vivo"	carbohydrate and lipid	Bracken (1965)
	Exeristes roborator	larva	in vitro & in vivo	carbohydrate	Thompson (1976a) (1979)
		larva	in vitro	lipid	Thompson (1977)

[a] Studies listed as in vitro are those in which the organism was maintained or growing during the course of the experiment outside the host in physiological saline solutions or in artificial nutrient media. In most cases with parasitic helminths the parasite was reared in the host and removed by dissection immediately prior to the experiment.

Table 2. The effects of varying the level of amino acids and deletion of carbohydrate and lipid on larval survival and development time Exeristes roborator reared in vitro on artificial media.[a,b] After Thompson (1976a).

| | Amino acid level | | | | | |
| | 1% | | 3% | | 6% | |
	% survival	Development Time	% survival	Development Time	% survival	Development Time
Control	89 ± 8*	5.8 ± 0.4a	87 ± 12a*	4.6 ± 0 ab	89 ± 6a*	3.5 ± 0.4a
-- D-glucose	0	---	43 ± 10b	6.2 ± 1.3a*	78 ± 18a	4.3 ± 0.5a*
-- triolein	71 ± 6	6.3 ± 0.3a	96 ± 4a*	4.1 ± 0.3b*	87 ± 8a*	3.9 ± 0.3a*
-- (D-glucose and triolein)	0	---	51 ± 16b	6.1 ± 0.6a	91 ± 9a	4.2 ± 0.7a

[a] Each experiment was repeated three times with 25 larvae per replicate.

[b] Numbers in a single column followed by the same letter and numbers in a single row followed by the same symbol are not significantly different statistically at the 95% level.

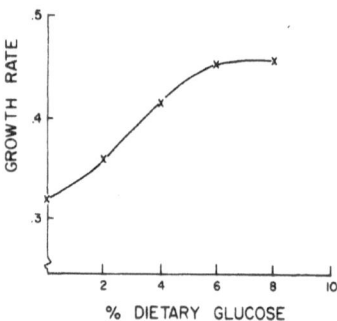

Fig.1 Effect of dietary glucose level on the growth rate of
       <u>Exeristes</u> <u>roborator</u> reared <u>in vitro</u> on chemically defined
       media. After Thompson (1979).

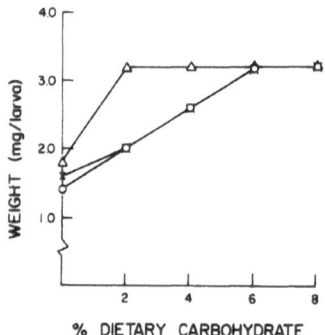

Fig.2 Effects of selected dietary carbohydrates on wet weight of
       <u>Exeristes</u> <u>roborator</u> reared <u>in vitro</u> on chemically defined
       media. Sucrose (Δ), glucose (0), fructose (X). After
       Thompson (1979).

Although the utilization of specific sugars varies considerably
between species (Read, 1959), glucose is utilized by most helminths,
and in contrast to the parasite, <u>E</u>. <u>roborator</u> above, sucrose is
seldom utilized. The carbohydrate and lipid requirements of
helminth species has been reviewed by Metterick and Jackson (1977),
Hieb (1977), and Crompton (1977). The direct quantification of
nutritive value of sugars in parasitic helminths has been difficult
due to the limited growth obtained with <u>in vitro</u> techniques. In
many cases, an assumption of nutritive value is made on the basis
of absorption, metabolic assimilation and/or the maintenance of

specific growth stages. However, examination of the data from the
in vivo studies of Read (1959) and others on H. diminuta demonstrates
specific relationships between dietary carbohydrate and growth.
Read et al. (1958) demonstrated that body weight was a linear
function of the quantity of carbohydrate ingested by the host. Although the authors did not correlate a specific physiological or
biological event to the time period of development, calculations
made from their data allowed the determination of relative growth
rates (Table 3). Rates were calculated using the weight of worms
maintained on hosts fed starch-free diets for the 9-day feeding time
as the initial weight for gain calculations. This weight, however,
was corrected by a multiplication factor of three determined from the
data of a previous study (Read and Rothman, 1957), which demonstrated
that carbohydrate deficiency caused a decrease in worm weight.
Starch was the carbohydrate substrate used as glucose is an ineffective in vivo substrate presumably because it is absorbed rapidly
in the first part of the intestine. In a similar fashion relative
growth rates were determined for E. roborator using the weight of
parasite larvae maintained on carbohydrate-free diets as the initial
weight rather than the weight of newly hatched larvae as in the calculation of absolute growth rate (Fig. 1, Table 3). The studies cited
overruled caloric deficiency as a factor limiting growth of either
species under the nutritional regimes present at the time of the
experiments. Rates thus obtained are shown in Fig. 3. Considering
the different fates of dietary carbohydrate in the groups represented by these two species, it is of interest that the growth rates
obtained are of the same order of magnitude.

The rate of utilization of carbohydrate by parasitic helminths
is largely a reflection of their fermentative capacity and adaptation to facultative anaerobiosis. Indeed, carbohydrate is the
primary substrate for energy production. The pattern of oxidative
metabolism is for the incomplete oxidation of carbohydrate substrates with the formation and excretion of partially oxidized end
products. The two major fermentation pathways are the glycolytic
sequence resulting in the accumulation of lactate and its subsequent excretion and the production and excretion of succinate.
The latter are produced by the activity of phosphoenolpyruvate
carboxykinase and malate dehydrogenase catalyzing the terminal
reactions following glycolytic oxidation and resulting in the
production of oxaloacetate and malate which is then converted to
succinate.

With some modifications this metabolic scheme has been found
in species of each group of parasitic helminths listed in Table 1.
The inhibition of succinic dehydrogenase by high ratios of succinate
to fumarate has been shown to be absent in Ascaris lumbricoides
(Kmetec and Bueding, 1961) and presumably in other species as
well. Lipid does not appear to play a significant role in energy

Table 3. Calculation of relative growth rates of _Hymenolepsis diminuta_ reared _in vivo_ and _Exeristes roborator_ reared _in vitro_ under varying dietary carbohydrate levels.

Species	Percent dietary carbohydrate	Mean weight (mg)	Mean Development period (days)	Mean Weight gain during development period (mg)	Mean weight during development period (mg)	Mean relative growth rate
Hymenolepsis diminuta (Extracted from Read and Rothman, 1957; Read et al., 1958)	0	100	9.0			0.145
	1	286	9.0	186	143	0.168
	6	429	9.0	329	218	0.183
	12	560	9.0	460	280	0.194
	21	800	9.0	700	400	
Exeristes roborator (Extracted from Thompson, 1979)	0	1.20	5.00	----	----	----
	2	1.50	4.60	0.30	0.75	0.09
	4	1.80	4.10	0.60	0.90	0.16
	6	2.10	3.80	0.90	1.05	0.23
	8	2.40	3.80	1.20	1.20	0.26

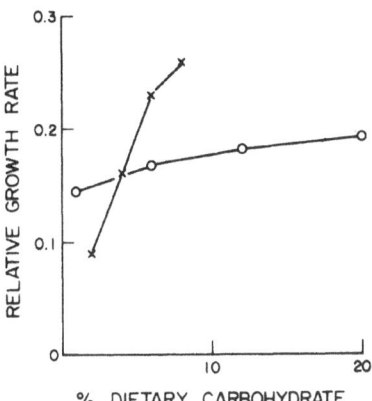

Fig.3 Effect of dietary carbohydrate level of the relative growth
       rates of the parasities Exeristes roborator in response to
       dietary glucose (X) and of the cestode Hymenolepsis diminuta,
       in response to dietary starch (o).

metabolism in the parasitic stages of helminths. In addition to
the extensive research literature on this topic, several excellent
reviews are available including those of Read and Simmonds (1963);
Read (1968); Saz (1969); Saz (1972); and Kurelac (1975).

     Parasitic Hymenoptera, in contrast to the above groups,
exhibit patterns of oxidative metabolism more typical of obligate
aerobic organisms and are similar to higher animals. Thus carbohydrate provides energy primarily by complete aerobic oxidation.
Excess carbohydrate is converted to storage lipid which may be
rapidly mobilized for the production of energy substrate intermediates in response to high metabolic demand. The energetics of
these various metabolic strategies have been discussed by Hochachka
and Somero (1973).

LIPIDS

     Little information is available on the nutrition of lipids in
metazoan parasites. Certain lipids are, however, required. Parasitic helminths (Hieb, 1978; Metterick and Jackson, 1978) as well
as Hymenoptera (Thompson, unpublished) have an absolute requirement
for dietary sterol, although their capacity for sterol modification
differs considerably. Parasitic Hymenoptera also require polyunsaturated fatty acid for pupal development (Yazgan, 1972; Thompson, 1977). The uptake and/or uitlization of a variety of free
fatty acids and/or glycerides has been demonstrated in several
parasities including H. diminuta (Bailey and Fairbairn, 1968),

Schistosoma mansoni (Meyer et al., 1970), and I. conquisitor (Yazgan,
1972), but quantification of their nutritive value for development
has not been made. Dietary lipids have been investigated in E.
roborator and the results emphasize the difficulty in studying
nutrition in vitro. Initial studies carried out on E. roborator related to nutritional balance demonstrated that dietary triglyceride
had little value as an energy source for larval development (Table
2). Deletion of triolein from diets containing amino acids at the
3% level had no effect on survival or development time and from
diets containing 1% amino acid had only a small effect on survival.
Furthermore, when both glucose and triolein were deleted at the 3%
amino acid level the decrease in survival from the control was not
greater than that observed when carbohydrate alone was deleted.
Additional studies with dietary supplements of triglycerides of
all the major synthesized fatty acids of the parasite did not im-
prove development with or without dietary carbohydrate (Table 4).
Thus dietary triglyceride did not replace the nutritional value
of carbohydrate and indeed when parasities were reared on diets
lacking carbohydrate some of the triglycerides were detrimental.
For example, survival was decreased nearly 50% by inclusion of
tripalmitin in artificial media lacking carbohydrate. Considering
the nature of metabolism in E. roborator it is likely that dietary
lipid has an important nutritional role during development. The
above results possibly reflect the artificial situation created
when naturally complex foodstuffs are replaced by simplified and
defined components. In the case of lipids, for example, the natural
dietary lipid consists not only of triglyceride but of lipoprotein,
lipid-protein complexes, partial glyceride, phospholipid, etc. The
problem of formulation and emulsification of the lipid component of
the artificial media may also be partially responsible for the difficulty in assessing the nutritional value of lipids. Although free
fatty acids were highly toxic to E. roborator when formulated into
the diet without emulsifier or with sodium dodecyl sulphate, which
is similar to the naturally occurring emulsifiers of invertebrate
guts, they were not toxic when emulsified in polyoxyethylene sorbi-
tan monoleate (Tween) (Thompson, 1977). Although similar toxicity
was not obtained with triglycerides, the emulsifiers failed to
demonstrate their nutritional value.

Extensive studies have been made on characterizing the lipids
of parasitic animals and determining the effects of nutrition on
their distribution and metabolism. Metazoan parasities contain
substantial amounts of lipid, generally 1 to 10% of their wet
weight, and composed largely of triglyceride and phospholipid
(Fairbairn, 1969; Thompson, 1979). Dietary fatty acids are readily
incorporated into the esterified lipids of parasitic helminths such
as H. diminuta (Bailey and Fairbairn, 1968) and A. lumbricoides
(Jezyk, 1968), and this capability has been shown to be due to the
glyceride-phospholipid biosynthetic pathway found in most animals

Table 4. The effects of triglyceride supplements with and without carbohydrate on larval survival and development time of Exeristes roborator reared in vitro on artificial media.[a,b] After Thompson (1977).

Supplement	With carbohydrate		Without carbohydrate	
	% larvae surviving to 4th instar ± S.D.[b]	Development time to 4th instar (days) ± S.D.	% larvae surviving to 4th instar ± S.D.	Development time to 4th instar (days) ± S.D.
None--fatty acid free	87 ± 8a	3.9 ± 0.3a	91 ± 9a	4.2 ± 0.7a
Tripalmitin	87 ± 7a	3.6 ± 0.5a	48 ± 13	5.1 ± 0.3a
Tripalmitolein	83 ± 9a	3.7 ± 0.2a	79 ± 18a	4.4 ± 0.8a
Tristearin	86 ± 7a	3.4 ± 0.5a	78 ± 17a	5.1 ± 0.3a

[a] Each experiment was repeated three times with 30 larvae per replicate.

[b] Numbers are in a single column followed by the same letter and numbers in a single row followed by the same symbol are not significantly different statistically at the 95% level.

(Buteau and Fairbairn, 1969). The fatty acids of helminth lipids are composed largely of unsaturated C18 acids and the fatty acid patterns of the esterified lipid fractions reflect to some extent the fatty acid patterns of the host milieu. Thus the fatty acid pattern of Trichinella spiralis is similar to the composition of a skeletal muscle (Castro and Fairbairn, 1969) and those of H. diminuta resemble the contents of the rat intestine (Ginger and Fairbairn, 1966). Further investigation demonstrated that many parasitic helminths including H. diminuta (Jacobsen and Fairbairn, 1967), Spirometra mansonoides (Meyer et al., 1966), and S. mansoni (Smith et al., 1970), lack both the ability to synthesize saturated fatty acids de novo or desaturate dietary fatty acids (Meyer and Meyer, 1972). In addition, the parasitic stages of helminths fail to oxidize fatty acids (Fairbairn, 1969; Castro and Fairbairn, 1969; Ward and Fairbairn, 1970a,b). Presumably this lack of metabolic capacity coupled with the rapid hydrolysis of absorbed dietary glyceride (Bailey and Fairbairn, 1968), creates a rather static composition from which substrate is provided for glyceride and phospholipid synthesis. Despite this lack of synthetic ability, however, many parasitic helminths do have a limited capacity to chain elongate existing fatty acids. Fairbairn (1970) suggests that additional control of the fatty acid composition of the helminth's lipids is achieved by selective acylation. However, no studies are available on the substrate specific nature of helminth acyltrans-ferase enzymes and these systems must be reasonably non-specific for the parasites to exhibit similar compositions to those of their hosts.

The nutritional role of fat in parasitic helminths remains obscure. In parasites such as H. diminuta their primary function may be structural and related to their role as components of phos-pholipid. Although triglyceride is readily synthesized from fatty acid of dietary origin substantial amounts are lost presumably as waste when the proglottids are shed. Fairbairn (1970) has suggested that " . . . in the overall economy of the parasite it is less trouble to convert excess fatty acids to triglyceride than it is to curtail their absorption and excretion."

In vivo studies with the insect parasite, Exeristes comstockii resulted in the finding of a similar reflection of parasite and host fatty acid compositions as that described above (Bracken and Barlow, 1967), but the similarity was more striking than in the case with parasitic helminths. The non host-species specific nature of E. comstockii allowed its development on a number of foreign hosts with markedly different compositions. In each case, the fatty acid patterns of the total lipids of parasite and host were almost identical. Further study resulted in the finding that such "dupli-cation" occurs in many parasitic Hymenoptera and appears particularly common in the family Ichneumonidae (Thompson and Barlow, 1974).

For example, the fatty acid compositions of three species, $\underline{E}$. comstockii, $\underline{E}$. roborator and $\underline{I}$. conquisitor and selected hosts are shown in Fig. 4. "Duplication" was later shown to be characteristic of the esterified lipid classes (Thompson and Barlow, 1973; Thompson and Adams, 1976).

In contrast to the parasitic helminths, $\underline{I}$. conquisitor, $\underline{E}$. roborator and $\underline{E}$. comstockii were all shown to rapidly synthesize fatty acid and incorporate a variety of radioactive lipid precursors including glucose and acetate into saturated and unsaturated fatty acids, phospholipid and triglyceride (Barlow and Bracken, 1971; Thompson and Barlow, 1972; Thompson and Johnson, 1978). Indeed, the pattern of radioactive incorporation into fatty acid corresponded well with accepted de novo pathways of synthesis. Fatty acid synthetase activity, that enzyme system which catalyzes the de novo synthesis of saturated fatty acid by condensation of acetyl- and malonyl-coenzyme A, was later isolated from developing larvae (Thompson and Johnson, 1978). Furthermore, quantitative studies demonstrated that the fractional turnover rates of fatty acids in $\underline{E}$. comstockii greatly exceeded those of two free-living host species, Galleria mellonella and Lucilia sericata used for comparison (Thompson and Barlow, 1972).

Tracer experiments with $\underline{E}$. comstockii reared in vivo on the above two species of host demonstrated that radioactivity was incorporated from ^{14}C-1-acetate into palmitic, palmitoleic, stearic and oleic acids. The relative amounts of these acids expressed as a percentage of the total differed considerably between the parasites reared on the two hosts and were nearly identical to those of the specific hosts. In the case of palmitoleic acid, which shows the greatest difference between these two host species, the concentration changes 10 fold, between 2 and 20%, in the parasite reared on one as compared to the other. However, despite the marked alterations in fatty acid composition, the specific activity did not change significantly. Specific activity may be considered as a ratio of the amount of metabolite synthesized and into which radioactivity has been incorporated less the amount degraded divided by the total amount present. On examining palmitoleic acid, this ratio is almost the same regardless of the 10-fold difference in amount and thus the amount of synthesis less degradation must also have changed 10 fold. The experiment demonstrates that the level of palmitoleic acid and other fatty acids is controlled by active metabolic processes and not by deposition of host fat. The above studies with parasitic Hymenoptera were reviewed more thoroughly by Barlow (1972).

Recent studies on lipid metabolism have been carried out in vitro with $\underline{E}$. roborator following development of artificial culture techniques and chemically defined media (Thompson, 1975). Initial in vitro investigations were aimed at characterizing the nature of

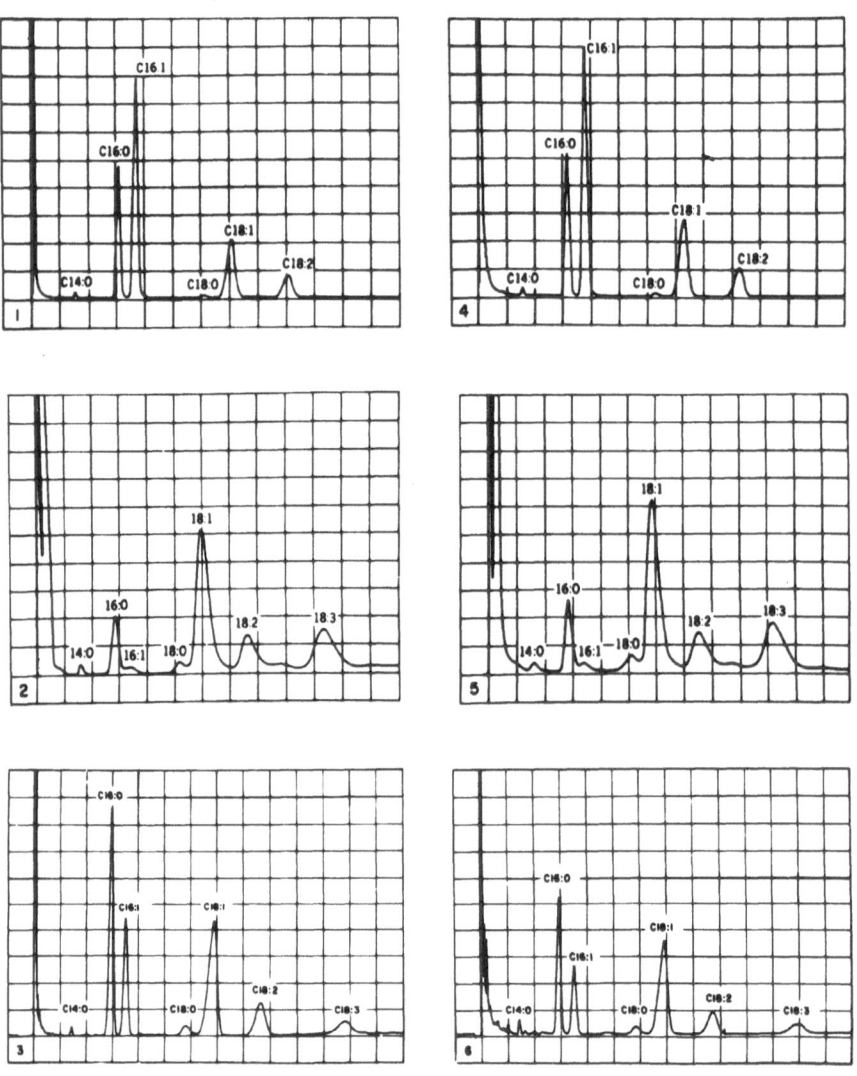

Fig.4 "Duplication" of fatty acid patterns of Exeristes comstockii,
Itoplectis conquisitor, and Exeristes roborator on selected
hosts. 1. Ostrinia nubilalis. 2. Neodiprion sertifer.
3. Gnormoshema operculella. 4. I. conquisitor reared on
O. nubilalis. 5. E. comstockii reared on N. sertifer. 6.
E. roborator reared on G. operculella.

fatty acid metabolism in the absence of dietary fatty acids and on
the effect of dietary fatty acid composition on that of the parasite in the absence of host metabolism. The fatty acid pattern of
the total lipids of E. roborator reared on fatty acid-free artificial media was similar to that of the parasite reared on Gnormoshema
operculella (Fig. 4) except for the absence of polyunsaturates which
are not synthesized and are only present if these are dietary components (Thompson, 1976b). Furthermore, the specific activities of
fatty acids following incorporation of radioactivity from ^{14}C-2-
acetate were similar to parasites reared on host material. It is
apparent, therefore, that the parasite has the ability to regulate
fatty acid levels in the absence of dietary lipid. However, when
the diet was supplemented with lipid extracted from two host species
the resulting fatty acid composition of the parasite was substantially
different from that of the host fats and except for the presence
of polyunsaturates, the fatty acid composition was similar to that
when the parasite was reared on the fatty acid-free media (Table 5).
It was concluded that the proportion of individual fatty acids in
the host ultimately determines the levels in the parasite, but that
the host plays a more complex and active role in affecting the regulation of fatty acid metabolism in the parasite in addition to
serving as a nutrient source of fatty acids.

In contrast to the above studies with dietary host lipid supplements, experiments involving the supplementation of fatty acid-
free diets with individual triglycerides did result in marked
laterations in fatty acid composition in the parasite (Table 6).
The addition of tripalmitin, tripalmitolein, trisstearin and triolein
resulted in increases in the proportions of the corresponding fatty
acids in the parasite. However in this case, the alterations were
accompanied by decreases in their specific activities. Thus, altered fatty acid composition appears to be due to dietary fat
deposition.

The significance of the influence of the host species on fatty
acid metabolism in the above parasites remains unknown. It may
play some role in the parasite's biology even though the parasite
can be reared in the absence of the host on artificial media. It
was previously suggested that a lack of control over fatty acid
composition by the parasite may impart an advantage to survival, and
that the degree to which various parasites regulate their compositions may be one factor influencing host suitability (Thompson and
Barlow, 1974). However, the rather strict maintenance of the
synthetic capacity of parasites reared in vitro on triglyceride
supplemented diets and of the composition of parasites reared on
diets containing host lipids suggest that under the specific nutritional regime presented by artificial media, the fatty composition
observed may be of importance in maintaining the physiological
integrity of the parasite.

Table 5. Fatty acid composition of Exeristes roborator reared in vitro on fatty acid free artificial media supplemented with host lipids. After Thompson (1976b).

	Percent Composition						
	Myristate	Palmitate	Palmitoleate	Stearate	Oleate	Linoleate	Linolenate
1. E. roborator reared on fatty acid-free media	2	23	10	9	57		
2. E. roborator reared on media supplemented with L. sericata lipids	1	19	8	10	48	3	13
L. sericata lipids	2	17	14	4	20	36	7
3. E. roborator reared on media supplemented with G. mellonella lipids	1	16	9	16	52	2	6
G. mellonella lipids	Trace (<1)	36	4	3	44	10	3
	Percent composition excluding polyunsaturates						
4. E. roborator reared on media supplemented with L. sericata lipids	1	22	9	12	56		
5. E. roborator reared on media supplemented with G. mellonella lipids	1	17	10	17	55		

Table 6.   Fatty acid composition and relative specific activity
(% cpm/% composition) of fatty acids of <u>Exeristes</u> <u>roborator</u>
reared <u>in</u> <u>vitro</u> on artificial media supplemented with
triglycerides and 2-^{14}C-acetate.   After Thompson and
Johnson (1978).

Supplement	Percent Composition Specific Activity			
	Palmitate	Palmitoleate	Stearate	Oleate
None (Control	23 / 0.9	16 / 0.8	10 / 1.4	51 / 1.0
Tripalmitin	70 / 0.4	4 / 2.5	9 / 1.1	17 / 2.5
Tripalmitolein	16 / 1.1	31 / 0.4	10 / 1.9	43 / 1.1
Tristearin	25 / 0.8	3 / 1.0	37 / 0.8	35 / 1.3
Triolein	25 / 1.4	6 / 1.3	13 / 1.5	56 / 0.7

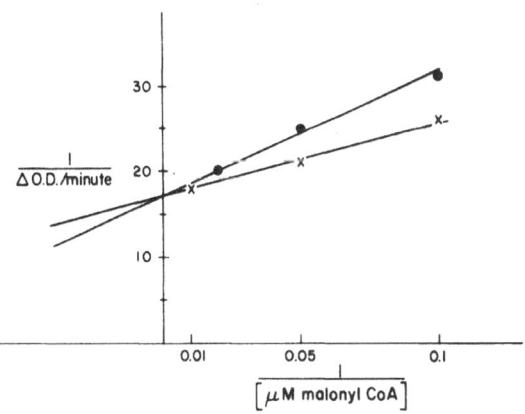

Fig.5   Linweaver-Burk plot of velocity - substrate concentration
relationship of fatty acid synthetase activity from <u>Exeristes</u>
<u>roborator</u> to malonyl-CoA, with (o) and without(x) phenyl-
methylsulphonyl fluoride.   After Thompson and Johnson (1978).

Table 7. The effect of phenylmethylsulphonyl fluoride on larval
          survival and development time of <u>Exeristes</u> <u>roborator</u>
          reared on fatty acid-free artificial media. After
          Thompson and Johnson (1978).

PMSF level	Percent survival	Development time
None (Control)	83 ± 5	3.4 ± 0.3
1 mM	85 ± 2	3.8 ± 0.2
2.5 mM	67 ± 2	4.7 ± 0.2
5 mM	0	---

        Studies were, therefore, carried out to determine if fatty
acid composition could be altered in parasites reared in vitro
on fatty acid-free diets and if parasite development was effected
by such changes. The hypolipidemic agent phenylmethylsulphonyl
fluoride (PMSF) was used in attempts to inhibit lipogenesis. Low
dietary levels of PMSF did indeed alter development and fatty acid
metabolism (Thompson and Johnson, 1978). Survival was decreased
and development time increased with increasing concentrations of
PMSF from 1 to 5 mM (Table 7) and fatty acid synthetase activity
was inhibited in an apparent competitive manner (Fig. 5). Accompanying the above effects were changes in the relative proportions of fatty acids; the saturates, palmitic and stearic acids
increased and their corresponding monounsaturates decreased (Table
8). This effect presumably resulted from a decreased level of
saturated fatty acid synthesis and thus a shortage of substrate
for desaturation, although PMSF may also have a direct effect on
desaturase activity, since the total level of lipid was not significantly different between mature larvae reared with and without
PMSF. This lack of difference suggests that although de novo
synthesis is depressed, the increased development time of larvae
fed PMSF allows accumulation of an equal quantity of fatty acid as
in control larvae by the time development is complete. A direct
demonstration that the altered fatty acid composition brought about
by PMSF was responsible for the accompanying changes in development
was not possible since the effects on fatty acid composition and
development could not be reversed in the presence of PMSF by inclusion of dietary fats (Tables 9, 10). Therefore, it may be that
other unknown effects of PMSF are responsible for the changes in
development noted.

Table 8. The fatty acid composition of <u>Exeristes</u> <u>roborator</u> reared
          on fatty acid-free artificial media with and without
          phenylmethylsulphonyl fluoride. After Thompson and
          Johnson (1978).

| Fatty acid | Control | Percent composition | | |
		1 mM	2 mM	3 mM
Palmitate	19	19	40	39
Palmitoleate	51	48	21	26
Stearate	5	6	8	15
Oleate	25	27	31	19

Table 9. The fatty acid composition of <u>Exeristes</u> <u>roborator</u> reared
          on fatty acid-free and triglyceride supplemented artifi-
          cial media containing 2.5 mM phenylmethylsulphonyl
          fluoride. After Thompson and Johnson (1978).

| Fatty acid | Control | Percent composition | | |
		2.5 mM PMSF	+ triolein	+ tristearin
Palmitate	22	41	46	46
Palmitoleate	18	9	15	10
Stearate	8	13	9	11
Oleate	52	37	30	33

Table 10. The effect of dietary lipid supplements on larval survival
and development time of Exeristes roborator reared on
artificial media with and without 2.5 mM phenylmethylsulphonyl fluoride. After Thompson and Johnson (1978).

	Without PMSF		With PMSF	
	Percent survival	Development time	Percent survival	Development time
Control	87 + 8	3.9 + 0.3	43 + 13	4.9 + 0.6
Galleria mellonella lipids	89 + 10	3.9 + 0.8	0	---
Tristearin	86 + 7	3.4 + 0.5	23 + 12	4.7 + 0.6
Triolein	89 + 6	3.5 + 0.4	21 + 13	4.9 + 0.5

Since the inclusion of dietary triglyceride in artificial media
resulted in alteration in the proportions of fatty acids in E.
roborator (Table 5) and such effects were not noted when fats were
included in diets containing PMSF, radiotracer studies were performed to determine if PMSF, in addition to its effects on fatty
acid synthesis, had an effect on the synthesis of the esterified
lipid fractions. Because triglyceride and phospholipid contain the
bulk of fatty acids in animal tissues, selective alterations in
acyltransferase activity could have resulted in the alterations in
the fatty acid composition of the total lipids noted above and
brought about by administration of PMSF. Radioactivity from ^{14}C-
2-acetate was incorporated into all the major lipid classes (Table
11). However, no difference in the distribution of radioactivity
was noted between the fractions of parasites reared on fatty acid-
free media and those of parasities reared on that media with PMSF
(Table 12).

Despite the failure to determine the mechanism of dietary
PMSF activity in E. roborator reared in vitro, it was noted during
the above investigations that proportionately small amounts of
radioactivity were incorporated into triglyceride (Table 12) suggesting low levels of that lipid fraction in parasites reared in
vitro. Triglyceride levels of E. roborator reared in vivo and in
vitro were, therefore, compared. The results demonstrated the
parasites reared in vivo on host insects contained substantially
greater amounts of triglyceride (Fig. 6).

Table 11. The percent distribution of radioactivity from dietary radiolabeled precursors incorporated into the major neutral lipid classes of Exeristes roborator reared on fatty acid-free diets. After Thompson and Johnson (1978).

Percent radioactivity incorporated

	Monglyceride	1, 2 diglyceride	1, 2 diglyceride	Free fatty acid	Triglyceride	Hydrocarbon sterol ester and wax
2-^{14}C-acetate	9	2	6	8	17	58
U-^{14}C-palmitic acid	8	10	4	source	24	54
^{14}C (carboxyl) tripalmitin	21	20	15	25	source	19

Table 12. The percent distribution of radioactivity from dietary 2-^{14}C-acetate incorporated into the major neutral lipid classes of Exeristes roborator reared on Galleria mellonella and fatty acid-free diets with and without 2.5 mM phenylmethylsulphonyl fluoride. After Thompson and Johnson (1978).

Percent radioactivity incorporated

	Monglyceride	1, 2 diglyceride	1, 2 diglyceride	Free fatty acid	Triglyceride	Hydrocarbon sterol ester and wax
Host reared	2	3	1	6	78	11
Fatty acid-free	9	2	5	5	13	65
Fatty acid-free + 2.5 mM PMSF	7	Trace	7	8	14	64

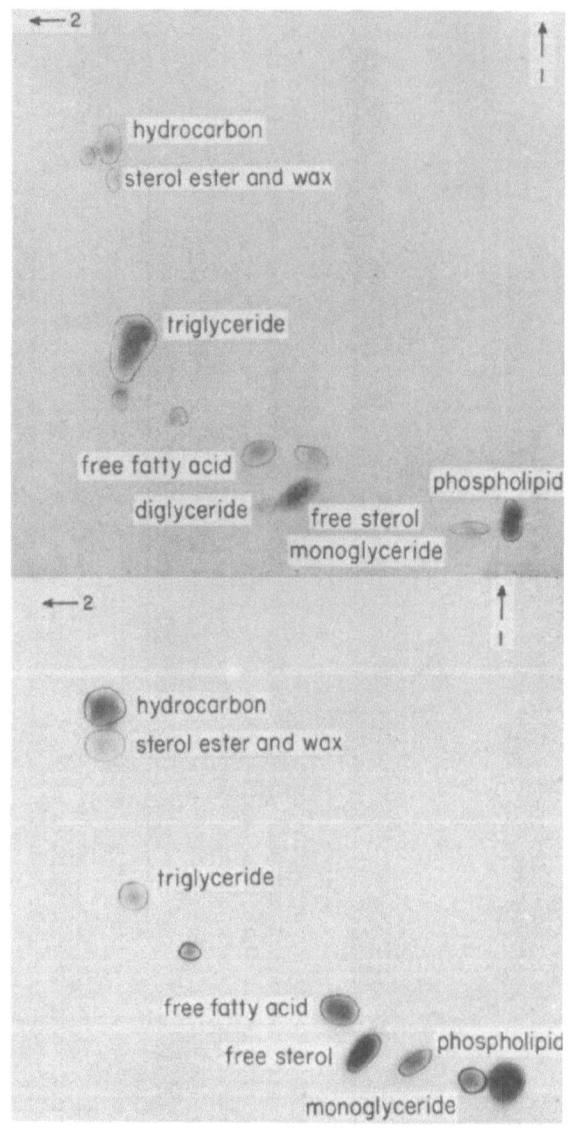

Fig.6   Thin layer chromatograms of lipids from Exeristes roborator
        reared in vitro on chemically defined media (bottom), and
        in vivo on Galleria mellonella (top). Developed thin layers
        were charred with H_2SO_4. After Thompson and Johnson (1978).

Although the findings of Barlow and Bracken (1971) demonstrate the metabolic activity was largely responsible for the alterations in saturated and monounsaturated fatty acid levels in E. comstockii reared on different host species with varying fatty acid patterns, the "duplication" of fatty acid composition in parasitic Hymenoptera includes the polyunsaturates as well, and these are not synthesized by the insects. Deposition of dietary fat, therefore, does occur, and, indeed, radioactive fatty acid was shown to be rapidly incorporated into complex lipid (Table 11). The phenomenon also appears to occur in the complex lipids, but not in the free fatty acid fraction (Thompson and Adams, 1976). If regulation of esterified fatty acid levels was determined during the turnover of the fatty acids themselves, the free fatty acid fraction of the parasite would be expected to be similar to that of the host since the free fatty acid pool is considered the source of fatty acids for complex lipid synthesis. Furthermore, isolated fatty acid synthetase activity was shown to be depressed by exogenous fatty acid rather than stimulated (Fig. 7). Considering these findings it appears likely that the mechanism of regulation over the fatty acid levels in parasites reared in vivo lies at the level of esterification and complex lipid formation. Considering this conclusion it is of interest that parasities reared in vitro, in which case dietary fats fail to influence the parasite's fatty acid composition in the same manner as occurs in vivo, have low levels of triglyceride.

The regulation of fatty acid levels in vivo through altered substrate specificity during glyceride synthesis is consistent with the findings of Barlow and Bracken (1971) described above.

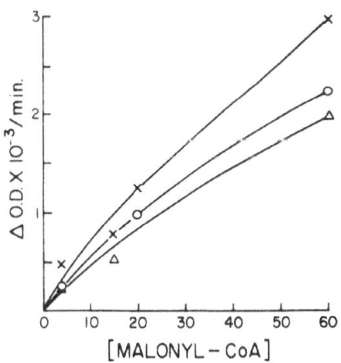

Fig.7   Effect of dietary palmityl-CoA (o) and oleyl-CoA (Δ) on the velocity-substrate concentration (μM) relationship of fatty acid synthetase activity (X), from Exeristes roborator reared on artificial media, to malonyl-CoA. After Thompson and Johnson (1978).

Following fatty acid synthesis, an increased acyltransferase specificity for one fatty acid would result in an increased level of that
fatty acid in the esterified lipids but the specific activity of
that fatty acid would be unaltered. However, since the individual
classes of esterified lipid all have somewhat different fatty acid
compositions, the substrate specificity of the acyltransferase
system responsible for the synthesis of each type in the parasite
must be affected by the proportion of fatty acids in the cor-
esponding complex lipid in the host. It is difficult to visualize
how such a mechanism could be operative and how the enzyme systems
might be exposed to the host lipids considering digestive lipase
activity and lipid hydrolysis which presumably occurs in the gut.

CARBOHYDRATE-LIPID INTERACTIONS

    The metabolic and nutritional interactions of carbohydrate and
lipid in parasitic animals is to some degree predictable considering
the different metabolic strategies described above. Although the
fate of nutritional fat in most parasitic helminths remains in
question, lipid-carbohydrate interactions have been observed. Passey
and Fairbairn (1957) found evidence in A. lumbricoides eggs indicating that lipid was being converted to carbohydrate, which was
then available as reserve substrate for energy metabolism. That
study demonstrated that the utilization of fat during embryonic
development occurred when oxygen consumption was inadequate to be
accounted for by complete oxidation. Furthermore, carbon dioxide
production accounted for only 72% of the fatty acid which was
utilized, although oxidation is functional in the eggs of this
species (Ward and Fairbairn, 1970b). As development proceeded and
the period of infectivity reached the continued loss of triglyceride
was accompanied by the resynthesis of glycogen and trehalose reserves which had also been depleted during embryonic development.
Indeed, Barrett et al. (1970) later demonstrated the net conversion
of triglyceride to carbohydrate in A. lumbricoides eggs via the
glyoxylate cycle. The activity of both isocitrate lyase and malate
synthetase corresponded well with changes in the rate of carbohydrate synthesis and incorporation of radioactivity from fatty acid
into glycogen and trehalose. These metabolic interrelationships
have recently been reviewed by Greichus and Greichus (1975).

    In higher animals, dietary carbohydrate, in addition to its
role in aerobic energy metabolism is the major precursor of fat
synthesis, providing the substrate acetyl-CoA via glycolysis.
Numerous in vivo and in vitro studies concerning the effects of
nutritional state on fat metabolism in vertebrates have been carried
out (MacDonald, 1966; Romos and Leveille, 1974), and this capacity
for lipogenesis has also been described in a few "free-living"
insects (Van Handel, 1965; Horie and Nakasone, 1971). In vitro
studies with parasitic Hymenoptera, specifically E. roborator, have
demonstrated the net conversion of dietary carbohydrate to lipid,

and fatty acid synthetase activity was readily isolated from larval tissue extracts (Thompson, 1980).

Increased growth rates observed in E. roborator in relation to dietary carbohydrate levels were due largely to increases in tissue protein and fat levels (Fig. 8). A four-fold greater concentration of lipid was observed in larvae reared on media containing 8% glucose as compared to those reared on carbohydrate-free media, representing a seven-fold increase in total lipid level per individual. Similar results were obtained with dietary sucrose and fructose (Fig. 9). These findings, however, are in contrast to studies with other animals. Varied lipid response to different dietary carbohydrates particularly glucose and sucrose have been well documented in higher animals (Zakim, 1973). Generally, sucrose feeding results in a greater accumulation of lipid than occurs with glucose, an effect which results from the fructose component of the disaccharide. Thus, sucrose and fructose feeding generally result in the same effect. Although fructose has differential effects on several lipogenic enzyme systems, the effect of increasing rates of lipid synthesis when compared to glucose is largely attributed to its role in glycolysis. The rate of fructose oxidation via the glycolytic pathway is greater than that of glucose due to the higher affinity of fructokinase for fructose as compared to that of glucokinase for glucose, and the fact that fructose oxidation does not involve the phosphofructokinase step, a key mediator in glycolysis. As a result, fructose provides substrate for lipogenesis at

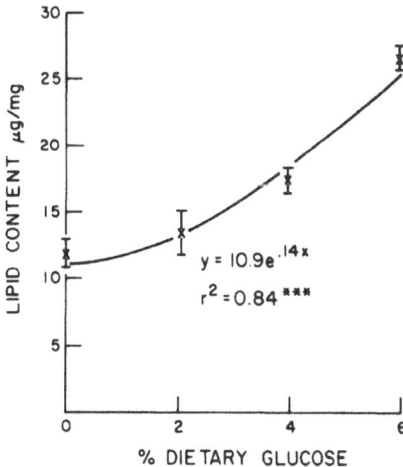

Fig.8 Effects of dietary glucose level on lipid concentration (µg/mg wet weight) in Exeristes roborator reared in vitro on artificial media.

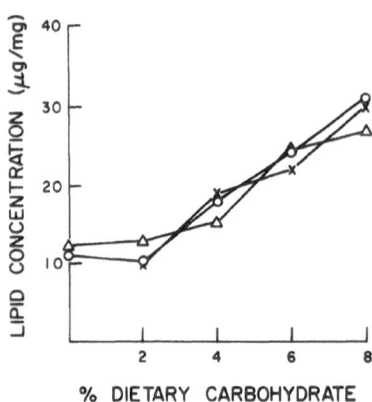

Fig.9 Effects of selected carbohydrates on the lipid concentration
        (μg/mg wet weight) in <u>Exeristes</u> <u>roborator</u> reared <u>in</u> <u>vitro</u>
        on artificial media. Sucrose (Δ), glucose (o), fructose (x).
        After Thompson (1979).

a greater rate than does glucose. Although parasite larvae reared
on media containing low levels of sucrose attained a higher weight
than those reared on equal levels of glucose (Fig.2), and therefore have a higher absolute lipid level, the concentration of lipid
was the same in both cases. No such difference was found between
glucose and fructose, and therefore, the differential effect of
sucrose on weight was attributed to its possible role as a phago-
stimulant.

    Further investigation demonstrated that increased lipid levels
in parasites reared on diets containing increasing levels of glucose
were due to increased triglyceride (Thompson, 1979). The lipid of
larvae reared on carbohydrate-free media as well as on media containing 2% glucose was composed largely of phospholipid with small
amounts of sterol and little triglyceride (Fig. 10). However as
the proportion of dietary carbohydrate increased, the concentration
of triglyceride increased as well. The results were supported by
those obtained from radioactive tracer studies (Fig. 11).

    The fatty acid synthetase complex was markedly affected by
dietary carbohydrate (Fig. 12). Isolated synthetase activity was
consistently greater in parasities reared on media containing carbohydrate than those reared on carbohydrate-free media and synthetase
activity appeared to be directly related to dietary glucose level.
These findings were similar to those reported with tissues from
vertebrate animals in which synthetase activity also fluctuates in
response to dietary carbohydrate (Romos and Leveille, 1974). In

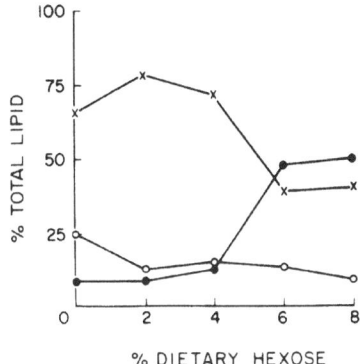

Fig.10 Effects of dietary glucose on the relative amounts of
       phospholipid (x), triglyceride (o) and sterol (o) in
       Exeristes roborator reared in vitro on artificial media.
       After Thompson (1979).

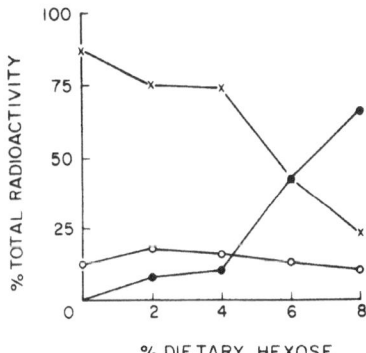

Fig.11 Effects of dietary glucose on the relative amount of radio-
       activity from ^{14}C-2-acetate incorporated into phospholipid
       (x), triglyceride (o) and hydrocarbons (o) in Exeristes
       roborator reared in vitro on artificial media. After
       Thompson (1980).

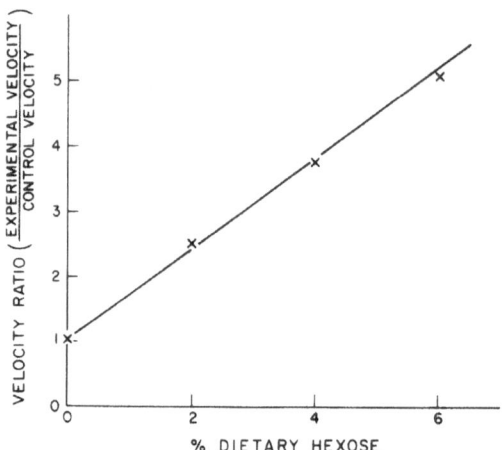

Fig.12 Effects of dietary glucose level on fatty acid synthetase
       activity in _Exeristes_ _roborator_ reared on artificial media.
       After Thompson (1980).

studies with rat liver, Craig et al. (1972) demonstrated that
such alterations in response to dietary carbohydrate were related
to the amount of synthetase protein present and that dietary carbohydrate stimulated native enzyme synthesis. Although estimates were
not made of the synthetase protein levels present in E. roborator,
kinetic investigations of crude preparations demonstrated that
dietary glucose had no effector action on isolated synthetase
(Fig. 13). The lipogenic response to dietary carbohydrate in E.

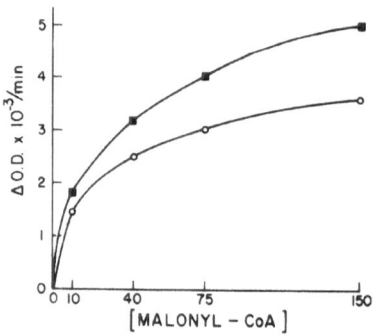

Fig.13 Effects of dietary glucose (□) and palmityl-CoA (o) on the
       velocity-substrate concentration (μM) relationship of malonyl
       CoA and fatty acid synthetase activity (x) isolated from
       Exeristes _roborator_ reared on artificial media. After
       Thompson (1980).

roborator, therefore, presumably results from stimulation of enzyme synthesis as well.

In addition to the above effects of dietary glucose on trigly-ceride and de novo fatty acid synthesis, substantial alterations occurred in the fatty acid compositions of the triglycerides as well as phospholipids in response to dietary carbohydrate (Fig. 14). The fatty acids of parasites reared on carbohydrate-free media were composed primarily of saturated acids, which probably reflects the low level of de novo fatty acid synthesis under such conditions and results in a lack of substrate for desaturation. The proportion of unsaturates in both lipid fractions increased with increasing levels of dietary glucose and these changes appeared to be synchronized with growth rate. To determine more precisely the metabolic factors which influence the fatty acid composition as mediated by dietary carbohydrate, the rates of increase and decrease in the absolute amount of saturated and unsaturated fatty acids in the phospholipid and triglyceride fractions synthesized in response to increased dietary glucose were calculated from the data in Table 13. Rates were calculated as:

$$\text{Rate} = \frac{\text{Difference in fatty acid concentration from control}}{\text{(Control development time) (Mean fatty acid concentration)}}$$

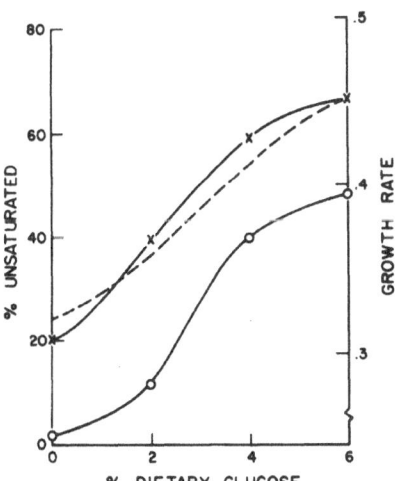

Fig.14 Effect of dietary glucose on the saturation of fatty acids in the triglycerides (o) and phospholipids (x) of Exeristes roborator reared in vitro on artificial media. After Thompson (1980).

Table 13.   Data summarizing the mean increases in triglyceride and phospholipid levels and the degree of unsaturation in 4th instar E. roborator reared on fatty acid-free artificial media in response to increasing dietary carbohydrate.

% Dietary glucose	Total lipid (µg/mg wet weight)	Total lipid % Tgl.	Total lipid % Ppl.	Lipid (µg/mg wet weight) Tgl.	Lipid (µg/mg wet weight) Ppl.	% unsaturation Tgl.	% unsaturation Ppl.	Fatty acid (µg/mg wet weight) Tgl. sat.	Tgl. unsat.	Ppl. sat.	Ppl. unsat.	Development time ratio
0 Control	12	9	66	1.1	7.9	1	20	1.09	.01	6.32	1.58	
2	14	9	78	1.3	10.9	10	38	1.17	.13	6.76	4.14	.92
4	18	13	72	2.3	13.0	40	60	1.38	.92	5.20	7.80	.82
6	28	48	39	13.4	10.9	45	65	7.40	6.03	3.82	7.10	.76

where the numerator was the difference between the fatty acid concentration in the experimental groups reared on diets containing
glucose and the level in the control insects reared in carbohydrate-
free media. The denominator was the product of the ratio of the
development time of the experimental groups to the control group
times the mean fatty acid concentration during the development period
of the experimental group for which the rate was being calculated.
Rates thus obtained are shown in Fig. 15. In most cases the rates
increased in response to dietary glucose levels. Of interest are
the lower and negative rates for the saturated fatty acids in the
phospholipid fraction indicating that the absolute amount of saturated fatty acids in that lipid fraction decreased in response to
dietary carbohydrate. Since a similar result was not obtained in
the triglyceride fraction, these lower rates are likely the result
of altered esterification patterns rather than altered fatty acid
synthesis per se. The increase in rate for saturated fatty acids
in the triglyceride fraction which accompanied the decreased rate
in phospholipid was presumably due to a shift of fatty acid substrate into triglyceride. The specific fatty acid compositions of
the triglyceride and phospholipid fraction of E. roborator reared
in vitro appear to be influenced by altered fatty acid substrate
specificity during glyceride synthesis. In addition, since the
increase in unsaturation was accompanied by an increase in the total
amount of lipid as described above, the absolute amount of unsaturated fatty acids increased and desaturation, therefore, must have
been stimulated in response to dietary glucose.

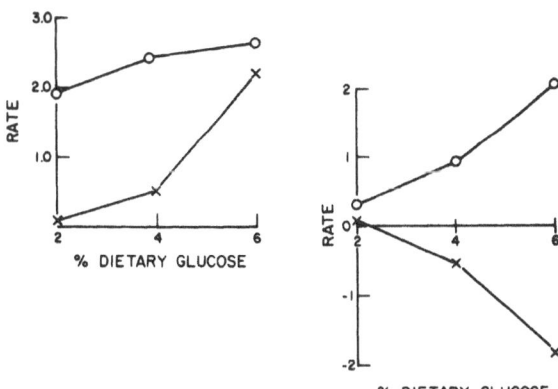

Fig.15 Effects of dietary glucose on the rate of incorporation of
       saturated (x) and unsaturated (o) fatty acids in the tri-
       glycerides (left) and phospholipids (right) of Exeristes
       roborator reared on chemically defined media. After
       Thompson (1980).

The above investigation demonstrates the effects of dietary carbohydrate on fatty acid metabolism in E. roborator and indicates that the previously described influence of dietary fats on fatty acid composition in vitro may be dramatically affected by dietary carbohydrate. Few studies have been carried out on the nutritional interaction of dietary carbohydrate and lipid. Carroll (1963, 1964), however, demonstrated in rats that the metabolic responses to different types of dietary carbohydrate can be modified by alterations in dietary lipid composition and that the effect of dietary fat substitution on liver lipids is influenced by the dietary carbohydrate component. It was later demonstrated that alterations in the relative proportions of dietary carbohydrate and fat modified the metabolic response to different carbohydrate-fat combinations, and it was concluded that the effects of altering the ratio of carbohydrate to fat in the diet involves more than the direct variation observed in relation to the absolute amount of each nutrient, individually (Carroll and Bright, 1965). Without additional studies on E. roborator fed dietary lipid supplements in conjunction with varying levels of carbohydrate it is not possible to assess the importance of possible lipid-carbohydrate interactions in influencing fatty acid metabolism.

CONCLUSION

The nature of carbohydrate and lipid nutrition in metazoan parasites may reflect to some degree adaptation attained during the development of the parasitic life strategy, but in most cases it is difficult to make conclusive assessments. Nutritional characteristics of parasitic helminths as contributions to fitness through adaptation or genetic loss have been discussed by Fairbairn (1970), but our present knowledge of parasitic Hymenoptera is too limited for such an attempt. However, the duplication of dietary fatty acid composition observed in vivo and the regulation of fatty acid levels presents an interesting case for examination. Although the compositions of parasite and host are similar, a lack of regulation is not responsible and the character represents metabolic diversity rather than metabolic absence. Genetic loss, therefore, cannot be considered since the parasities regulate these levels strictly, although the specific levels are in some manner influenced directly by the corresponding levels in the host. In considering such a metabolic characteristic as an adaptation it is assumed that the trait is goal-directed and that it has arisen as a consequence of natural selection. The latter of necessity, therefore, implies purpose which should be detected if sufficient information is available. Such is not the case in the present situation with parasitic Hymenoptera. Although a relationship was previously suggested between the above trait and host numbers as described above, the contribution of the trait to a non host-specific strategy is not intuitively reasonable with the present state of knowledge. The expression of the character may be the result of accidental

selection and have appeared as a result of genetic association with
other adaptive genes, which indeed, may be related to host specificity. A contribution to reproductive fitness must be evident in
order to conclude that adaptation has occurred.

LITERATURE CITED

Bailey, H. H. and Fairbairn, D., 1968, Lipid metabolism in helminth
    parasites. V. Absorption of fatty acids and monoglycerides
    from micellar solution by Hymenolepis diminuta (Cestoda),
    Comp. Biochem. Physiol. 26:819-836.
Barlow, J. S., 1972, Some host-parasite relationships in fatty acid
    metabolism, in: "Insect and Mite Nutrition," J. G. Rodriguez,
    ed., North Holland, Amsterdam.
Barlow, J. S. and Bracken, G. K., 1971, Incorporation of Na-1-[14]C-
    acetate into the fatty acids of two insect parasities (Hymenop-
    tera) reared on different hosts, Can. J. Zool. 49:1297-1300.
Barrett, J., Ward, C. W., and Fairbairn, D., 1970, The glyoxylate
    cycle and the conversion of triglycerides to carbohydrates in
    developing eggs of Ascaris lumbricoides, Comp. Biochem. Physiol.
    35:577-586.
Bartlett, B. R., 1964, Patterns in the host-feeding habit of adult
    parasitic Hymenoptera, Ann. Entomol. Soc. Amer. 57:344-350.
Bracken, G. K., 1965, Effects of dietary components on fecundity
    of the parasitoid Exeristes comstockii (Cress.) (Hymenoptera:
    Ichneumonidae), Can. Ent. 97:1037-1041.
Bracken, G. K. and Barlow, J. S., 1967, Fatty acid composition of
    Exeristes comstockii (Cress.) reared on different hosts, Can.
    J. Zool. 45:57-61.
Bruce, J. I., Ruff, M. D., Davidson, D. E., and Crum, J. W., 1974,
    Shistosoma mansoni and Shistosoma japonicum: Comparison of
    selected aspects of carbohydrate metabolism, Comp. Biochem.
    Physiol. 49B:157-164.
Bueding, E., 1952, Carbohydrate metabolism, in: "Chemical Physiology
    of Endoparasitic Animals," T. Von Brand, ed., Academic Press,
    New York.
Buteau, G. H. and Fairbairn, D., 1969, Lipd metabolism in helminth
    parasites. VIII. Triglyceride synthesis in Hymenolepis dimi-
    nuta (Cestoda), Expl. Parasitol. 25:265-275.
Carroll, C., 1963, Influence of dietary carbohydrate-fat combina-
    tions on various functions associated with glycolysis and
    lipogenesis in rats. Effects of substituting sucrose for rice
    starch with unsaturated and with saturated fat, J. Nutrition
    79:93-100.
Carroll, C., 1964, Influence of dietary carbohydrate-fat combinations
    on various functions associated with glycolysis and lipo-
    genesis in rats. Glucose vs. sucrose with corn oil as two
    hydrogenated oils, J. Nutrition 82:103-172.

Carroll, C. and Bright, E., 1965, Influence of carbohydrate-to-fat ratio on metabolic changes induced in rats by feeding different carbohydrate-fat combinations. J. Nutrition 87:202-210.

Castro, G. A. and Fairbairn, D., 1969, Carbohydrates and lipids in Trichinella spiralis larvae and their utilization in vitro, J. Parasitol. 55:51-58.

Castro, G. A. and Roy, S. A., 1974, Disaccharides and the nutrition of Trichinella spiralis, J. Parasitol. 60:887-889.

Clegg, J. A. and Smyth, J. D., 1968, Growth, development and culture methods: Parasitic Platyhelminths, in: "Chemical Zoology," M. Florkin and B. T. Scherr, eds., Academic Press, New York.

Craig, M. C., Nepokroeff, C. M., Lakshmanan, M. R., and Porter, J. W., 1972, Effect of dietary change on the rates of synthesis and degradation of rat fatty acid synthetase, Arch. Biochem. Biophys. 152:619-630.

Crompton, D. W. T., 1977, Qualitative requirements and utilization of nutrients: Acanthocephala, in: "Handbook of Nutrition and Food," Section D: Nutritional Requirements, Vol. I, M. Rechcigl, Jr., ed., Chemical Rubber Co., Cleveland.

Crompton, D. W. T. and Nesheim, M. C., 1973, Relationship between Moniliformis dubius (Acanthocephala) and the carbohydrate intake of rats, Parasitology 67:ii.

Dadd, R. H., 1977, Qualitative requirements and utilization of nutrients: Insects, in: "Handbook of Nutrition and Food," Section D: Nutritional requirements, Vol. I, M. Rechcigl, Jr., ed, Chemical Rubber Co., Cleveland.

Dunagen, T. T., 1962, Studies on in vitro survival of Acanthocephala, Proc. Helminthol. Soc. Wash. 29:131-135.

Dunkley, L. C. and Metterick, D. F., 1969, Hymenolepis diminuta: Effect of quality of host dietary carbohydrate on growth, Exp. Parasitol. 25:146-161.

Fairbairn, D., 1969, Lipid components and metabolism of Acanthocephala and Nematoda, in: "Chemical Zoology," M. Florkin and B. T. Scheer, eds., Academic Press, New York.

Fairbairn, D., 1970, Biochemical adaptation and loss of genetic capacity in helminth parasites, Biol. Rev. 45:29-72.

Ferguson, J. D. and Castro, G. A., 1973, Metabolism of intestinal stages of Trichinella spiralis, Amer. J. Physiol. 225:85-89.

Ginger, C. D. and Fairbairn, D., 1966, Lipid metabolism in helminth parasites. II. The major origins of the lipids of Hymenolepis diminuta (Cestoda), J. Parasitol. 52:1097-1107.

Graff, D. J., 1964, Metabolism of C^{14}-glucose by Moniliformis dubius (Acanthocephala), J. Parasitol. 50:230-234.

Greichus, A. and Greichus, Y. A., 1975, Lipid metabolism in the hog roundworm, Int. J. Biochem. 6:1-7.

Hammond, R. A., 1968, Some observations on the role of the body wall of Acanthocephalus ranae in lipid uptake, J. Exp. Biol. 48:217-225.

Hieb, W. F., 1977, Qualitative requirements and utilization of
    nutrients: Nematoda, in: "Handbook of Nutrition and Food,"
    Section D: Nutritional Requirements, Vol. I, M. Rechcigl, Jr.,
    ed., Chemical Rubber Co., Cleveland.
Horie, Y. and Nakasone, S., 1971, Effects of the levels of fatty
    acids and carbohydrates in a diet on the biosynthesis of fatty
    acids in larvae of the silkworm, Bombyx mori, J. Insect
    Physiol. 17:1441-1450.
Hochachka, P. W. and Somero, G. N., 1973, The influence of oxygen
    availability, in: "Strategies of Biochemical Adaptation,"
    W. B. Saunders, ed., Philadelphia.
House, H. L., 1956, Nutritional studies with Pseudosarcophaga affinis
    (Fall.), dipterous parasite of the spruce budworm, Choristoneura
    fumiferana (Clem.), V. Effects of various concentrations of
    the amino acid mixture, dextrose, potassium, the salt mixture,
    and lard on growth and development; and a substitute for lard,
    Can. J. Zool. 35:182-189.
House, H. L., 1974, Nutrition, in: "The Physiology of Insecta,"
    M. Rockstein, ed., Academic Press, New York.
House, H. L., 1977, Nutrition of natural enemies, in: "Biological
    Control by Augmentation of Natural Enemies," R. L. Ridgway and
    S. B. Vinson, eds., Plenum Press, New York.
House, H. L. and Barlow, J. S., 1960, Effects of oleic and other
    fatty acids on the growth rate of Agria affinis (Fall.)
    (Diptera:Sarcophagidea), J. Nutrition 72:409-414.
Jacobsen, N. S. and Fairbairn, D., 1967, Lipid metabolism in helminth
    parasites, III. Biosynthesis and interconversion of fatty acid
    by Hymenolepis diminuta (Cestoda), J. Parasitol. 53:355-361.
Jezyk, P. F., 1968, Glycerolipid metabolism in Ascaris lumbricoides,
    Can. J. Biochem. 46:1167-1173.
Kmetec, E. and Bueding, E., 1961, Succinic and reduced diphosphopy-
    ridine nucleotide oxidase systems of Ascaris muscle, J. Biol.
    Chem. 236:584-591.
Kurelac, B., 1975, Molecular biology of helminth parasites, Int.
    J. Biochem. 6:375-380.
Laurie, J. S., 1957, The in vitro fermentation of carbohydrates by
    two species of cestodes and one species of Acanthocephala,
    Exp. Parasitol. 6:245-260.
Lumsden, R. D. and Harrington, G. W., 1966, Incorporation of
    linoleic acid by the cestode Hymenolepis diminuta (Rudolph,
    1819), J. Parasitol. 52:695-700.
MacDonald, F., 1966, Lipid responses to dietary carbohydrates, Adv.
    Lip. Res. 4:39-67.
Mellini, E., 1975, Possibilità de allevamento di insetti entomofagi
    parassiti su diete artificiali (Possibilities of breeding
    parasitic entomophagous insects on artificial diets), Bolletino
    dell'Istituto di Entomologia della Università di Bologna,
    XXXIII:257-290.

Metterick, D. F. and Jackson, D. J., 1977, Qualitative requirements and utilization of nutrients: Platyhelminthes, in: "Handbook of Nutrition and Food," Section D: Nutritional Requirements, Vol. I, M. Rechcigl, Jr., ed., Chemical Rubber Co., Cleveland.

Meyer, F. and Meyer, H., 1972, Loss of fatty acid biosynthesis in flatworms, in: "Comparative Biochemistry of Parasites," H. Van den Bossche, ed., Academic Press, New York.

Meyer, F., Kimura, S., and Mueller, J. F., 1966, Lipid metabolism in the larval and adult forms of the tapeworm, Spirometra mansonoides, J. Biol. Chem. 241:4224-4232.

Meyer, F., Meyer, H., and Bueding, E., 1970, Lipid metabolism in the parasitic and free living flatworms, Schistosoma mansoni and Dugesia dorotocephala, Biochim. Biophys. Acta. 210: 257-266.

Passey, R. F. and Fairbairn, D., 1957, The conversion of fat to carbohydrate during embryonation of Ascaris eggs, Can. J. Biochem. Physiol. 35:511-525.

Rathbone, L. and Rees, K. R., 1954, Glycolysis in Ascaris lumbricoides from the pig, Biochim. Biophys. Acta. 15:126-133.

Read, C. P., 1959, The role of carbohydrates in the biology of cestodes. VIII. Some conclusions and hypotheses, Exp. Parasitol. 8:365-382.

Read, C. P., 1966, Nutrition of intestinal helminths, in: "The Biology of Parasites," E. J. L. Soulsby, ed., Academic Press, New York.

Read, C. P., 1968, Intermediary metabolism of flatworms, in: "Chemical Zoology," M. Florkin and B. T. Scheer, eds., Academic Press, New York.

Read, C. P. and Rothman, A. H., 1957, The role of carbohydrates in the biology of cestodes. IV. Some effects of host dietary carbohydrates on growth and reproduction of Hymenolepis diminuta, Exp. Parasitol. 6:294-305.

Read, C. P., Schiller, E. L., and Phifer, K., 1958, The role of carbohydrates in the biology of cestodes. V. Comparative studies on the effects of host dietary carbohydrate on Hymenolepis spp., Exp. Parasitol. 7:198-216.

Read, C. P. and Simmons, J. E., 1963, Biochemistry and physiology of tapeworms, Physiol. Rev. 43:263-305.

Rothstein, M. and Nicholas, G. L., 1969, Culture methods and nutrition of nematodes and acanthocephala, in: "Chemical Zoology," M. Florkin and B. T. Scheer, eds Academic Press, New York.

Romos, D. R. and Leveille, G. A., 1974, Effect of diet on activity of enzymes involved in fatty acid and cholesterol synthesis, Adv. Lip. Res. 12:97-145.

Saz, H. J., 1969, Carbohydrate and energy metabolism of nematodes and acanthocephala, in: "Chemical Zoology," M. Florkin and B. T. Scheer, eds., Academic Press, New York.

Saz, H. J., 1972, Comparative biochemistry of carbohydrates in nematodes and cestodes, in: "Comparative Biochemistry of Parasites," H. Van den Bossche, ed., Academic Press, New York.

Silverman, P. H., 1965, In vitro cultivation procedures for parasitic helminths, Adv. Parasitol. 3:159-222.

Silverman, P. H. and Hansen, E. L., 1971, In vitro cultivation procedures for parasitic helminths: Recent advances, Adv. Parasitol. 9:227-258.

Smith, T., Brooks, T., and Lockard, V., 1970, In vitro studies on cholesterol metabolism in the blood fluke, Schistosoma mansoni, Lipids 5:854-856.

Thompson, S. N., 1975, Defined meridic and holidic diets and aseptic feeding procedures for artificially rearing the ectoparasitoid Exeristes roborator (Fabricius), Ann. Entomol. Soc. Amer. 68:220-226.

Thompson, S. N., 1976a, Effects of dietary amino acid level and nutritional balance on larval survival and development of the parasite, Exeristes roborator, Ann. Entomol. Soc. Amer. 69: 835-838.

Thompson, S. N., 1976b, Regulation of lipid metabolism in the insect parasite Exeristes roborator (Fabricius), J. Parasitol. 62: 303-306.

Thompson, S. N., 1977, Lipid nutrition during larval development of the parasitic wasp, Exeristes, J. Insect Physiol. 23:579-583.

Thompson, S. N., 1979, Effects of dietary carbohydrate on larval development and lipogenesis in the parasite Exeristes roborator (Fabricius) (Hymenoptera: Ichneumonidae), J. Parasitol., in press.

Thompson, S. N., 1980, Effects of dietary glucose on in vivo fatty acid metabolism and in vitro synthetase activity in the parasite Exeristes roborator (Fabricius), Insect Biochem., in press.

Thompson, S. N. and Adams, J. D., 1976, Characterization of selected lipids of the parasite Exeristes roborator (Fabricius), Comp. Biochem. Physiol. 55B:591-593.

Thompson, S. N. and Barlow, J. S., 1972, Synthesis of fatty acids by the parasite Exeristes comstockii (hymenoptera) and two hosts, Galleria mellonella (Lep.) and Lucilia sericata (Dip.), Can. J. Zool. 50:1105-1110.

Thompson, S. N. and Barlow, J. S., 1973, The inconsistent phospholipid fatty acid composition in an insect parasitoid, Itoplectis conquisitor, Comp. Biochem. Physiol. 44B:59-64.

Thompson, S. N. and Barlow, J. S., 1974, The fatty acid composition of parasitic Hymenoptera and its possible biological significance, Ann. Entomol. Soc. Amer. 67:627-632.

Thompson, S. N. and Johnson, J., 1978, Further studies on lipid metabolism in the insect parasite Exeristes roborator (Fabricius), J. Parasitol. 64:731-740.

Van Handel, E., 1965, The obese mosquito, J. Physiol. 181:478-486.

Waldbauer, G. P., 1968, The consumption and utilization of food by
    insects, Adv. Insect Physiol. 5:229-288.
Ward, C. W. and Fairbairn, D., 1970a, Enzymes of Beta- oxidation and
    the tricarboxylic acid cycle in adult Hymenolepis diminuta
    (Cestoda) and Ascaris lumbricoides (Nematoda), J. Parasitol.
    56:100-1012.
Ward, C. W. and Fairbairn, D., 1970b, Enzymes of oxidation and
    their function during development of Ascaris lumbricoides
    eggs, Devl. Biol. 22:366-387.
Weinstein, P. P., 1966, The in vitro cultivation of helminths with
    reference to morphogenesis, in: "The Biology of Parasites,"
    E. L. J. Soulsby, ed., Academic Press, New York.
Yazgan, S., 1972, A chemically defined synthetic diet and larval
    nutritional requirements of the endoparasitoid Itoplectis
    conquisitor (Hymenoptera), J. Insect Physiol. 18:2123-2142.
Zakim, D., 1973, Influence of fructose on hepatic synthesis of
    lipids, Prog. Biochem. Pharmacol. 8:161-188.

DIETARY MODULATION OF GLUCOSE-6-PHOSPHATE DEHYDROGENASE AND 6-PHOSPHOGLUCONATE DEHYDROGENASE IN DROSOPHILA

B. W. Geer, J. H. Williamson, D. R. Cavener and B. J. Cochrane

Department of Biology
Knox College
Galesburg, Illinois 61401

Department of Biology
University of Calgary
Calgary, Alberta T2N 1N4 Canada

Department of Molecular and Population Genetics
University of Georgia
Athens, Georgia 30602

Department of Zoology
University of North Carolina
Chapel Hill, North Carolina 27514

INTRODUCTION

The pentose phosphate cycle (pentose shunt) derives substrate from glycolysis and the initial portion of the pathway consists of reactions that convert glucose-6-phosphate to ribulose-5-phosphate. This portion of the pathway, termed the oxidative pentose shunt pathway, generates NADPH in the reactions catalyzed by glucose-6-phosphate dehydrogenase (G6PD; EC 1.1.1.49) and 6-phosphogluconate dehydrogenase (6PGD; EC 1.1.1.44). Flux through the oxidative shunt pathway is closely correlated with the rate of lipogenesis (Wise and Ball, 1964) and the primary purpose of the pathway ostensibly is the formation of NADPH for lipid synthesis (Fig. 1).

Early studies of the enzymes of the pentose shunt in Drosophila melanogaster were concentrated on gene dosage effects

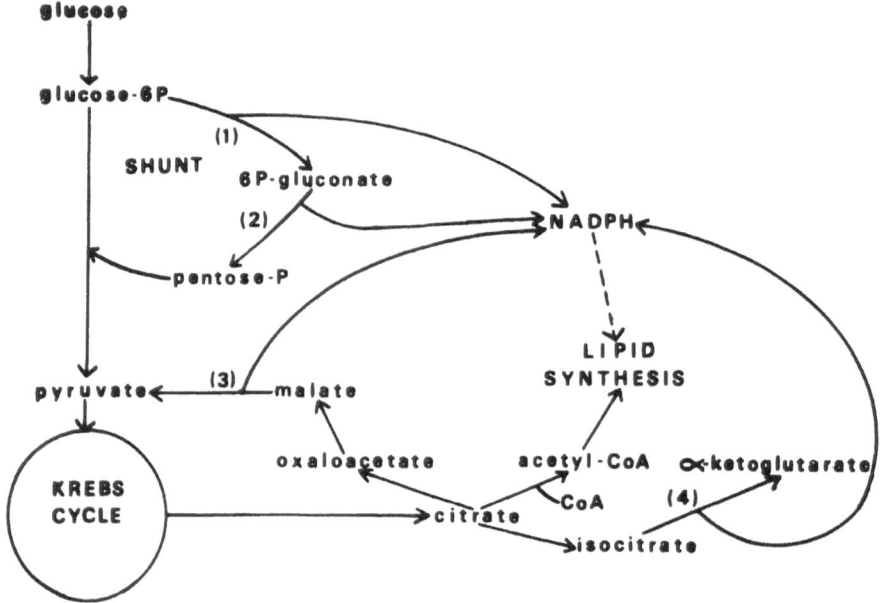

Fig.1   A metabolic scheme showing the contribution of glucose-6-
        phosphate dehydrogenase (1), 6-phosphogluconate dehydro-
        genase (2), malic enzyme (3) and NADP-isocitrate dehydro-
        genase (4) to the NADPH pool that is required for lipid
        synthesis (from Geer et al., 1976).

on tissue activities (Seecof et al., 1969; Bowman and Simmons,
1973) and allozyme variation (Young et al., 1964; Young, 1966).
The studies of the dietary modulation of the pentose shunt
enzymes in D. melanogaster were initiated because of this research
and a knowledge that the levels of the oxidative NADP-enzymes were
known to fluctuate in the mammalian liver in response to dietary
variations in carbohydrate and lipid (Tepperman and Tepperman,
1958, 1965; Fitch and Chaikoff, 1960; Potter and Ono, 1961;
Johnson and Sassoon, 1967).

     The nutrition of D. melanogaster is relatively well known
(Sang, 1956, 1979; Geer, 1963, 1966, 1972; Geer and Vovis, 1965)
and the availability of extensive genetic tools suggests that D.
melanogaster may represent a model system with which to study
the regulation of the pentose shunt enzymes.

MODIFICATION OF THE TISSUE ACTIVITIES OF THE NADP-ENZYMES BY
DIETARY SUCROSE AND LIPID

     D. melanogaster was initially examined to determine whether
dietary carbohydrate and lipid modulate the pentose shunt enzymes

in a manner similar to the pattern observed in the mammalian liver
(Geer et al., 1976). Larvae of the Canton-S wild-type strain were
grown for 4 days on modified Sang's medium C with 29 mM sucrose
(Table 1) and transferred to a series of diets constituting a
gradient of sucrose concentrations. Individuals of the Canton-S
strain possess the A-allozyme of 6PGD and the B-allozyme of G6PD.
The tissue activities of G6PD and 6PGD and NADP-malic enzyme
(EC 1.1.1.40) increased in the larvae in response to increased
concentrations of dietary sucrose, whereas the tissue activity of
NADP-isocitrate dehydrogenase (EC 1.1.1.42) declined (Table 2).
The tissue activities of the pentose shunt enzymes were particularly sensitive to alterations in dietary sucrose. The activities
of the pentose shunt enzymes paralleled the dietary sucrose
concentration from 9 to 290 mM, being more than 2-fold greater in
larvae fed 290 mM sucrose than the levels observed in larvae fed
a sucrose-free medium. The tissue activity of NADP-malic enzyme
was not different within the concentration range of 0 to 145 mM

Table 1. The composition of modified Sang's Medium C, altered
          as indicated in the text for experimental purposes.

	g
Agar	1.5
Casein	3.5
Sucrose	1.0
RNA	0.1
Cholesterol	0.03
Choline	0.008
Ca pantothenate	0.0016
Nicotinic acid	0.0012
Folic acid	0.001
Riboflavin	0.001
Pyridoxine·HCl	0.00025
Thiamine·HCl	0.002
Biotin	0.00002
$NaHCO_3$	0.140
$MgSO_4 \cdot 7H_2O$	0.0246
KH_2PO_4	0.183
Na_2HPO_4	0.189
Water to	100 ml

Table 2.   Tissue levels of oxidative NADP-enzymes in larvae fed diets containing different sucrose concentrations[a].

| | | | | Enzymes | |
Sucrose concentration	Soluble protein per larva	Glucose-6-phosphate dehydrogenase[b]	6-Phosphogluconate dehydrogenase[b]	NADP-isocitrate dehydrogenase[b]	NADP-Malic enzyme[b]
. mM	µg				
0	52.1	24.5±1.9	26.2±4.8	346.0±8.7	156.3±24.7
9	72.4	29.1±2.4[c]	30.9±3.2	298.8±23.2	144.9±23.6
29	71.6	38.2±2.8[c]	32.6±2.4	317.9±15.0	160.4±27.4
73	75.6	43.6±2.2	43.8±2.7[c]	244.9±36.0[c]	167.1±27.8
145	65.6	51.4±4.5	47.3±4.7	202.0±24.3	173.9±26.9
218	62.0	54.3±3.4	56.7±2.9	219.5±38.6	203.9±24.4[c]
290	49.4	66.4±3.2	61.2±2.3	227.3±55.1	223.3±28.3
435	44.8	63.5±2.4	60.2±4.6	193.9±29.8	218.1±17.0

[a]From Geer et al. (1978).
[b]The mean ± S.D. based on nine determinations. The activities are the nanomoles of cofactor reduced per mg of larval protein per minute at 30°C.
[c]Compared with the activity for larvae fed a sucrose-free diet, larvae fed sucrose at this concentration and higher concentrations had an enzyme activity that was different at at least the 0.05 level.

sucrose, but higher sucrose concentrations significantly altered
the tissue activity of this enzyme. The decline of the tissue
activity of NADP-isocitrate dehydrogenase in response to high
levels of dietary sucrose was perhaps due to the instability of
this enzyme in a low $NADP^+$ environment. The activity of 6PGD increased in the larval fat body in response to dietary carbohydrate,
whereas the G6PD activity increased in the larval fat body, gut and
other tissues. A number of laboratories have repeated these observations. As indicated in a later section, there is genetic variation between strains in the degrees to which the tissue activities
of G6PD and 6PGD are modulated by dietary sucrose.

Rocket immunoelectrophoresis performed using antiserum prepared with purified G6PD indicated that the diet-induced differences
in the tissue activity of this enzyme in wild-type larvae are
accompanied by differences in the concentration of cross-reacting
material (CRM) (Table 3). Consequently, the modulation of G6PD
by dietary sucrose is due to differential synthesis and/or degradation of the enzyme.

The tissue activities of the oxidative NADP-enzymes were
significantly modified by the addition of lipid or fatty acid to
the diet. The tissue activities of G6PD and 6PGD were 75-100%
higher in crude homogenates of 6-day old fatty acid-deprived
larvae than in homogenates of palmitate-sucrose fed larvae (Table
4). Dietary glycerol had little, if any, influence on the tissue
activities of the NADP-enzymes, consequently fatty acid is the
apparent active modulating ingredient of dietary lipid for larvae.

Table 3. The G6PD tissue activity and CRM concentration in 6-day
old larvae fed diets containing different sucrose
concentrations.

Sucrose concentration	G6PD	
	activity	CRM
mM	nmole/mg protein/min	ng enzyme/mg protein
9	22.7 ± 2.9[a]	575 ± 82[a]
29	38.4 ± 1.3	780 ± 60
145	76.8 ± 8.8	1562 ± 345

[a]The mean $\pm$ S.D. of 6 samples. Each sample was a homogenate of
7 male Canton-S larvae.

Table 4.   G6PD and 6PGD tissue levels in 6-day old Canton-S larvae
            fed the fat-free sucrose diet supplemented with different
            lipid components[a].

Enzyme	Diet[b]		
	Fat free-sucrose plus glycerol	Fat free-sucrose plus palmitate	Fat free-sucrose
Glucose-6-phosphate dehydrogenase	26.7 ± 3.0	14.8 ± 3.6[c]	30.1 ± 3.4
6-Phosphogluconate dehydrogenase	23.6 ± 2.5	14.9 ± 2.3[c]	26.1 ± 4.3

[a]From Geer et al. (1976).

[b]The fat free-sucrose diet was supplemented with 5.3 mM glycerol
or 11.5 mM palmitic acid. All diets contained 29 mM sucrose. The
mean ± S.D. is given for 12 samples. The activities are nanomoles
of NADPH formed/mg protein/min at 30°C.

[c]Different at the 0.01 level from the corresponding mean for larvae
fed the fat free-sucrose diet.

A MODEL OF THE CARBOHYDRATE-LIPID REGULATORY SYSTEM

     The pentose shunt enzymes in D. melanogaster are probably
regulated in two different ways. A "coarse" regulatory mechanism
acts to roughly synchronize the concentrations of the pentose
shunt enzymes in larval tissues with the demand for lipid synthesis, whereas allosteric control of the functional enzyme molecules
precisely coordinates NADPH production with the requirement of the
reduced cofactor for lipid synthesis. The relative concentrations
of NADPH and NADP$^+$ in the mammalian liver are important to the
allosteric regulation of the oxidative NADP-enzymes (Gumaa et al.,
1971). G6PD activity is limited by the availability of NADP$^+$
(Gumaa et al., 1971) but the enzyme is inhibited by high concentrations of NADPH (Krebs and Eggleston, 1974).

     A model of the coarse regulatory system of D. melanogaster
was postulated (Geer et al., 1976). In the model (Fig. 2) the
tissue levels of G6PD and 6PGD increase in response to an inducing
metabolite generated by glycolysis and decrease in response to a
metabolite supplied by lipid degradation. Both the inducing and
repressing regulatory systems are proposed to include a modulating
protein that interacts specifically with an inducing or repressing
metabolite. Presumably, the regulatory regions adjacent to each

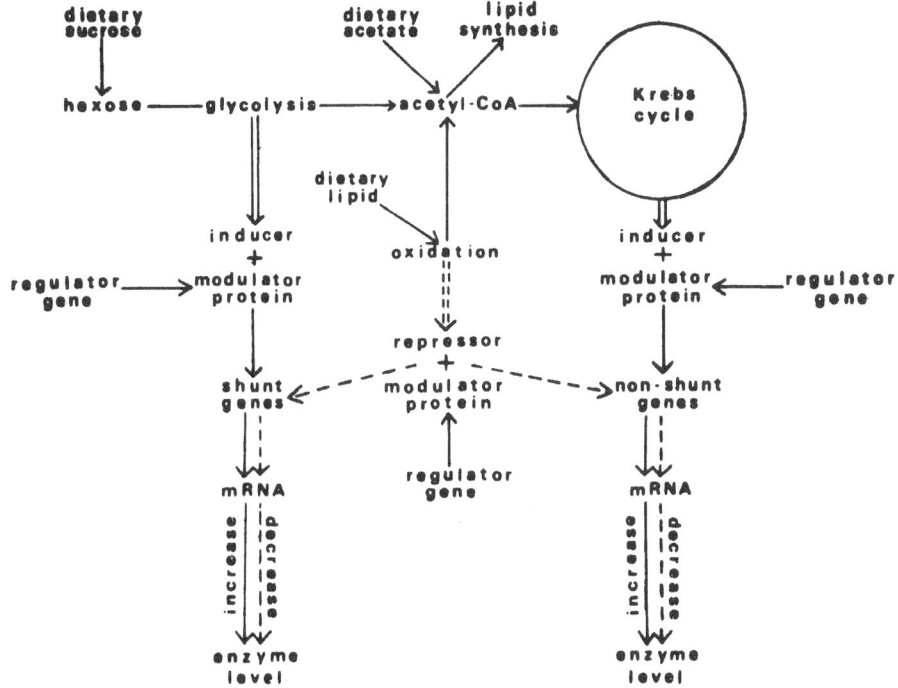

Fig.2 A model of the regulatory systems that control the oxidative NADP-enzyme tissue levels (from Geer et al., 1976).

of the oxidative NADP-enzyme structural genes contain sites for
binding the inducing and modulating proteins. Interaction of the
regulatory proteins with the binding sites could influence transcription, maturation or the stability of the messenger RNA.
Alternatives to the proposed model are regulatory systems that
act by controlling the rate of messenger RNA translation or enzyme
degradation.

THE EFFECTS OF DIFFERENT CARBON SOURCES ON THE PENTOSE SHUNT
ENZYME LEVELS

    To test the hypothesis that the oxidative shunt enzymes are
induced by a glycolytic derivative, larvae were fed diets containing compounds convertible to glycolytic intermediates (Geer
et al., 1978). It was anticipated that glucose and fructose would
be converted to hexose phosphate, ribose to ribose-5-phosphate,
gluconate to 6-phosphogluconate, glyceraldehyde to glyceraldehyde-
3-phosphate, and D-glycerate to 3-phosphoglycerate. The assumption

was that compounds not convertible to the molecule that induces
higher tissue levels of the shunt enzymes could be identified in
this way.

Larvae fed diets containing glucose or fructose had tissue
levels of G6PD as high as larvae fed an equi-carbon concentration
of sucrose. Dietary ribose and gluconate induced G6PD levels that
were slightly, but insignificantly, lower than the level found in
sucrose-fed larvae. Dietary pyruvate did not alter the shunt
enzyme level from that observed in the larvae fed an unsupplemented
diet, whereas dietary D-glycerate was found to be a potent nutritional inducer of G6PD.

Supplementing the diet with both D-glycerate and sucrose had
a spectacular effect on the tissue activities of the pentose shunt
enzymes (Table 5). G6PD and 6PGD were elevated to tissue activities
greater than those observed with the highest dietary sucrose
concentrations. An assessment of the rate of lipid synthesis indicated that the D-glycerate-sucrose combination increased the tissue
activities of the pentose shunt enzymes without elevating the rate
of synthesis above that noted in larvae fed sucrose alone (Geer
et al., 1978).

THE EFFECTS OF DIETARY SUCROSE AND D-GLYCERATE ON METABOLITE LEVELS

The tissue levels of intermediates of the pentose shunt and
glycolysis were determined to assess the in vivo influences of
dietary sucrose and D-glycerate (Geer et al., 1978). The glycolytic
intermediates - glucose-6-phosphate, fructose-6-phosphate, triose
phosphate and pyruvate - increased in the larval tissues as the
dietary sucrose concentration was raised from 9 mM to 290 mM.
Glucose-6-phosphate and fructose-6-phosphate were both increased
about 85%, triose phosphate 39% and pyruvate 54% by the dietary
sucrose elevation (Fig. 3). In contrast, the tissue level of 3-
phosphoglycerate was inversely related to the dietary concentration
of sucrose, declining by 61% as the dietary sucrose level was increased from 9 to 290 mM. The tissue level of the shunt intermediate, 6-phosphogluconate, was influenced much more markedly by
an increase in dietary sucrose concentration than were the tissue
levels of the intermediates of glycolysis. The increase of 6-
phosphogluconate was 542% when dietary sucrose was raised from
9 mM to 290 mM. Thus, the carbon-flow through the oxidative
pentose shunt pathway was greatly increased by the increase in
dietary sucrose concentration.

Because pentose shunt activity in D. melanogaster is
positively correlated with lipid synthesis (Geer et al., 1976),
the tissue concentrations of α-glycerophosphate, a lipid substrate, and long chain fatty acyl-CoA, a product of lipid

Table 5. The influences of dietary sucrose and D-glycerate on the rates of lipid and protein synthesis and the tissue activities of G6PD and 6PGD[a].

Supplement	Lipid[b]/Protein	Label into lipid[c]	Label into protein[c]	Glucose-6-phosphate dehydrogenase[d]	6-Phospho-gluconate dehydrogenase[d]
29 mM sucrose	1.29	1,163+29	572+74	30.7+3.1	31.8+3.6
145 mM sucrose	1.62	3,746+46[e]	1,204+64[e]	41.2+2.9[e]	36.5+3.0
290 mM sucrose	1.64	5,157+63[e]	1,074+53[e]	60.1+6.1[e]	44.3+5.8[e]
174 mM D-glycerate and 29 mM sucrose	1.06	925+49[f]	566+79	76.3+6.9[f]	50.7+5.2[f]

[a]From Geer et al (1978). The data is for 6-day old Canton-S larvae.

[b]μg of total lipid per larva/μg of soluble protein per larva.

[c]The nanomoles of labelled glucose incorporated into tissue protein or lipid per mg of larval protein during the final 2-day feeding period. The mean ± S.D. for three determinations.

[d]The mean ± S.D. for six samples. The activities are nanomoles of cofactor reduced per mg of larval protein per minute at 30°C.

[e]Different at the 0.01 level from the corresponding mean for larvae fed a 29 mM sucrose diet.

[f]Different at the 0.05 level from the corresponding mean for larvae fed a 29 mM sucrose diet.

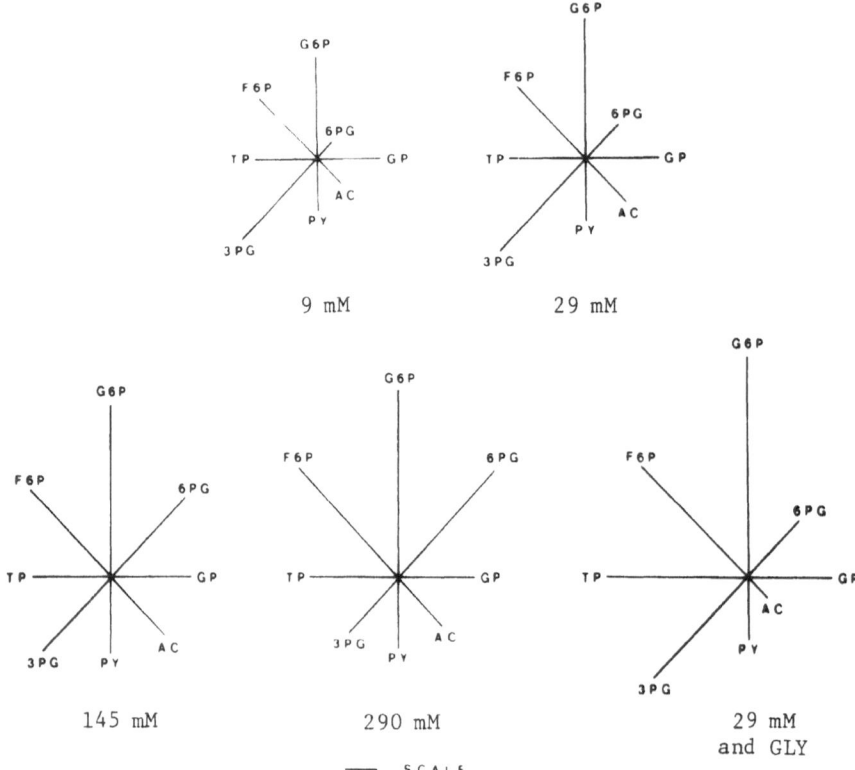

Fig.3   The concentrations of metabolites in Canton-S larvae fed
        diets containing 9, 29, 145 or 290 mM sucrose, or a mix-
        ture of 29 mM sucrose and 174 mM D-glycerate (GLY). The
        concentration of each metabolite is indicated according to
        scale by the length of the line from the point of intersec-
        tion to the tip. The length of the scale unit line is equal
        to 100 nanomoles of glucose-6-phosphate (G6P), fructose-6-
        phosphate (F6P), 3-phosphoglycerate (3GP), α-glycerophosphate
        (GP) or 6-phosphogluconate (6PG) per g fresh weight; 50
        nanomoles of pyruvate (PY) per g fresh weight; 10 nanomoles
        of triose phosphate (TP); or 5 nanomoles of long chain acyl-
        CoA (AC) per g fresh weight. From Geer et al (1978).

degradation, were examined (Fig. 3). The concentration of α-glycerophosphate increased as the dietary sucrose concentration was
elevated but peaked at 145 mM. The α-glycerophosphate concentration
paralleled the tissue lipid content in larvae over the range of the
dietary sucrose gradient. The tissue level of long chain acyl-CoA
was relatively low in all of the test groups but was highest in
larvae fed high concentrations of sucrose.

Dietary D-glycerate influenced the metabolite concentrations in the test larvae in different ways than did sucrose (Fig. 3). Glucose-6-phosphate, fructose-6-phosphate, triose phosphate and 3-phosphoglycerate were boosted by the D-glycerate-sucrose combination to levels beyond those promoted by diets containing sucrose. A dietary D-glycerate-sucrose combination caused the tissue levels of pyruvate and α-glycerophosphate to be high, but the levels were only those expected from the additive input of the two carbon sources. Thus, dietary D-glycerate caused a dramatic accumulation of the metabolites involved in the initial steps of glycolysis. Feeding D-glycerate with sucrose increased the tissue level of 6-phosphogluconate only slightly above that observed when 29 mM sucrose was fed alone. Hence high tissue levels of the shunt enzymes and a high concentration of glucose-6-phosphate did not significantly increase the pentose shunt activity when D-glycerate was fed with sucrose (Geer et al., 1978).

These observations are consistent with the hypothesis that the tissue levels of G6PD and 6PGD are increased by a metabolite derived from the initial portion of glycolysis.

INACTIVATION OF G6PD AND 6PGD BY LONG CHAIN FATTY ACYL-CoA

Because of the possible role of long chain fatty acyl-CoA in the regulation of the tissue levels of G6PD and 6PGD, purified G6PD and 6PGD from D. melanogaster were examined for inactivation by palmityl-CoA for 5 min. at 30°C in a Tris-HCl buffer (Table 6). Supplementation of the buffer with ions, cofactor or substrate did not prevent the inactivation process. Rat liver G6PD has been reported to be inactivated by palmityl-CoA (Taketa and Pogell, 1966) and Kelly et al. (1975) have suggested that the irreversible inactivation of G6PD by long chain fatty acyl-CoA may occur when lipid is fed. Our observations suggest that this could also occur in Drosophila. Nevertheless, the concurrent increases in long chain fatty acyl-CoA content and G6PD and 6PGD activities that are noted when the dietary sucrose level is elevated evidence a separate carbohydrate induction system.

THE ELECTROPHORETIC FORMS OF G6PD AND 6PGD

Two electrophoretic forms of 6PGD have been found in D. melanogaster. The presence of the faster A-form and the slower B-form in the tissues of the fly are determined by Pgd^A and Pgd^B alleles that map at 1-0.64 of the recombination map (Gvozdev et al., 1970). This locus is situated at 2D3-5 of the cytological map (Gerasimova and Ananjev, 1972). An intermediate form of 6PGD is present in individuals heterozygous for Pgd^A and Pgd^B (Kazazian et al., 1965), indicating that the polypeptides that comprise the A and B-forms can associate into a functional molecule.

Table 6. Drosophila G6PD and 6PGD incubated 5 min at 30°C in 0.1
          M Tris-HCl, pH 7.9[a].

Supplement	Δ O.D./min	% inactivation
G6PD		
none, not incubated	0.072	–
none	0.047	35
15 µM palmityl-CoA	0.001	98.4
6PGD		
none, not incubated	0.068	–
none	0.060	12
15 µM palmityl-CoA	0.002	97.1

[a]The A allozyme of G6PD and the A allozyme of 6PGD of Oregon-R
adults were purified to homogeneity according to acrylamide gel
electrophoresis using the methods of Williamson et al (1980a,c).

Williamson et al. (1980c) found that the A-form is a dimer with a
native molecular weight of 105,000 daltons and subunit molecular
weights of 55,000 and 53,000.

The two electrophoretic variants of G6PD, the fast A-form
and the slow B-form (Young et al., 1964), differ in size. The A-
form is a dimer with a native molecular weight of about 147,000
daltons (Steele et al., 1969; Williamson et al., 1980a), whereas
the B-form is apparently a tetramer of about 300,000 molecular
weight. The reported estimates of the molecular weight of the B-
form of G6PD differ markedly (Steele et al., 1969; Lee et al.,
1978), and a re-examination is warranted. The A-form of G6PD
consists of subunits of 66,800 and 61,000 daltons (Williamson et al.,
1980a). Steele et al. (1968) noted that under certain conditions
the B-form dissociates to produce the A-form, suggesting that the
subunits of the two forms of the enzyme are the same.

Zw^A and Zw^B alleles of a gene at 1-63 (Young, 1966) determine
the electrophoretic forms exhibited by the individual. The Zw
locus is located at 18D1-2-18E1-2 of the cytological map. Giesel
(1976) has speculated that the gene at the Zw locus is a regulator
rather than a structural gene.

ALLOSTERIC REGULATION OF G6PD AND 6PGD

The kinetic properties of the A-form of 6PGD (Krochko and Williamson, 1980; Williamson et al., 1980c) and the A-form of G6PD (Williamson et al., 1980a) are indicated in Table 7. Drosophila G6PD is competitively inhibited by NADPH in respect to $NADP^+$; whereas 6-phosphogluconate, fructose-1,6-diphosphate, ribulose-5-phosphate, glyceraldehyde-3-phosphate and ATP do not influence the binding of $NADP^+$ or glucose-6-phosphate by the enzyme. In contrast, Drosophila 6PGD is inhibited by NADPH, but it is also inhibited by fructose-1,6-diphosphate in a manner that is competitive with 6-phosphogluconate and by ribulose-5-phosphate in a way that is mixed in respect to 6-phosphogluconate (Krochko and Williamson, 1980). According to the regulatory properties of 6PGD and G6PD, the activity of the pentose shunt usually would be determined by the $NADPH/NADP^+$ ratio which is influenced by the rate of lipogenesis. Because of the inhibitory effect of ribulose-5-phosphate on 6PGD, rapid flux through the reductive pentose phosphate pathway to ribulose-5-phosphate would inhibit the oxidative pathway. This

Table 7. Properties of the A-form of G6PD and the A-form of 6PGD of Drosophila[a].

	G6PD	6PGD
Apparent gene locus	1-63	1-0.64
Native MW	146,000	105,000
Subunit MW	61,000 and 66,800	53,000 and 55,000
K_M substrate	1.2×10^{-4}M	8.1×10^{-5}M
K_M NADP$^+$	1.7×10^{-5}M	2.2×10^{-5}M
K_I NADPH	4.1×10^{-5}M	7.2×10^{-5}M
Inhibition by		
ribulose-5-phosphate	No	Yes
fructose-1,6-diphosphate	No	Yes
6-phosphogluconate	No	–

[a]Compiled from the results of Williamson et al. (1980a,c).

would re-enforce the inhibition of the pentose shunt enzymes by
NADPH when the rate of lipogenesis is low but would allow pentose
phosphate to be formed for RNA synthesis.

Administration of a fat-free, high carbohydrate diet to D.
melanogaster would effect a rapid rate of lipogenesis and reduce
the NADPH/NADP$^+$ ratio resulting in a more rapid rate of 6-phospho-
gluconate formation. The accumulation of 6-phosphogluconate in
turn would negate the inhibition of 6PGD by ribulose-5-phosphate
and fructose-1,6-diphosphate and drive the oxidative phase in a
forward direction. Since 6-phosphogluconate inhibits the phosphoglucose isomerase from plant and animal tissues (Parr, 1956),
an accumulation of this compound would slow the formation of
fructose-1,6-diphosphate and indirectly enhance the flux through
the oxidative pentose phosphate pathway.

In brief, the activity of the oxidative pentose phosphate
cycle in D. melanogaster is usually a function of the intracellular
NADPH/NADP$^+$ ratio. The dietary modulation of G6PD and 6PGD by
carbohydrate and lipid amplifies or diminishes the allosteric
affects of metabolites, depending upon the dietary regime being
followed.

GENETIC LESIONS OF THE PENTOSE SHUNT

Both low activity and null-activity mutations of 6PGD and
G6PD have been isolated for the Pgd and Zw genes respectively
(see Lucchesi et al., 1979). Deficiencies for G6PD have little
or no effect on the viability of the fly (Geer et al., 1974), but
deficiencies for 6PGD are semi-lethal or lethal and a complete
absence of 6PGD is lethal (Bewley and Lucchesi, 1975; Gvozdev et
al., 1975). Individuals carrying a lethal Pgd-null mutation may
be rescued by introducing a Zw-null mutation into their genome
(Gvozdev et al., 1976; Hughes and Lucchesi, 1977); that is, a
deficiency for G6PD negates the lethal effect of 6PGD.

An examination of the allosteric properties of Drosophila
G6PD and 6PGD sheds light on the lethal nature of Pgd-null
mutations. A genetic block in the oxidative pentose phosphate
cycle reduces the capacity to form NADPH by about 40% in Drosophila
(Geer et al., 1979b). Since the activity of G6PD is a function of
the NADPH/NADP$^+$ ratio and not the concentration of 6-phosphogluco-
nate, a 6PGD deficiency simultaneously eliminates the inhibition of
G6PD and the route for 6-phosphogluconate metabolism. In Pgd-
null individuals possessing a normal complement of G6PD, the un-
inhibited enzyme tentatively forms 6-phosphogluconate at a more
rapid rate than in normal individuals having an operative pentose
shunt pathway. Theoretically, 6-phosphogluconate would reach a
lethal concentration in individuals expressing a Pgd-null mutation

because of the indirect influence of the mutation on the activity
of glucose-6-phosphate dehydrogenase. As indicated, this hypothesis
has not been confirmed.

6-Phosphogluconate may be lethal at abnormally high intracellular concentrations because it inhibits a step in glycolysis
(Parr, 1956; Gvozdev et al., 1976; 1977). The lethal condition
is relieved by the inclusion of fatty acid in the diet (Hughes
and Lucchesi, 1978), apparently because the dietary fatty acid
lowers the G6PD level in the Pgd-null larvae and reduces the
accumulation of 6-phosphogluconate.

GENETIC EVIDENCE FOR THE PHYSIOLOGICAL ROLE OF THE PENTOSE SHUNT

Wild-type and Pgd$^-$ Zw$^-$ larvae were examined to determine the
physiological impact of deficiencies for 6PGD and G6PD (Geer et
al., 1979b). The NADPH concentration in early third instar Pgd$^-$
Zw$^-$ larvae was 60% of the concentration in wild-type larvae, but
this resulted in a 6-fold difference in the NADPH/NADP$^+$ ratio
(Table 8). One effect of the lower NADPH/NADP$^+$ ratio on the
individual was evident when the lipid contents of the test larvae
were examined. The triglyceride content of 6-day old wild-type
larvae fed a 290 mM sucrose diet was significantly greater than
that of larvae fed a 9 mM sucrose diet (Table 9). The difference
in triglyceride content may be attributed to the capacity of the
larvae to respond to the added sucrose. A comparison of pentose
shunt proficient and deficient larvae showed that the triglyceride
contents of Pgd$^-$ Zw$^-$ larvae were not increased as much by a high
dietary sucrose concentration as were the triglyceride contents
of their shunt proficient counterparts. The differences in

Table 8. The NADP$^+$ and NADPH concentrations in the tissues of early
third instar larvae[a].

Genotype	$\dfrac{10^{-10} \text{ g NADPH}}{\text{μg protein}}$	$\dfrac{10^{-10} \text{ g NADP}^+}{\text{μg protein}}$	$\dfrac{\text{NADPH}}{\text{NADP}^+}$
Pgd$^+$ Zw$^+$	11.56[b]	1.26	9.2
Pgd$^-$ Zw$^-$	6.97	4.57	1.5

[a] From Geer et al. (1979b).
[b] The average of 3 determinations. The larvae were cultured on an
agar-cornmeal-yeast-sucrose-glucose medium (Lewis, 1960) and
selected by size. Pgd$^-$ Zw$^-$ larvae possess less than 10% of the
normal tissue activities of 6PGD and G6PD. Pgd$^+$ Zw$^+$ was the
Oregon-R strain.

Table 9. The lipid composition of pentose shunt deficient and proficient larvae fed low and high sucrose diets[a].

Genotype	Dietary sucrose level (mM)	Lipid content (%)				% Normal triglyceride increase
		Phospholipid	Mono- & Diglyceride	Fatty acids	Triglyceride	
Pgd+ Zw+	9	41.7[b]	3.6	0.6	54.1	
	290	12.6	3.4	0.5	84.4	
Pgd- Zw-	9	39.2	4.5	0.5	55.8	
	290	16.4	4.5	1.4	77.6	72[c]
C(1)DX, Pgd+ Zw+	9	55.5	3.9	0.2	40.4	
	290	14.4	1.6	0.5	83.4	
Pgd^{n2} Zw-	9	52.1	4.9	0.3	42.7	
	290	19.5	2.6	0.8	77.1	80
C(1)DX, Pgd+ Zw+	9	36.8	2.6	0.3	60.3	
	290	17.9	1.5	0.4	80.2	
Pgd^{n3} Zw-	9	39.7	2.4	0.6	57.6	
	290	26.4	2.4	0.5	70.7	66

[a] From Geer et al. (1979b). Individuals of the Pgd- Zw-, Pgd^{n2} Zw- and Pgd^{n3} Zw- strains were deficient for G6PD and 6PGD. Each strain had the autosomal background of the control strain listed directly above it.

[b] The average of three determinations.

[c] (The % triglyceride content increase in the Pgd- Zw- strain/the % triglyceride content increase in the Pgd+ Zw+ strain) X 100.

triglyceride content between pentose shunt deficient larvae fed low and high sucrose diets were on the average only 73% as great as the corresponding differences in triglyceride content observed in shunt proficient larvae. In the case of a pentose shunt deficiency, the availability of NADPH apparently limits the rate that carbohydrate is converted to lipid when a high sucrose diet is fed to larvae.

INTERACTION OF THE PENTOSE SHUNT WITH OTHER PATHWAYS

Lipid and Carbohydrate Metabolism

The activity of the oxidative pentose shunt pathway is tightly linked to lipogenesis. The triglyceride content of larvae is related to the carbohydrate content of the diet; the capacity of the individual to respond to dietary carbohydrate is limited by the activity of the pentose shunt; and dietary lipid represses the tissue activities of the pentose shunt enzymes. Acetyl-CoA carboxylase (EC 6.4.1.2), the rate-limiting enzyme of fatty acid synthesis, was somewhat higher in larvae fed a 29 mM sucrose diet than in larvae fed a sucrose-free medium. Dietary lipid effectively repressed acetyl-CoA carboxylase (Table 10). The tissue activity of acetyl-CoA carboxylase was essentially the same in 4-day old adult males raised on a fat-free 29 mM sucrose diet as in third instar larvae fed the same diet.

The glycogen content of the third instar larvae of D. melanogaster was influenced by the dietary carbohydrate concentration. The glycogen content of larvae fed a sucrose-free medium was only 59% of that of larvae fed a 290 mM sucrose diet (Table 10). Nevertheless, the highest glycogen level in larvae was only 46% as high as the level in adult tissue. Adult Drosophila depend upon glycogen as an energy source for flight and the flight muscle comprises much of the body mass of the adult.

Phosphoenolpyruvate (PEP) carboxykinase (EC 4.1.1.32), an enzyme important to the regulation of gluconeogenesis, was most prominent in larvae fed a sucrose-free diet (Table 10). Addition of sucrose to the diet progressively repressed PEP-carboxykinase over a range of 0 to 290 mM sucrose. When fed with sucrose, lipid did not exert a significant effect on PEP-carboxykinase. The tissue activity of PEP-carboxykinase declined during the pupal period and was at an extremely low level in 4-day old adult males.

Repression of PEP-carboxykinase by dietary sucrose in larvae ensures that glycogen will not be synthesized from non-carbohydrate substrates when carbohydrate is available. Similarly, the repression of acetyl-CoA carboxylase inhibits fatty acid synthesis when fatty acids are components of the diet. Lipogenesis, glycogen

Table 10. The tissue levels of glycogen and enzymes of carbohydrate metabolism and fatty acid synthesis in Canton-S larvae.

Sucrose level (mM)	Glycogen content[a] mg/100 mg wet wt	Acetyl-CoA carboxylase[b]	PEP carboxykinase[c]
Larva			
0	2.7	0.94+0.22[d]	14.82+0.12[d]
9	3.0	1.60+0.28	6.52+0.75
29	3.6	1.85+0.28	4.55+0.64
145	4.2	1.54+0.24	3.74+0.50
290	4.6	---	2.05+0.42
29 + 0.1% phosphatidylcholine	---	0.07+0.05	4.07+0.66
Adult			
29	10.0	1.71+0.52	2.08+0.37

[a] Determined by the procedure of Keepler and Decker (1974).
[b] Assayed by the method of Arinze and Mistry (1971).
[c] Assayed by the method of DeZwann and Marrewijk (1973).
[d] Mean + S.D. nanomoles of product formed/mg protein/min at 30°C.

synthesis and the pentose shunt are linked to a common substrate, dietary carbohydrate, and the rates of these processes are integrated with the rate of glycolysis.

Ethanol Metabolism

Fermenting fruit, which contains moderate to high concentrations of ethanol, is a major source of food for natural populations of Drosophila and larvae are attracted to high concentrations of ethanol (Parsons and King, 1977). The frequencies of allozyme-determining alleles of alcohol dehydrogenase have been shown to change in response to dietary ethanol in populations of D. melanogaster (Gibson, 1970; Van Delden et al., 1975; Cavener and Clegg, 1978). The in vitro kinetic properties of the alternate allozymes of alcohol dehydrogenase appear to be consistent with the direction of gene frequency change observed in cage populations.

Enzyme polymorphisms of 6PGD, malate dehydrogenase, α-glycerophosphate dehydrogenase and alcohol dehydrogenase were of apparent adaptive value in cage populations exposed to dietary ethanol for more than 50 generations (Table 11). An electrophoretically fast form of 6PGD was close to fixation in replicate ethanol cages, whereas the slow form was at much higher frequencies in control cages. Specific allozymes of malate dehydrogenase and α-glycerophosphate dehydrogenase were also of benefit to D. melanogaster in an ethanol-rich environment, and a fitness interaction between alcohol dehydrogenase and α-glycerophosphate dehydrogenase genotypes was evident.

Metabolic relationships provide a reasonable explanation for the correlated responses of different enzymes to ethanol selection pressure. Alcohol dehydrogenase is instrumental in the conversion of ethanol to acetate but also forms NADH. Acetyl-CoA, the product that is subsequently formed, may be used for mitochondrial oxidation but is most likely used for fatty acid synthesis. Two shuttle systems, the α-glycerophosphate shuttle and the malate-aspartate

Table 11. Adh^S, $\alpha Gpdh^S$, Pgd^S and Mdh^S mean gene frequencies for control and ethanol selection populations after more than fifty generations[a].

	Adh^S	$\alpha Gpdh^S$	Pgd^S	Mdh^S
Control	67.5	41.6	46.3	85.0
10% Ethanol	5.5	72.0	4.5	100.00

[a]Adapted from Cavener and Clegg (1980). The gene symbols signify genes for alcohol dehydrogenase (Adh), α-glycerophosphate dehydrogenase ($\alpha Gpdh$), malate dehydrogenase (Mdh) and 6-phosphogluconate dehydrogenase (Pgd). $Pgd^S = Pgd^B$.

shuttle, transport NADH equivalents into the mitochondrion (Cavener
and Clegg, 1980). Also, malate may be converted to pyruvate to
consumate a cytoplasmic NADH to NADPH transhydrogenase system. This
reaction provides NADPH for lipid synthesis.

The utilization of dietary ethanol requires alcohol dehydrogenase for the initial degradative step, α-glycerophosphate dehydrogenase for the NADH shuttle system of the mitochondrion, and malate
dehydrogenase for NADH-NADPH transhydrogenation and for a second NADH
shuttle system. Because of the formation of NADPH by malate dehydrogenase and NADP-malic enzyme, low activity forms of 6PGD and G6PD may
have positive selection values when ethanol is at a high concentration
in the diet. That is, the pentose shunt enzymes compete for $NADP^+$
with NADP-malic enzyme and slow the removal of cytoplasmic NADH via
the malate dehydrogenase-NADP malic enzyme system. The Pgd^A allele
increased in frequency in the ethanol environment and, in addition,
flies expressing Pgd^A had an in vitro 6PGD activity that was lower
than that of flies exhibiting the Pgd^B phenotype. Consequently,
the low activity form of 6PGD had an apparent selective advantage
under conditions favoring a low oxidative pentose shunt activity.

Differences in fitness have been attributed to different
alleles at the Zw and Pgd loci in D. melanogaster in cage populations not exposed to ethanol (Bijlsma and Van Delden, 1977;
Bijlsma, 1978). Although the oxidative pentose shunt pathway is
non-essential for survival in D. melanogaster, it is necessary for
the maintenance of optimal physiological conditions for growth.
The capacity for lipid synthesis may limit the fitness potential,
but the availability of different dietary substrates for lipid synthesis such as ethanol or acetate may influence the selection of
alleles of the pentose shunt genes.

Alternate Pathways

The non-essential nature of the oxidative pentose shunt pathway provides fuel for speculation. Although alleles of the Pgd
and Zw loci impart apparent advantages in fitness, deficiencies for
G6PD, 6PGD and even NADP-malic enzyme can simultaneously be
tolerated by D. melanogaster (Geer et al., 1979a). Under non-stress
conditions D. melanogaster apparently is able to form sufficient
NADPH for survival despite the absence of three major sources of
reduced cofactor. The reductive phase of the pentose shunt ostensibly operates in a reverse direction to supply pentose phosphate
for nucleic acid synthesis when the oxidative phase is non-func-
tional.

It is possible that one or more metabolic pathways exist that
by-pass the oxidative pentose shunt pathway. A NAD(P)-glucose
dehydrogenase and an NAD-gluconate dehydrogenase have been detected
in D. melanogaster by electrophoretic methods (Cavener, 1980). A

gluconate kinase would have to exist to complete a by-pass pathway. Only indirect evidence of the existence of this enzyme has been gathered. Dietary gluconate partially replaces sucrose for growth (Geer et al., 1978), suggesting that this intermediate enters the pentose shunt after being phosphorylated.

There are other potential alternate pathways to the oxidative pentose shunt pathway. One of the possibilities is a "sorbitol by-pass" that includes the reactions catalyzed by sorbitol dehydrogenase (Bischoff, 1978) and glucose oxidase (Cavener, 1980). This potential pathway also has gluconate as an intermediate and may play an important role in the male reproductive tract. Another possible pathway includes a reaction catalyzed by β-hydroxyacid dehydro-genase, an enzyme recently isolated from <u>Drosophila</u> (Cannistraro et al., 1979). This pathway would channel the carbon flow through glucuronate to xylulose and would involve the phosphoryla-tion of that intermediate.

All of the hypothetical pathways are tentatively coupled to the reductive phase of the pentose shunt and allow the reductive shunt pathway to operate in a forward direction. Seemingly, each alternate pathway could operate without being intimately coupled to lipogenesis. The existence and importance of each pathway in metabolizing dietary carbohydrate remains to be determined.

GENETIC VARIATION IN THE RESPONSE TO DIETARY CARBOHYDRATE

Seven wild-type strains established from single wild-caught females in the Galesburg, Illinois area were placed in axenic culture on Sang's modified C medium and tested to determined the extent of the response to dietary sucrose (Table 12). The activities of G6PD and 6PGD were 2 to 3-fold greater in larvae fed 290 mM sucrose than in larvae fed 9 mM sucrose for most of the strains. Strain 1 possessed extremely high tissue activities of the pentose shunt enzymes on both test diets, whereas strain 4 ex-hibited a weak G6PD response and strain 7 showed a weak 6PGD re-sponse. Clearly, both the tissue activity and the degree of the dietary modulation of the pentose shunt enzymes exhibit genetic variation in strains established from natural populations.

A similar comparison was performed for lines that were coiso-genic for the X and second chromosomes and homozygous for different third chromosomes. The X-chromosome of the strains carried $\underline{Zw}^A$ and $\underline{Pgd}^A$. Two of the 5 test strains, NC19 and NC25, exhibited weak G6PD responses to a high dietary sucrose concentration (Table 13). Although lower in magnitude, the 6PGD responses of the two strains were not significantly different from those of the other test strains. Crosses between a high G6PD response strain, CL55, and the two low strains gave F_1 progeny with diverse nutritional

Table 12. Variation of the dietary modulation of G6PD and 6PGD in
           strains of D. melanogaster established from single wild-
           caught females[a].

| Strain | G6PD | | 6PGD | |
| | mM sucrose | | mM sucrose | |
	29	290	29	290
1	56.5+7.7	124.3+15.7	29.9+3.1	82.5+9.4
2	21.3+2.3	57.7+2.6	18.8+3.4	47.0+5.5
3	40.6+4.5	70.5+8.9	27.0+2.4	55.7+9.5
4	27.4+2.7	31.8+2.9	20.0+3.7	57.7+8.3
5	20.8+3.2	64.7+6.7	21.7+2.8	52.5+9.1
6	27.2+2.6	56.0+8.6	18.1+5.9	38.2+2.9
7	21.8+2.3	57.9+4.6	21.9+1.9	31.3+5.4

[a]Mean ± S.D. for 6 homogenates of 7 larvae each. The activities
are nanomoles of NADPH formed/mg protein/min at 30°C.

traits. The CL55/NC19 progeny exhibited a weak G6PD dietary response, but the CL55/NC25 progeny showed a G6PD response intermediate to those of the parental strains. Apparently the dominance relationships of the low response chromosomes to the CL55
chromosomes are different, suggesting different genetic bases for
each. No G6PD induction was seen in the F_1 progeny of the two non-
inducing lines, evidencing the apparent dominance of the genetic
determinants situated on the NC19 third chromosome. The non-inducing
strains had reductions in G6PD activity in both the larval fat
body and gut.

    These experimental results indicate that genetic determinants
are situated on the third chromosome of D. melanogaster that
influence the response of G6PD to dietary carbohydrate. The data
on the Chapel Hill and Galesburg strains of D. melanogaster also
suggest that variation exists in the 6PGD response to dietary
carbohydrate.

    An examination of 5 lines coisogenic for the X and third
chromosomes and homozygous for different second chromosomes did

Table 13.   Sucrose induction of G6PD and 6PGD in isogenic lines.

Line	G6PD mM sucrose		6PGD mM sucrose	
	10	300	10	300
RI42	13.3+5.0[a]	35.1+11.2	24.7+4.1	37.1+10.3
RI08	16.0+4.8	26.2+4.4	23.2+5.0	35.8+9.0
CL55	12.3+3.1	30.6+6.9	24.7+1.0	39.7+6.8
NC19	15.8+3.2	19.4+4.3	25.5+4.0	30.5+5.5
NC25	20.5+1.2	23.4+3.9	30.9+1.0	38.1+5.0
F1 progeny				
NC19/CL55	13.5+1.8	15.4+1.7	17.4+6.9	28.9+3.5
NC25/CL55	13.2+3.2	21.3+4.2	19.2+9.9	39.2+4.9
NC19/NC25	17.6+3.3	17.1+7.8	31.1+5.4	30.6+11.6

[a]Mean ± S.D. for 5-7 replicates of 6-10 third instar larvae cultured on modified Sang's Medium C.   The activities are nanomoles of NADPH formed/mg protein/min at 28°C.

not detect any significant differences in the dietary induction of G6PD and 6PGD.

The X-chromosome appears to contain an important determinant of the G6PD response to dietary sucrose. Four experimental strains were established from a minimum of 16 genomes which in turn arose from a single population collected in nature near Athens, Georgia in 1978. Thus, the genetic backgrounds of the four test strains were randomized and the differences that are evident most likely are due to the G6PD and 6PGD genotypes. In each of the test strains alleles that determine a fast or a slow allozyme of G6PD and 6PGD were expressed. The test strains constituted the four possible double homozygous combinations of the A and B-forms of G6PD and 6PGD present in the natural population (Table 14). The strains possessing the A-form of G6PD expressed a very weak sucrose response as well as a low G6PD activity. In contrast, strains

Table 14.   Modulation of G6PD and G6PD by dietary sucrose in selected
             strains possessing different allozymes[a].

| | G6PD | | 6PGD | |
| | mM sucrose | | mM sucrose | |
Genotype	9	290	9	290
Pgd^{BB} Zw^{BB}	10.3+0.3[b]	26.7+0.6	16.2+0.8	34.9+0.4
Pgd^{BB} Zw^{AA}	6.4+0.4	6.8+2.0	18.0+1.0	35.1+1.3
Pgd^{AA} Zw^{BB}	11.1+1.0	25.7+1.9	14.4+0.6	27.6+1.3
Pgd^{AA} Zw^{AA}	7.0+1.0	8.8+2.0	16.4+0.9	27.6+1.6

[a] Pgd^{BB} denotes that larvae possessed only the B-allozyme and Pgd^{AA}
indicates the A-allozyme of 6PGD. Zw^{BB} indicates that larvae
possessed the B-allozyme of G6PD and Zw^{AA} denotes the A-allozyme.

[b] Mean ± S.D. for 6 samples. The activities are nanomoles of NADPH
formed/mg wet weight/min at 30°C.

homozygous for the B-form of G6PD exhibited a higher tissue activity
and a strong sucrose response. The strains with the A and B-forms
of 6PGD had similar sucrose responses but the 6PGD tissue activities
of the two forms were different in larvae fed a 290 mM sucrose diet.
These data implicate the Zw locus as the X-linked determinant for
the G6PD sucrose response.

There is a need to isolate and purify the G6PD and 6PGD from
strains with different regulatory traits in reference to carbohydrate and other dietary ingredients. The adaptive value of a
gene might reside either in the kinetic properties of its enzymic
product or the influence of the gene on the tissue concentration of
the enzyme.

The genetic region containing the structural gene for NADP-
malic enzyme includes an apparent regulatory element (Williamson
et al., 1980b). Circumstantial evidence presented in this paper
indicates that regulatory determinants for G6PD reside on the X
and third chromosomes of D. melanogaster. Furthermore, the Pgd
locus may consist of structural gene and regulatory components.
A recent examination of Pgd mutations indicated that 10 of the
mutations are null activity, CRM-negative mutations and one
mutation produces a full complement of CRM but exhibits a low

6PGD activity. Conceivably the Pgd region consists of DNA comprising the structural gene component and DNA that controls the synthesis and maturation of the transcript. The latter DNA could be subject to regulatory modulation that alters the tissue concentration of the enzyme. A mutation in the regulatory DNA could yield a CRM-negative mutation or a regulatory site with modified properties.

The study of the genetic determinants associated with the dietary modulation of G6PD and 6PGD is only beginning. The promise remains that much can be learned about the regulation of the oxidative pentose shunt enzymes through nutritional and genetic studies of D. melanogaster.

SUMMARY

1. The tissue activities of G6PD and 6PGD are induced in the larvae of D. melanogaster by a derivative of an early step of glycolysis and repressed by a fatty acid derivative.

2. G6PD and 6PGD are subject to allosteric regulation. Both are inhibited by NADPH. 6PGD is also inhibited by ribulose-5-phosphate and fructose-1,6-diphosphate. Consequently, the rate of lipogenesis usually determines the current rate of pentose shunt activity by influencing the NADPH/NADP$^+$ ratio. However, under certain physiological conditions other intermediates may limit the pentose shunt activity.

3. A deficiency for the oxidative pentose shunt modifies the NADPH/NADP$^+$ ratio and limits the capacity of D. melanogaster to synthesize triglyceride from dietary carbohydrate.

4. Because of pentose shunt is integrated with other metabolic pathways associated with the utilization of dietary energy sources, enzyme polymorphisms for G6PD and 6PGD appear to be of adaptive value to D. melanogaster for coping with diverse dietary conditions.

5. Strains established from natural populations exhibit varied G6PD and 6PGD sucrose responses. Genetic determinants for the G6PD sucrose response are situated in the X and third chromosomes of D. melanogaster.

REFERENCES

Arinze, J. C. and Mistry, S. P., 1971, Activities of some biotin enzymes and certain aspects of gluconeogenesis during biotin deficiency, Comp. Biochem. Physiol. 38B:285-294.

Bewley, G. C. and Lucchesi, J. C., 1975, Lethal effects of low and "null" activity alleles of 6-phosphogluconate dehydrogenase in Drosophila, Genetics 79:451.

Bijlsma, R., 1978, Polymorphism at the G6PD and 6PGD loci in Drosophila melanogaster. II. Evidence for interaction in fitness, Genet. Res. Camb. 31:227.

Bijlsma, R. and Van Delden, W., 1977, Polymorphism at the G6PD and 6PGD loci in Drosophila melanogaster. I. Evidence for selection in experimental populations, Genet. Res. Camb. 30: 221.

Bischoff, W. L., 1978, Ontogeny of sorbitol dehydrogenases in Drosophila melanogaster, Biochem. Genet. 16:485.

Bowman, J. T. and Simmons, J. R., 1973, Gene modulation in Drosophila: dosage compensation of Pgd^+ and Zw^+ genes, Biochem. Genet. 10:319.

Cannistraro, V. J., Borack, L. I., and Chase, T., 1979, Subunit structure and kinetic properties of L-β-hydroxyacid dehydrogenase of Drosophila, Biochim. Biophys. Acta 569:1

Cavener, D. R., 1980, The genetics of male specific glucose oxidase and the identification of other unusual hexose enzymes in Drosophila melanogaster, Biochem. Genet., in press.

Cavener, D. R. and Clegg, M. T., 1978, Dynamics of correlated genetic systems. IV. Multi locus effects of ethanol stress environments, Genetics 90:629.

Cavener, D. R. and Clegg, M. T., 1980, Multigenic adaptation to ethanol in Drosophila melanogaster, Evolution, in press.

DeZwann, A. and Marrewijk, W. J. A., 1973, Intercellular localization of pyruvate carboxylase, phosphoenol pyruvate carboxylase, phosphoenol pyruvate carboxykinase and malic enzyme, and the absence of glyoxylate cycle enzymes in the sea mussel (Mylitus endulis L.), Comp. Biochem. Physiol. 44B:1057.

Fitch, W. W. and Chaikoff, I., 1960, Extent and patterns of adaptation of enzyme activities in livers of normal rats fed diets high in glucose and fructose, J. Biol. Chem. 235:554.

Geer, B. W., 1963, A ribonucleic acid-protein relationship in Drosophila nutrition, J. Exp. Zool. 154:353.

Geer, B. W., 1966, Utilization of D-amino acids for growth by Drosophila melanogaster larvae, J. Nutr. 90:31.

Geer, B. W., 1972, The functions of choline and carnitine during spermatogenesis in Drosophila, in: "Insect and Mite Nutrition," J. G. Rodriguez, ed., North-Holland Publishing Co., Amsterdam.

Geer, B. W., Bowman, J. T., and Simmons, J. R., 1974, The pentose
    shunt in wild-type and glucose-6-phosphate dehydrogenase
    deficient Drosophila melanogaster, J. Exp. Zool. 187:77.
Geer, B. W., Kamiak, S. N., Kidd, K. R., Nishimura, R. A., and Yemm,
    S. J., 1976, Regulation of the oxidative NADP-enzyme tissue
    levels in Drosophila melanogaster. I. Modulation by dietary
    carbohydrate and lipid, J. Exp. Zool. 195:15.
Geer, B. W., Krochko, D., and Williamson, J. H., 1979a, Ontogeny
    cell distribution and the physiological role of NADP-malic
    enzyme in Drosophila melanogaster, Biochem. Genet. 17:867.
Geer, B. W., Lindel, D. L., and Lindel, D. M., 1979b, The relation-
    ship of the oxidative pentose shunt pathway to lipid synthesis
    in Drosophila melanogaster, Biochem. Genet. 17:881.
Geer, B. W. and Vovis, G. F., 1965, The effects of choline and re-
    lated compounds on the growth and development of Drosophila
    melanogaster, J. Exp. Zool. 158:223.
Geer, B. W., Woodward, C. G., and Marshall, S. D., 1978, Regulation
    of the oxidative NADP-enzyme tissue levels in Drosophila
    melanogaster. II. The biochemical basis of dietary carbohy-
    drate and D-glycerate modulation, J. Exp. Zool. 203:391.
Gerasimova, T. I. and Ananjev, E. V., 1972, Cytogenetical localiza-
    tion of structural gene Pgd for 6-phosphogluconate dehydro-
    genase in Drosophila melanogaster., Drosophila Inform. Serv.
    48:93.
Gibson, J., 1970, Enzyme flexibility in Drosophila melanogaster,
    Nature 227:959.
Giesel, J. T., 1976, Biology of a duplicate gene system with glucose
    6-phosphate dehydrogenase activity in Drosophila melanogaster:
    genetic analysis and differences in fitness components and
    reaction to environmental parameters among Zw genotypes,
    Biochem. Genet. 14:823.
Gumaa, K. A., McLean, P., and Greenbaum, A. L., 1971, Compartment-
    alization in relation to metabolic control in liver, Essays
    in Biochemistry 7:39.
Gvozdev, V. A., Birstein, F. J., and Faizullin, L. Z., 1970, Gene
    dependent regulation of 6-phosphogluconate dehydrogenase
    activity of Drosophila melanogaster, Drosophila Inform.
    Serv. 45:163.
Gvozdev, V. A., Gerasimova, T. I., Kogan, G. L., and Braslavskaya,
    O. Y., 1976, Role of the pentose phosphate pathway in the
    metabolism of Drosophila melanogaster elucidated by mutations
    affecting glucose 6-phosphate and 6-phosphogluconate dehydro-
    genases, FEBS Lett. 64:85.
Gvozdev, V. A., Gerasimova, T. I., Kogan, G. L., and Rosovsky,
    J. M., 1977, Investigations on the organization of genetic
    loci in Drosophila melanogaster: lethal mutations affecting
    6-phosphogluconate dehydrogenase and their suppression,
    Molec. Gen. Genet. 153:191.

Gvozdev, V. A., Gostimsky, S. A., Gerasimova, T. I., Dubrovskaya,
    E. S., and Braslavskaya, O. Y., 1975, Fine genetic structure
    of the 2D3-2F5 region of the X-chromosome of Drosophila
    melanogaster, Molec. Gen. Genet. 141:269.
Hughes, M. B. and Lucchesi, J. C., 1977, Genetic rescue of a lethal
    "null" activity allele of 6-phosphogluconate dehydrogenase in
    Drosophila melanogaster, Science 197:1114.
Hughes, M. B. and Lucchesi, J. C., 1978, Dietary rescue of a lethal
    "null" activity allele of 6-phosphogluconate dehydrogenase in
    Drosophila melanogaster, Biochem. Genet. 21:1.
Johnson, B. C. and Sassoon, H. R., 1967, Studies on the induction
    of liver glucose-6-phosphate dehydrogenase in the rat, Ad-
    vances in Enzyme Regulation 5:93.
Kazazian, H. H., Jr., Young, W. J., and Childs, B., 1965, X-linked
    6-phosphogluconate dehydrogenase in Drosophila: Subunit
    associations, Science 150:1601.
Keepler, D. and Decker, K., 1974, Glycogen determination with
    amyloglucosidase, in: "Methods of Enzymatic Analysis,"
    Vol. 2., H. U. Bergmeyer, ed., Academic Press, New York.
Kelly, D. S., Watson, J. J., Mack, D. O., and Johnson, B. C.,
    1975, Glucose-6-phosphate dehydrogenase is not induced in the
    mammalian liver by dietary carbohydrate, Nutr. Rep. Int. 12:
    121.
Krebs, H. A. and Eggleston, L. V., 1974, The regulation of the
    pentose phosphate cycle in rat liver, Advances in Enzyme
    Regulation 12:421.
Krochko, D. and Williamson, J. H., 1980, 6-Phosphogluconate dehydro-
    genase from Drosophila melanogaster: physical and kinetic
    properties in comparison with the yeast enzyme, Comp. Biochem.
    Physiol., in press.
Lee, C. Y., Langley, C. H., and Burkhart, J., 1978, Purification
    and molecular weight determination of glucose-6-phosphate
    dehydrogenase and malic enzyme from mouse and Drosophila,
    Analyt. Biochem. 86:697.
Lewis, E. G., 1960, A new standard food medium, Drosophila Inform.
    Serv. 34:117.
Lucchesi, J. C., Hughes, M. B., and Geer, B. W., 1979, Genetic
    control of the pentose phosphate pathway enzymes in Drosophila,
    Current Topics in Cellular Regulation 15:143.
Parr, C. W., 1956, Inhibition of phosphoglucose isomerase, Nature
    178:1401.
Parsons, P. A. and King, S. B., 1977, Ethanol: larval discrimination
    between two Drosophila sibling species, Experientia 33:898.
Potter, V. C. and Ono, T., 1961, Enzyme patterns in rat liver
    and Morris hepatoma 5123 during metabolic transition, Cold
    Spring Harbor Symp. Quant. Biol. 26:355.
Sang, J. H., 1956, The quantitative nutritional requirements of
    Drosophila melanogaster, J. Exp. Biol. 33:45.

Sang, J. H., 1979, The nutritional requirements of Drosophila,
    in: "The Genetics and Biology of Drosophila," Vol. 2a,
    M. Ashburner and T. R. F. Wright, eds., Academic Press,
    New York.
Seecof, R. L., Kaplan, W. D., and Futch, D. G., 1969, Dosage
    compensation for enzyme activities in Drosophila melanogaster,
    Proc. Nat. Acad. Sci. (U.S.A.) 62:528.
Steele, M. W., Young, W. J., and Childs, B., 1968, Glucose 6-phos-
    phate dehydrogenase in Drosophila melanogaster: starch gel
    electrophoretic variation due to molecular instability,
    Biochem. Genet. 2:159.
Steele, M. W., Young, W. J., and Childs, B., 1969, Genetic regula-
    tion of glucose-6-phosphate dehydrogenase activity in
    Drosophila melanogaster, Biochem. Genet. 3:359.
Taketa, K. and Pogell, B. M., 1966, The effect of palmityl coenzyme
    A on glucose-y-phosphate dehydrogenase and other enzymes, J.
    Biol. Chem. 241:720.
Tepperman, H. M. and Tepperman, J., 1958, The hexose monophosphate
    shunt and adaptive hyperlipogenesis, Diabetes 7:478.
Tepperman, H. M. and Tepperman, J., 1965, Effect of saturated fat
    diets on rat liver NADP-linked enzymes, Amer. J. Physiol.
    209:773.
Van Delden, W., Kamping, A., and van Dijk, M., 1975, Selection at
    the alcohol dehydrogenase locus in Drosophila melanogaster,
    Experientia 31:418.
Williamson, J. H., Geer, B. W., and Krochko, D., 1980a, Glucose-6-
    phosphate dehydrogenase from Drosophila melanogaster. I.
    Purification and properties of the dimeric form, Comp.
    Biochem. Physiol., in press.
Williamson, J. H., Krochko, D., Bentley, M. M., and Thwaites, T.,
    1980b, Purification and properties of NADP-malic enzyme from
    Men Drosophila melanogaster, in preparation.
Williamson, J. H., Krochko, D., and Geer, B. W., 1980c, 6-Phospho-
    gluconate dehydrogenase from Drosophila melanogaster. I.
    Purification and Properties of the A-isozyme, Biochem. Genet.
    18:87.
Wise, E. M. and Ball, E. G., 1964, Malic enzyme and lipogenesis,
    Proc. Nat. Acad. Sci. (U.S.A.) 52:1255.
Young, W. J., 1966, X-linked electrophoretic variation in 6-
    phosphogluconate dehydrogenase in Drosophila melanogaster,
    J. Heredity 57:58.
Young, W. J., Porter, J. E., and Childs, B., 1964, Glucose-6-
    phosphate dehydrogenase in Drosophila: X-linked electro-
    phoretic variants, Science 143:140.

RESOLUTION AND PARTIAL CHARACTERIZATION OF THE DIGESTIVE

PROTEINASES FROM LARVAE OF THE BLACK CARPET BEETLE[a]

J. E. Baker

Stored-Product Insects Research and Development Lab
AR, SEA, USDA
Savannah, Georgia 31403

INTRODUCTION

Keratins are the major protein components of wool, hair, horn, and feathers. The relatively high cystine content (8-18%) and the resultant extensive crosslinking of the protein chains via disulfide bonds makes native keratin relatively insoluble and resistant to biological degradation (Waterhouse, 1958). Because of its chemical stability, few organisms have been shown to be able to digest keratin. A few microorganisms, including several strains of fungi in the genus Streptomyces, produce keratinolytic enzymes (Morihara et al., 1967). However, among higher organisms, only the insects, some 30 species of Lepidoptera in the family Tineidae and 15 species of Coleoptera in the family Dermestidae, have been shown to digest this protein (Waterhouse, 1958).

Among the dermestids, larvae of the black carpet beetle, Attagenus megatoma (F.), are one of the most important fabric pests. These larvae will feed on clean woolen test cloth with a consumption index (C.I.) of 0.06 mg dry wt food consumed/mg fresh body wt per day and with an approximate digestibility (A.D.) coefficient of 42.9% (Baker, 1974). However, these larvae do not grow. The addition of selected nutrients (vitamins, minerals, and a sterol) to the clean woolen fabric increased the C.I. to 0.09 and the A.D. to 50.5% and allowed growth, although slow, to occur.

---

[a]Mention of a commercial product in this paper does not constitute a recommendation by the USDA.

These results corroborated earlier behavioral (Mallis et al.,
1962) and nutritional (Fraenkel and Blewett, 1946) studies indicating that stained or soiled areas on fabrics were the preferred
feeding sites for wool feeding insects.

The digestion of wool by the dermestids is thought to be
enhanced by the strong reducing conditions (-190 to -230 mV) in
the larval midguts of these species (Waterhouse, 1952a). It was
postulated that this midgut environment may reduce the disulfide
bonds of keratin and allow hydrolysis by the gut proteinases to
occur. Similar reducing conditions (-230 to -280 mV) are present
in larvae of the webbing clothes moth, Tineola bisselliella (Hummel)
(Waterhouse, 1952b). However, there is also evidence for a
keratinase in T. bisselliella (Powning and Irzykiewicz, 1962) in
addition to a complex series of peptidases and proteinases (Ward,
1975a).

Although larvae of A. megatoma are used to bioassay mothproofing test chemicals on woolen cloth, they are commonly reared
in the laboratory on ground Purina Laboratory Chow[R] containing
5% (wt/wt) brewer's yeast. The larval development time on the
Purina diet is about 36 wk (Baker, 1977a). For studies on the
digestive enzymes of A. megatoma, 8-10 wk-old larvae were selected.
At this age the mean larval weight is about 6 mg for males and 10
mg for females. The larvae are feeding actively at this time
with a C.I. of 0.11 and an A.D. of 70.2% on the Purina diet (Baker,
1975). In addition, the secretion of the alkaline proteinase
present in the midguts (Baker, 1976; 1977b; 1978) is near maximum.

Using a combination of specific enzyme substrates, electrophoresis, and ion exchange, gel filtration, and affinity chromatographic techniques, the midgut digestive proteinases
were partially resolved and characterized. The following report
describes the procedures used and some of the properties of the
isolated enzymes.

MATERIALS AND METHODS

The abbreviations used in the text are noted below.[a]

---

[a]Abbreviations used: DMF, N,N-dimethylformamide; tris, tris
(hydroxymethyl)-aminomethane; SBTI, soybean trypsin inhibitor
(type IS); OVTI, ovomucoid trypsin inhibitor (type IIO); PMSF,
phenylmethylsulfonylfluoride; EDTA, ethylenediaminetetraacetic
acid - disodium salt; TLCK, N-α-tosyl-L-lysinechloromethyl ketone·
HCl; TPCK, L-1-tosylamide-2-phenylethyl-chloromethyl ketone. DEAE-
Sephacel, CM-Sephadex, and Sephadex G-100 were from Pharmacia. All
other substrates and inhibitors were purchased from Sigma. Agarose-
γ-aminocaproyl-D-tryptophane was from Miles.

Source of Insects

Larvae of A. megatoma were obtained from 8 to 10 wk-old cultures reared at 28°C and 45-55% RH on ground Purina Laboratory Chow[R] fortified with brewer's yeast (95:5, w/w).

Preparation of Homogenates

Midguts of larvae were dissected under ice-cold 1% NaCl, rinsed, and homogenized in 1% NaCl in a glass-glass homogenizer. The crude homogenate was centrifuged at 5300 g for 30 min (4°C) and the supernatant was stored at -20°C. Only midguts containing numerous food particles were selected.

Proteinase Activity

Total proteinase activity was measured with a modification of the casein digestion method of Kunitz (1947). Aliquots of enzyme were preincubated in 1.0 ml 50 mM tris-chloride pH 8.0 for 5 min at 37°C prior to the addition of 1.0 ml pre-warmed 1% casein in buffer. The reaction was terminated with 2.0 ml 10% trichloracetic acid. A proteinase unit was defined as a change of 1.0 absorbance unit at 280 nm per min per mg protein.

. In parts of this study, proteinase activity was also measured by using the substrate hide powder azure. Enzyme was added to a mixture of 10 mg of the powder in 3.0 ml 50 mM tris-chloride pH 8.0 at 30°C. The reaction was terminated with 0.6 ml 20% acetic acid and the mixture filtered. One hide powder azure unit was defined as an increase of 1.0 absorbance unit/min at 650 nm in a final volume of 3.62 ml.

Artificial Substrates

The change in molar extinction at the wavelength of maximum difference was used to monitor the hydrolysis of the artificial proteinase substrates. Trypsin-like activity was measured with p-tosyl-L-arginine methyl ester·HCl (TAME) $\Delta E_{247} = 540$ (Hummel, 1959), α-N-benzoyl-L-arginine ethyl ester (BAEE) $\Delta E_{255} = 808$ (Schwert and Takenaka, 1955), and α-N-benzoyl-DL-arginine-p-nitroanilide·HCl in 15% DMF (BApNA) $\Delta E_{410} = 8800$ (Erlanger et al., 1961). Chymotrypsin-like activity was measured with N-acetyl-L-tyrosine ethyl ester in 5% methanol (ATEE) $\Delta E_{237} = 400$ (Schwert and Takenaka, 1955) N-benzyol-L-tyrosine ethyl ester in 17% methanol (BTEE) $\Delta E_{256} = 964$ (Hummel, 1959), glutaryl-L-phenylalanine-p-nitroanilide (GPpNA) in 10% DMF $\Delta E_{410} = 8800$ (Erlanger et al., 1966), and N-benzoyl-L-tyrosine-p-nitroanilide (BTpNA) in 20% DMF $\Delta E_{410} = 8800$. Aminopeptidase activity was measured with L-leucine-p-nitroanilide·HCl (LpNA) in 10% DMF

ΔE_{410} = 8800 (Wachsmuth et al., 1966; Erlanger et al., 1961) and
L-leucyl-β-naphthylamide·HCl (LβNA) in 5% methanol and 10% DMF (in
a 5.0 ml volume, a unit change in optical density at 520 nm equals
0.11 μmole β-naphthylamine) (Marks et al., 1968). Carboxypeptidase
A activity was measured with hippuryl-DL-phenyllactic acid (HPLA)
in 0.1 M NaCl ΔE_{254} = 592 (McClure et al., 1964) and hippuryl-L-
phenylalanine (HP) in 0.1 M NaCl ΔE_{254} = 280 (Davies et al., 1968).
Carboxypeptidase B activity was measured with hippuryl-L-arginine
(HA) in 0.1 M NaCl ΔE_{254} = 360 (Folk et al., 1960). Assays were
conducted as follows: 3.0 ml of 1 mM substrate in 50 mM tris-
chloride buffer pH 8.0 were placed in a 1.0 cm cuvette. Changes
in absorbance at 30°C after addition of enzyme were determined with
a Gilford model 250 recording spectrophotometer equipped with an
automatic cuvette changer. Blank corrections were made with
cuvettes containing substrate and boiled enzyme.

Inhibitor Studies

The effect of selected trypsin and chymotrypsin inhibitors on
total proteinase activity and on the hydrolysis of selected substrates was studied. For the inhibitors of total proteinase
activity, enzyme was preincubated with inhibitor in 1.0 ml buffer
for 60 min at 37°C before the addition of casein. The inhibitor
concentrations tested are given in the results section. For the
artificial substrates, inhibitor and enzyme (in a total buffer
volume of 200 μl) were preincubated for 60 min at 30°C prior to
the addition of substrate. The following inhibitors were tested:
SBTI and OVTI, naturally occurring trypsin inhibitors; PMSF, a
selective inhibitor of serine proteinases (Gold, 1965); TLCK
and TPCK, site-directed inhibitors of trypsin and chymotrypsin,
respectively (Shaw, 1967); and the metal chelator, EDTA.

Ion Exchange Chromatography

DEAE-Sephacel cellulose anion exchanger (45 ml of the stock
ethanolic mixture) was equilibrated with 20 mM tris-chloride buffer
pH 7.5 and poured into a 1.6 cm i.d. column (final packed column
height was 13.6 cm). The sample was applied to the column through
a sample valve and a linear gradient of 250 ml 0.4 M NaCl in buffer
into 250 ml buffer was begun. The initial flow rate was 0.42 ml/
min and 4.0 ml fractions were collected. At tube 130, the gradient
was stepped to 1.0 M NaCl in buffer. The sample for this column
was prepared as follows: 200 midguts were dissected into a total
of 10 ml 1% NaCl. This sample was concentrated to 1.8 ml via
ultrafiltration over UM-10 membranes and passed through a Sephadex
G-25 column equilibrated with 20 mM tris-chloride pH 7.5. The
exchanged enzyme sample was again concentrated to 1.9 ml over
UM-10 membranes and 1.7 ml containing 7.8 mg protein was applied
to the column. Each fraction was checked for absorbance at 280

nm and 100 µl aliquots from every second tube were used to assay for caseinolytic activity and BApNA, GPpNA, HPLA, and LpNA hydrolysis.

Protein that was not bound to the anion exchange resin was applied to a cation exchange resin carboxymethyl-Sephadex (C-50-120). The CM-Sephadex (0.65 g) was equilibrated with 20 mM phosphate pH 7.5 buffer and poured into a 1.6 cm i.d. column (final packed height was 14.7 cm). The sample from the DEAE-Sephacel column was transferred to the phosphate buffer system by repeated ultrafiltration. Protein bound to the CM-Sephadex resin was eluted with a linear gradient of 250 ml 0.2 M NaCl in 20 mM phosphate pH 7.5 buffer into 250 ml of the same buffer. The column flow rate was about 0.5 ml/min and 4.0 ml fractions were collected. Aliquots of every other tube were analyzed for enzymatic activity.

Gel Filtration Chromatography

A Sephadex G-100 column (2.5 x 31 cm) equilibrated with 0.1 M NaCl in 5 mM tris-chloride buffer pH 7.8 was used to further purify the fractions from the ion exchange columns and to obtain an estimate of their molecular weight. The column was calibrated with ribonuclease A (MW 13,700), chymotrypsinogen A (MW 25,000), ovalbumin (MW 43,000), and bovine serum albumin (MW 67,000) (Andrews, 1964). The void volume was determined with blue dextran. Following sample application directly to the surface of the gel column, the first 50 ml of effluent was discarded and then 2.0 ml fractions were collected (flow rate 0.4 ml/min). Each tube was analyzed for absorbance at 280 nm and 100 µl aliquots of every second tube were analyzed for enzyme activity. Molecular weights of peaks were estimated by calculating their Kav (partition coefficient) values and reading from the plot of Kav against log molecular weight for the standard proteins.

Affinity Chromatography

The affinity gel agarose-γ-aminocaproyl-D-tryptophane was used to selectively remove chymotrypsin-like impurities from several of the purified trypsin-like fractions. The gel was equilibrated in 20 mM phosphate pH 7.8 buffer and a 0.9 x 3.5 cm column was prepared. Trypsin-like activity was not bound to this gel. Attempts were made to recover the chymotrypsin-like activity by changing the ionic strength and/or the pH of the buffer.

Ultrafiltration

Samples were concentrated via ultrafiltration over UM-10 or YM-10 membranes with an Amicon Stirred Cell[R] filtering chamber. The nitrogen pressure was 50 lb in^{-2}.

Electrophoresis

Polyacrylamide gel electrophoresis was carried out with a
Buchler Polyanalyst[R] apparatus using the procedures of Davis (1962)
and Ornstein (1962). Electrophoresis was carried out at 3 ma/tube
using a separating gel of 7.5%, a tris-glycine pH 8.6 buffer, and
bromophenol blue as the tracking dye. Protein bands were stained
with 0.05% aniline blue black in 7% acetic acid or with 0.25%
coomassie brilliant blue (R250) in a solution of 50% methanol:
acetic acid, 90:10 (v/v). Trypsin-like activity on the gels was
detected by incubating the gels in 0.5 mM $BA_\beta NA$ in 50 mM tris-
chloride pH 8.0 for 15-30 min at room temperature and staining
for 30 min with 0.15% fast garnet GBC in 0.33 M acetate buffer
pH 4.5 containing 5% Triton X-100 (Ward, 1975a). Chymotrypsin-
like activity was detected by incubating the gels for 15-30 min in
1 mM N-glutaryl-L-phenylalanine-β-naphthylamide in 50 mM tris-
chloride pH 8.0 (35% DMF in buffer was required to keep this substrate in solution). Bands were stained with fast garnet and
stored in 0.33 M acetate buffer pH 4.5. Aminopeptidase activity
was detected on the gels by incubating in 1 mM leucine-β-naphthyl-
amide in 50 mM phosphate pH 7.5 containing 5% MeOH and 10% DMF.
These gels were also stained with fast garnet. The gels were
scanned at selected wavelengths with a Gilford Model 2520 gel
scanner.

Cationic proteins were examined electrophoretically by using
the cationic gel system of Racusen (1967) with no stacking gel
(Ward, 1975a). The gel system has a running pH of 7.5 and uses
an imidazole-taurine buffer system. The tracking dye was methyl
green. The staining procedures used on these gels were similar
to those with the anionic gel system.

Protein Analysis

Total protein was estimated with the procedure of Lowry et al.
(1951). Bovine serum albumin was used as the standard.

Determination of Kinetic Constants

The values for K_m and V_{max} of selected trypsin, chymotrypsin,
and aminopeptidase substrates were determined by using the method
of Lee and Wilson (1971). Lineweaver-Burk plots of the average
velocity ($\bar{v}$) over a fixed time period against the average substrate concentrations ($\bar{S}$) over the same time period were prepared
and the slope and intercept values were obtained by using the
Statistical Analysis Systems (SAS) general linear models analysis
(Barr et al., 1976).

Amino Acid Composition

An anionic trypsin (ca. 700 µg) and a cationic trypsin (ca. 500 µg) from A. megatoma were passed through a Sephadex G-25 column equilibrated with 1 mM PMSF in 50 mM ammonium bicarbonate pH 7.8. The samples were concentrated over YM-10 membranes, lyophilized, and the amino acid composition of the PMSF-enzymes determined. The molar ratios of the amino acids were normalized to 1 histidine residue and the rounded, whole integer values for each amino acid were multiplied to approximate the estimated molecular weight of each enzyme.

RESULTS

The substrates specific for the endopeptidases trypsin and chymotrypsin were rapidly hydrolyzed by the midgut homogenates (Table 1). For each group of enzymes the rate of hydrolysis of the ester analogs was much faster than that of the amides. TAME

Table 1.   Hydrolysis of artificial proteinase substrates by the midgut homogenate prepared from larvae of Attagenus megatoma.

Substrate	Enzyme	Volume of homogenate used for reaction (µl)[a]	Rate of hydrolysis µmoles/min per mg protein
TAME	Trypsin	5[b]	77.3
BAEE	Trypsin	10[b]	18.1
BApNA	Trypsin	10[b]	1.3
ATEE	Chymotrypsin	5[b]	158.5
BTEE	Chymotrypsin	5[b]	47.4
GPpNA	Chymotrypsin	40-60[b]	0.23
BTpNA	Chymotrypsin	30-40[b]	0.42
LpNA	Aminopeptidase	20-50[b]	0.04
LβNA	Aminopeptidase	80-100[c]	0.01
HPLA	Carboxypeptidase A	20-40[c]	0.83
HP	Carboxypeptidase A	50-60[c]	0.05
HA	Carboxypeptidase B	50-60[c]	0.05
casein	Total proteinase	10-30[c]	80 x 10^{-2} PU

[a] In a total reaction volume of 3.0 ml.

[b] Homogenate concentration equivalent to 5 guts/ml.

[c] Homogenate concentration equivalent to 20 guts/ml.

and ATEE were the most rapidly hydrolyzed substrates. The substrates used to monitor exopeptidase activity were also hydrolyzed.
LpNA and HPLA were the most sensitive substrates for aminopeptidase
and carboxypeptidase A activity, respectively. Carboxypeptidase
B-like activity against the substrate HA was also demonstrated.

SBTI was a more effective proteinase inhibitor against
caseinolytic activity than was OVTI (Table 2). PMSF inhibited
ca. 88% of the proteinase activity and indicated the presence of
serine proteinases. The active-site directed inhibitors TLCK
and TPCK inhibited ca. 75 and 45% of the caseinolytic activity,
respectively. This suggests that trypsin-like enzymes may be
responsible for more of the proteinase activity in the midguts
than chymotrypsin-like enzymes. Caseinolytic activity was not
inhibited significantly when the homogenate was preincubated with
1, 2, or 5 mM EDTA.

Table 2.   Effect of inhibitors on caseinolytic activity of a crude
           midgut homogenate from larvae of Attagenus megatoma.

Inhibitor[a]	Preincubation concentration	Mean % inhibition[b]
SBTI	1 µg/ml	64.5
	5 µg/ml	74.8
OVTI	10 µg/ml	53.5
PMSF[c]	1 mM	88.9
EDTA	1 mM	10.4
	2 mM	4.3
	5 mM	9.3[d]
TLCK	0.1 mM	74.8
	0.2 mM	74.1
TPCK[e]	0.1 mM	45.6
	0.2 mM	42.3

[a]Inhibitors were preincubated with enzyme in a total volume of
1.0 ml for 60 min at 37°C prior to addition of substrate.

[b]Mean inhibition values determined from 3 replicates.

[c]Added as a solution in 2-propanol.

[d]Activated by this value over control.

[e]Added as a solution in methanol.

Fractionation on DEAE-Sephacel

Preliminary small-scale experiments indicated that ca. 50% of the protein and 80% of the caseinolytic activity of the midgut homogenate (following buffer exchange) was not bound to the DEAE-Sephacel equilibrated with 20 mM tris-chloride at pH 6.5, 7.0, 7.5, or 8.0. A linear gradient of NaCl to 0.4 M was used to elute the protein that did bind to the resin at pH 7.5 (Fig. 1). Two major peaks absorbing at 280 nm were eluted in the first 10 fractions (tubes 3-6 and 7-10). Small peaks were also noted at tubes 16-20, 48-50, and 116-122 and just after the step to 1.0 M NaCl in buffer (tubes 134-136).

Approximately 93% of the caseinolytic activity that was recovered was confined to tubes 3-6. The remaining activity was found in tubes 114-123. Essentially all of the aminopeptidase (LpNA) and carboxypeptidase (HPLA) activity was found in tubes 3-6. GPpNA activity was found in tubes 3-6 and in a peak eluted with 0.13 M NaCl (tubes 41-49) (Fig. 2). BApNA activity was associated

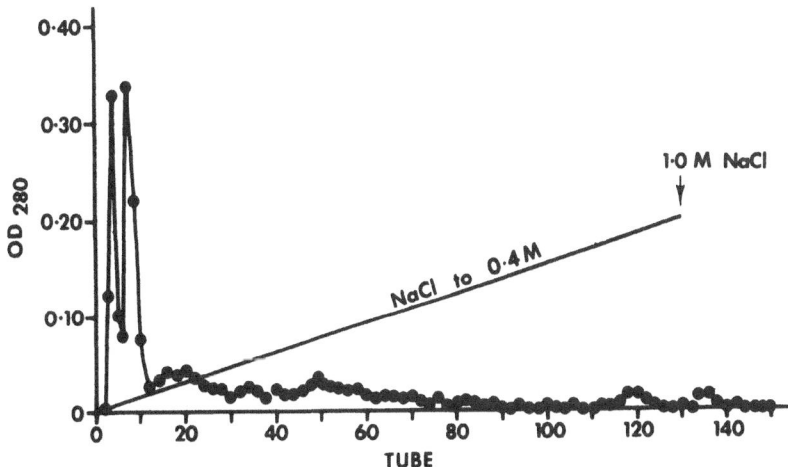

Fig.1 Pattern of digestive proteinases of larvae of Attagenus megatoma from a DEAE-Sephacel column equilibrated in 20 mM tris-chloride pH 7.5 and eluted with a NaCl gradient. The protein was monitored by measuring the absorbance at 280 nm of the first 10 tubes and of every second tube thereafter on a Gilford model 250 spectrophotometer.

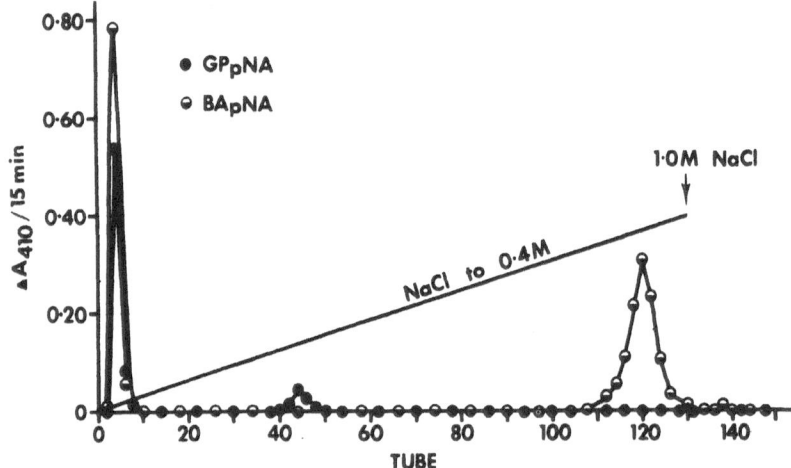

Fig.2   Pattern of digestive proteinases of larvae of <u>Attagenus</u>
        <u>megatoma</u> from a DEAE-Sephacel column equilibrated in 20 mM
        tris-chloride pH 7.5 and eluted with a NaCl gradient. Ali-
        quots (100 µl) of every second tube were analyzed for
        trypsin (BApNA) and chymotrypsin (GPpNA) activity.

with tubes 3-6, with a large symetrical peak eluted with 0.34 M
NaCl beginning at tube 110, and with a small peak that eluted in
2 tubes just after the step to 1.0 M NaCl. In additional tests
with the DEAE-Sephacel resin, the late emerging major peak that
had BApNA activity was eluted symetrically with 0.27 M NaCl from
a 1.0 x 8.0 cm column using a flow rate of 1.0 ml/min and with 0.35
M NaCl from a 1.6 x 14.0 cm column using a flow rate of 0.5 ml/min.
These columns could be used for samples prepared from as many as
400 midguts.

     The four groups of tubes with enzyme activity (peak I, tubes
3-6; peak II, tubes 7-10; peak III, tubes 41-49; and peak IV, tubes
110-129) were concentrated via ultrafiltration and tested against
the substrates shown in Table 3. Peak I contained activity against
all of the tested substrates. It was the only peak that contained
significant amounts of exopeptidase activity. Peak II contained
a small amount of tryptic and chymotryptic-like activity. This
activity may have been eluting on the shoulder of peak I. Peak III
contained activity against BTpNA and GPpNA, and ATEE (data for ATEE
were not tabulated). There was also a slight trace of LpNA and HPLA
activity in this peak but no activity against TAME, BAEE, or BApNA.
Peak IV contained activity against the trypsin substrates TAME,
BAEE, and BApNA but did not hydrolyze any of the other substrates.

Table 3. Hydrolysis of artificial substrates by combined fractions of a midgut homogenate of larvae of Attagenus megatoma eluted from a DEAE-Sephacel anion exchange column with a gradient of NaCl.

	Rate of hydrolysis μmoles/min per mg protein				
Substrate	Original sample	Peak I (Tubes 3-6)	Peak II (Tubes 7-10)	Peak III (Tubes 41-49)	Peak IV (Tubes 110-129)
TAME	86.5	177.8	21.0	0	211.3
BAEE	18.8	44.7	4.6	0	56.6
BApNA	1.37	3.43	0.27	0	2.57
BTpNA	0.27	0.67	T[a]	0.74	0
GPpNA	0.06	0.34	0.01	0.03	0
LpNA	0.04	0.11	T	T	0
HPLA	0.77	1.97	0	T	0

[a] T = <0.01 μmole/min per mg protein.

Electrophoresis of Peaks I, III, and IV

When peak I was subjected to electrophoresis in the anionic
gel system, 1 faint band migrated slightly towards the anode (Rm =
0.07). It was apparent that most of the proteins in peak I moved
to the cathode (upper buffer). When peak I was run with the current
reversed in the same gel system, a heavy band of protein was driven
into the top of the gel towards the cathode but no migration or
separation of sample occurred. Reducing the acrylamide concentration to 4.5% still did not allow the proteins present to migrate
into the gel.

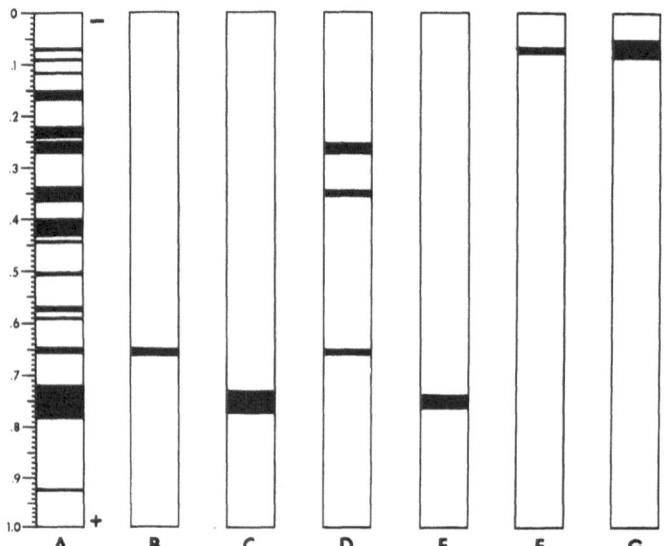

Fig. 3   Acrylamide gel zymograms of combined fractions from the DEAE-
Sephacel column using the anionic gel system. A - crude midgut homogenate (80 μg protein) stained for protein; B - crude
midgut homogenate (80 μg protein) incubated with GPβNA and
stained with fast garnet; C - crude midgut homogenate (80
μg protein) incubated with BAβNA and stained with fast
garnet; D - Peak III (25 μg protein) stained for protein;
E - Peak IV (25 μg protein) stained for protein; F - Aminopeptidase fraction (10 μg protein) stained for protein; and
G - Aminopeptidase fraction (10 μg protein) incubated with
LβNA and stained with fast garnet. Migration was toward
the anode and mobilities expressed relative to that of
bromophenol blue.

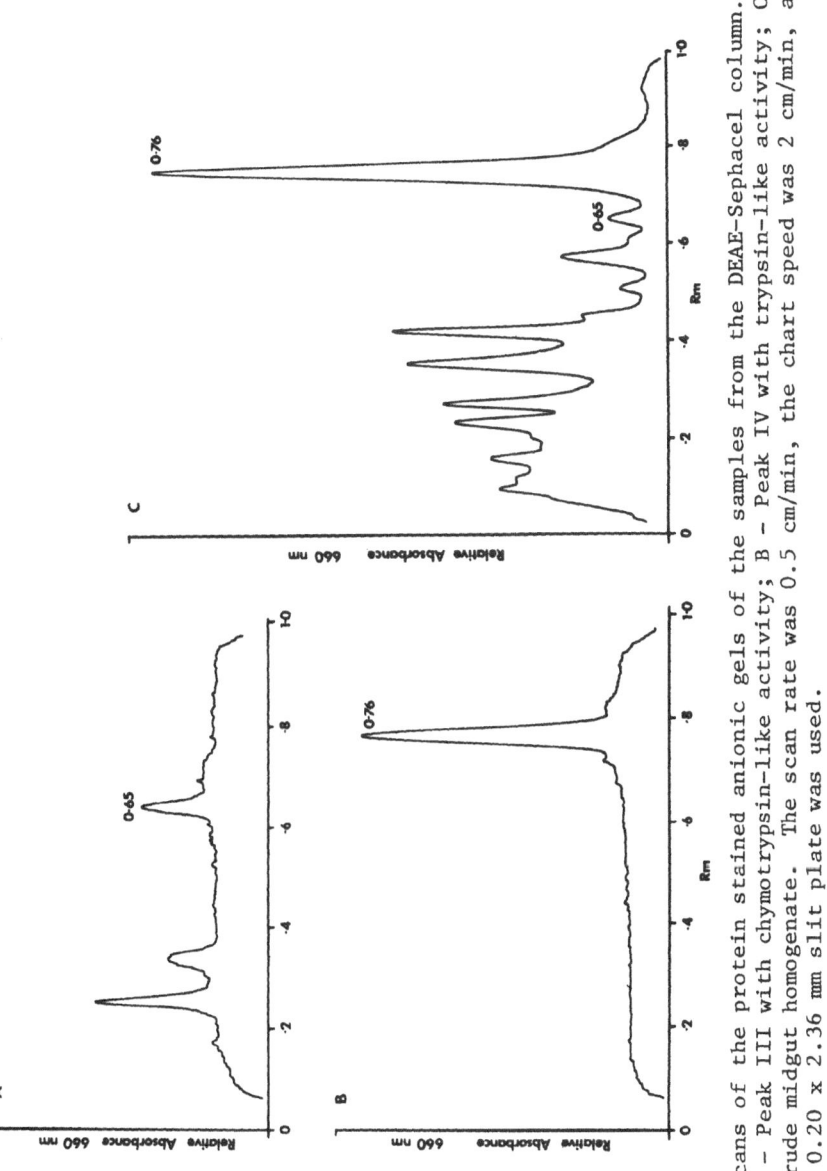

Fig.4 Scans of the protein stained anionic gels of the samples from the DEAE-Sephacel column. A – Peak III with chymotrypsin-like activity; B – Peak IV with trypsin-like activity; C – crude midgut homogenate. The scan rate was 0.5 cm/min, the chart speed was 2 cm/min, and a 0.20 x 2.36 mm slit plate was used.

Peak III was separated into 3 anionic protein bands (Fig. 3D, Fig. 4A). Bands were located at Rm 0.26, 0.35, and 0.65. However, when the gel was incubated with GPβNA, only the band at Rm 0.65 gave a positive reaction. When the crude midgut homogenate was separated on this anionic gel system (Fig. 3A) and the gel incubated with GPβNA, a single band of chymotrypsin-like activity at Rm 0.65 was observed (Fig. 3B). Thus, apparently only 1 anionic chymotrypsin-like enzyme is present in the larval midgut.

Electrophoresis of peak IV indicated a homogeneous, distinct anionic protein band at Rm 0.76 (Fig. 3E, Fig. 4B). This band gave a strong positive reaction when the gel was incubated with BAβNA and stained with fast garnet. When the crude midgut homogenate was separated on the anionic gel system and the gel incubated with BAβNA, only a single area of trypsin-like activity was seen (Fig. 3C). This activity zone corresponded in mobility to the purified trypsin at Rm 0.76. Thus, similar to the single anionic chymo-trypsin-like enzyme there apparently is only 1 anionic trypsin-like enzyme present. It was also apparent from the scan (Fig. 4C) of the protein-stained gel that the trypsin-like enzyme was by far the predominant anionic component in the crude larval midgut homogenate.

Molecular Weights of Anionic Enzymes

The anionic chymotrypsin-like fraction (peak III) and the trypsin-like fraction (peak IV) were passed through a calibrated Sephadex G-100 column to obtain an estimate of their molecular weights (Table 4). The enzymatic activity of both fractions eluted in symmetrical peaks with similar elution volumes. The estimated MW's were 20,900 and 21,600 for the trypsin-like and chymotrypsin-like enzyme, respectively.

Table 4.   Molecular weight estimation of the DEAE-Sephacel fractions by passage through a calibrated (2.5 x 31 cm)
Sephadex G-100 column.

Enzyme[a]	Ve (ml)	Kav	MW
Anionic trypsin	109	0.504	20,900
Anionic chymotrypsin	107	0.488	21,600

[a] BApNA and GPpNA were used to monitor the column effluent for tryp-
sin- and chymotrypsin-like activity, respectively.

Fractionation on CM-Sephadex

The protein fraction in the midgut homogenate that did not
bind to the DEAE-Sephacel resin (peak I, tubes 3-6) was applied to
the CM-Sephadex cation exchange resin. The patterns of enzyme
activity eluted from this column are shown in Fig. 5. The
aminopeptidase activity against LpNA was not bound to the CM-Sephadex and eluted in a distinct peak in tubes 3-5. No other tubes with
LpNA activity were found. Three peaks of chymotrypsin-like activity
against BTpNA were separated. The major cationic chymotrypsin
eluted in a sharp peak between tubes 10-20. Two smaller peaks of
chymotrypsin-like activity were found between tubes 30-45. Two peaks
of trypsin-like activity against BApNA were found. The first
cationic trypsin eluted with the two small chymotrypsin peaks
between tubes 32-42. The major cationic trypsin began eluting on
the shoulder of the second small chymotrypsin peak (tube 44) and
was completely eluted by tube 60. From this CM-Sephadex column,
tubes 3-6, 8-18, 30-41, and 46-56 were combined, concentrated via
ultrafiltration, and reassayed.

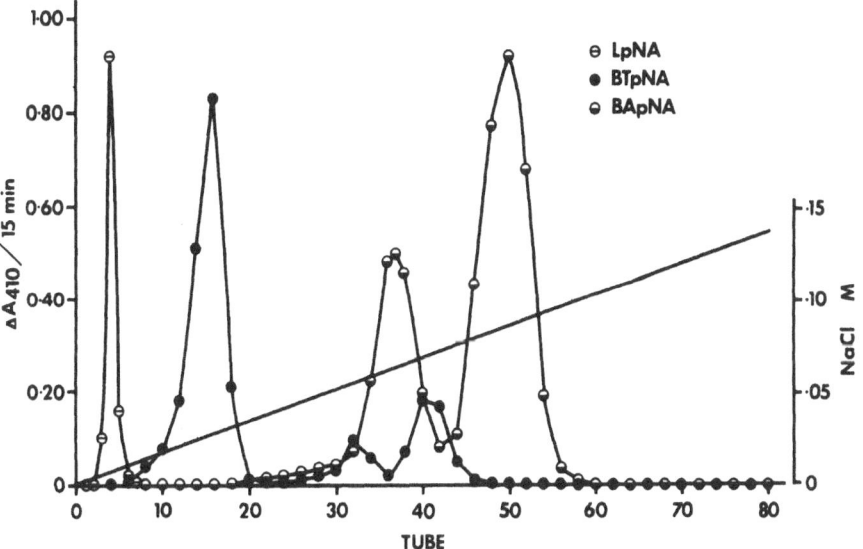

Fig.5 Pattern of digestive proteinases of larvae of Attagenus
       megatoma from a CM-Sephadex column equilibrated in 20 mM
       phosphate pH 7.8 and eluted with a NaCl gradient. Aliquots
       (100 µl) of every second tube were analyzed for aminopeptidase
       (LpNA), chymotrypsin (BTpNA), and trypsin (BApNA) activity.

The concentrated aminopeptidase peak (tubes 3-5) contained only LpNA activity (Table 5). The major cationic chymotrypsin (tubes 8-18) was extremely active against the ester substrates ATEE and BTEE and also hydrolyzed BTpNA and GPpNA quite rapidly. There was a trace amount of TAME activity in this sample, however, no other substrates were hydrolyzed. Tubes 30-41 contained activity against all of the substrates except LpNA. This was the only fraction that contained carboxypeptidase activity against HPLA. The major cationic trypsin peak (tubes 46-56) was very active against all of the trypsin substrates, TAME, BAEE, and BApNA. In the concentrated sample, ATEE and BTEE activity were also present.

The three cationic chymotrypsins from the CM-Sephadex column were designated CC-1, CC-2, and CC-3 in order of elution, CC-1 being the major cationic chymotrypsin. Similarly, the two cationic trypsins were designated CT-1 and CT-2, CT-2 being the major trypsin-like enzyme. Additional CM-Sephadex columns were run to collect CC-2 and CC-3 separately since they were both present in the CT-1 fraction.

Affinity Chromatography

The chymotrypsin activity in CT-1 and CT-2 could be completely removed by passing the samples through small columns of the affinity gel, agarose-γ-aminocaproyl-D-tryptophane. A typical elution

Table 5. Activity of combined tubes separated on the CM-Sephadex column and concentrated via ultrafiltration.

| Substrate | Activity[a] (μmoles/min per mg protein) | | | |
	[3-5]	[8-18]	[30-41]	[46-56]
TAME	0	5.3	192.7	700.7
BAEE	0	0	41.2	142.7
BApNA	0	0	2.8	9.2
ATEE	0	1096.1	666.6	50.7
BTEE	0	478.5	207.2	11.9
BTpNA	0	6.5	1.0	0
GPpNA	0	3.6	0.9	0
LpNA	3.6	0	0	0
HPLA	0	0	10.7	0

[a]The amount of protein in the 2.0 ml reaction volume for each substrate was 0.5 μg for [3-5], 1.3 μg for [8-18], 1.8 μg for [30-41], and 0.6 μg for [46-56].

pattern for CT-2 through this column is shown in Fig. 6. From this column tubes 4-13 were concentrated. No ATEE activity was found in the individual tubes or in the concentrated sample. Since the chymotrypsin was only a minor component in CT-2, there was no change in specific activity towards BApNA. However, when the chymotrypsin was removed from CT-1 with this affinity gel, a 2.5-fold increase in specific activity against BApNA could be obtained.

Attempts to recover the chymotrypsin activity from this affinity gel have met with only limited success. We have eluted with up to 1.0 M NaCl in the phosphate buffer at pH 7.8 and have lowered the pH to 4.0. ATEE activity has been recovered but not in a sharply eluted band.

Electrophoretic Analysis of the Cationic Enzymes

Electrophoretic comparisons were made of the purified cationic enzymes and the crude midgut homogenate by using the cationic gel system. When the gels with crude homogenate were stained for

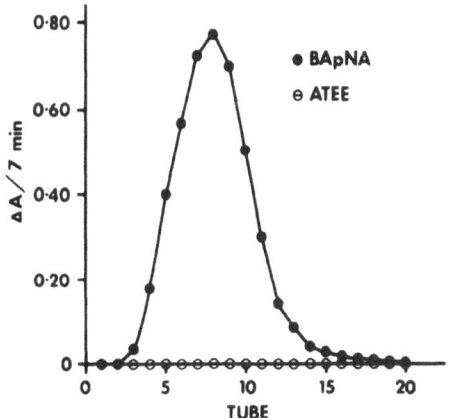

Fig.6 Elution pattern of CT-2 through a 0.9 x 3.5 cm column of agarose-γ-aminocaproyl-D-tryptophane equilibrated in 20 mM phosphate pH 7.8. Twenty 1.0 ml fractions were collected and 100 μl aliquots were analyzed for trypsin (BApNA) and chymotrypsin (ATEE) activity.

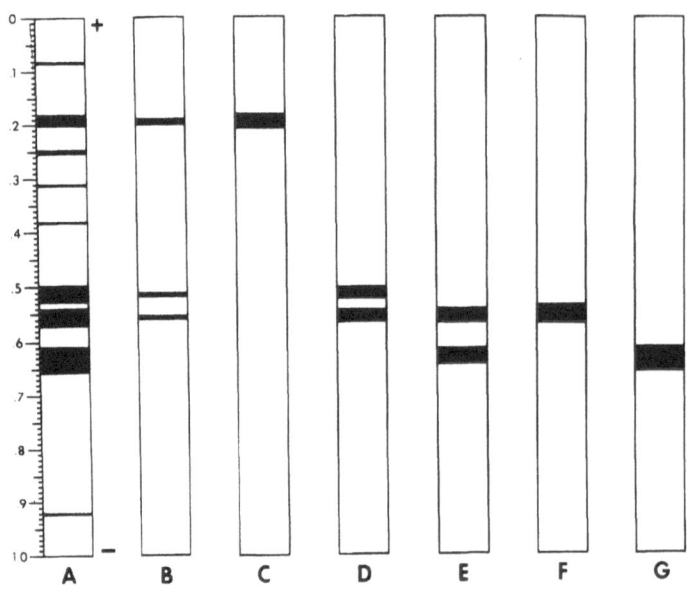

Fig. 7   Acrylamide gel zymograms of combined fractions from the
          CM-Sephadex column using the cationic gel system. A -
          crude midgut homogenate (80 μg protein) stained for pro-
          tein; B - crude midgut homogenate (80 μg protein) incubated
          with GPβNA and stained with fast garnet; C - CC-1 (25 μg
          protein) stained for protein; D - Mixed fraction (25 μg
          protein) of CT-1, CC-2, and CC-3 stained for protein; E -
          crude midgut homogenate (80 μg protein) incubated with
          BAβNA and stained with fast garnet; F - CT-1 (25 μg
          protein) stained for protein; and G - CT-2 (25 μg protein)
          stained for protein. Migration was toward the cathode and
          mobilities expressed relative to that of methyl green.

protein, four main bands could be seen at Rm 0.19, 0.51, 0.55, and
0.66 (Fig. 7A). An additional five faint bands could be seen and
were detected in the gel scan. When the gel of the crude homogenate was incubated with GPβNA and stained with fast garnet, three
areas of chymotrypsin-like activity were found (Fig. 7B). A large
band of activity was observed at Rm 0.19 and two faint bands of
chymotrypsin activity were observed close together in the area
between Rm 0.50-0.55. The large band corresponded to the slowest
moving protein band in the crude homogenate and was also identical
to the single band obtained when CC-1 was analyzed (Fig. 7C). The
weaker bands corresponded to the enzymes CC-2 and CC-3.

When the gels with crude homogenate were incubated with BAβNA and stained with fast garnet, trypsin-like activity was observed in bands at Rm 0.54 and 0.63 (Fig. 7E). These corresponded to the fastest moving major protein bands in the crude homogenate and also the purified enzymes CT-1 (Fig. 7F, Rm 0.55) and CT-2 (Fig. 7G, Rm 0.63). CT-1 and CT-2 appeared to be homogeneous in this gel system.

Molecular Weights of the Cationic Enzymes

The partially purified enzymes from the CM-Sephadex column were passed through a calibrated Sephadex G-100 column. The trypsins and chymotrypsins all had similar elution volumes (Table 6). The estimated molecular weights were similar and ranged from 18,800 for CC-2 to 22,600 for the two trypsins. The aminopeptidase eluted much earlier and had an estimated MW of about 126,000.

Comparisons of the Trypsin-like Enzymes

Several properties of the anionic trypsin and the two cationic trypsins from A. megatoma were compared. The pH activity curves for the hydrolysis of BApNA by the three enzymes are shown in Fig. 8A. The anionic trypsin was most active between pH 8.0 and 9.0 with maximum activity at pH 8.5. CT-1 and CT-2 had a slightly broader range of maximum activity, about pH 8.0 to 10.0. Below pH 6.5 and above pH 10.0 the activity of all three enzymes dropped rapidly.

Table 6. Molecular weight estimation of the CM-Sephadex fractions by passage through a calibrated Sephadex G-100 column.

Enzyme[a]	Ve (ml)	Kav	MW
CC-1	114	0.514	21,100
CC-2	117	.540	18,800
CC-3	114	.510	21,200
CT-1	112	.500	22,600
CT-2	112	.500	22,600
AP[b]	68	.051	126,000

[a] BApNA and GPpNA, and LpNA were used to monitor the column effluent for trypsin-, chymotrypsin-, and aminopeptidase-like activity, respectively.

[b] Aminopeptidase.

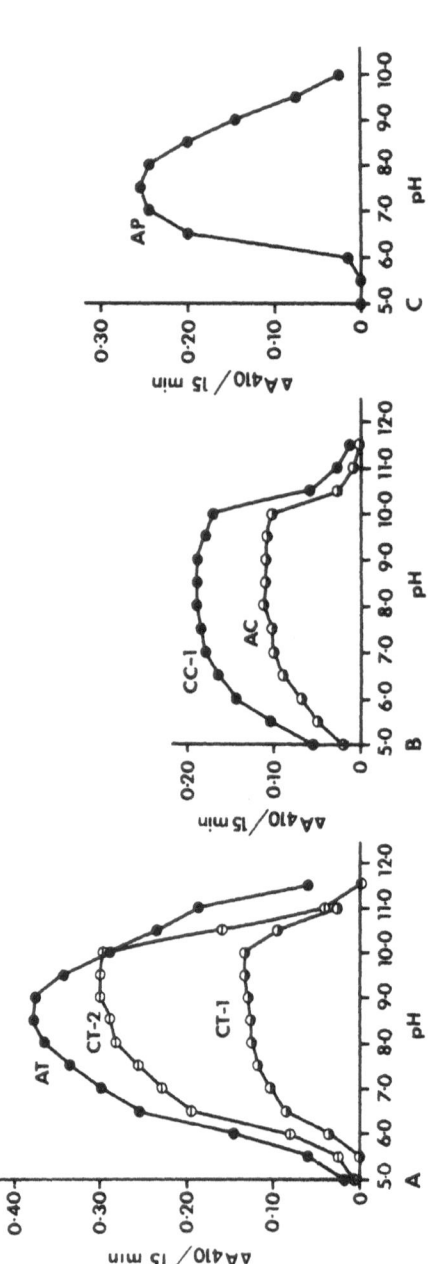

Fig. 8 The effect of pH on the hydrolysis of: A – BApNA by the trypsin-like enzymes; B – BTpNA by the chymotrypsin-like enzymes; and C – LpNA by the aminopeptidase fraction. The 0.1 M buffer systems used were: citrate – 5.0, 5.5, and 6.0; phosphate – 6.5 and 7.0; tris-chloride – 7.5 and 8.0; 2-amino-2-methyl-1,3-propanediol – 8.5, 9.0, 9.5, and 10.0; and carbonate – 10.5, 11.0, and 11.5.

Table 7. The effect of selected inhibitors on the hydrolysis of BApNA by the trypsin-like enzymes from larvae of A. megatoma.

Inhibitor	Final concentration[a]	Mean inhibition (%)		
		Anionic	CT-1	CT-2
PMSF	1 mM	91	97	95
TLCK	0.1 mM	99	91	98
TPCK	0.1 mM	0	5	6
SBTI	1 μg/ml	37	95	96
OVTI	1 μg/ml	33	54	13
EDTA	1 mM	21	7[b]	38[b]

[a] In 2.0 ml reaction volume. Enzyme was first preincubated with inhibitor in a total volume of 200 μl for 60 min at 30°C prior to addition of substrate.

[b] Activated by this value over control.

Table 8. Kinetic constants for the hydrolysis of BApNA by the three trypsin-like enzymes from larvae of A. megatoma.

Enzyme	Km^a moles/L	$Vmax^a$ moles/L-min	$kcat^b$ min^{-1}
Anionic trypsin	2.1×10^{-4} (2.0-2.3)	2.8×10^{-6} (2.7-2.9)	96
CT-1	3.3×10^{-3} (1.6-10.0)	8.3×10^{-6} (4.5-50.0)	280
CT-2	1.7×10^{-3} (1.2-2.5)	8.3×10^{-6} (6.6-11.1)	170

[a] Means from duplicate analyses at each of eight substrate concentrations. Values in parentheses are based on calculations of intercept ± standard error of estimate.

[b] Moles substrate hydrolyzed per mole enzyme per minute.

PMSF, a serine proteinase inhibitor, and TLCK, an active-site-directed inhibitor of trypsins, completely blocked the activity of all three trypsin-like enzymes (Table 7). TPCK, the chymotrypsin inhibitor, had little effect. SBTI strongly inhibited the hydrolysis of BApNA by CT-1 and CT-2 but was not nearly as effective against the anionic trypsin. OVTI was a less effective inhibitor than SBTI. The metal chelator, EDTA, had no inhibitory effect but appeared to slightly activate the enzymes.

The kinetic constants, K_m and V_{max}, for the hydrolysis of BApNA by the three trypsin-like enzymes from $\underline{A}$. $\underline{megatoma}$ were compared (Table 8). The anionic trypsin had a K_m of 2.1×10^{-4}M for BApNA. This was about 10-fold lower than the K_m's of the two cationic trypsins. The V_{max} values of CT-1 and CT-2 were identical (8.3×10^{-6}M - min^{-1}) and about 3-fold higher than that of the anionic trypsin. Similarly, the catalytic constants of the cationic enzymes were higher than that of the anionic enzyme.

The two cationic trypsins, CT-1 and CT-2, hydrolyzed the insoluble proteinase substrate, hide powder azure (HPA), quite rapidly (Table 9). The specific activity of CT-1 (7.6 units/mg protein) and CT-2 (5.6 units/mg protein) was greater than twice that of bovine trypsin. Little color release occurred when the anionic trypsin was incubated with HPA (0.8 units/mg protein) indicating that HPA was not a good substrate for this enzyme.

Table 9. Hydrolysis of hide powder azure (HPA) by the trypsin-like enzymes from larvae of $\underline{A}$. $\underline{megatoma}$ and comparison with bovine trypsin.

Enzyme[a]	ΔA_{650}/35 min[b]	Units/mg protein
Anionic trypsin (1.6 µg)	0.05 + .001	0.8
CT-1 (1.7 µg)	0.45 + .01	7.6
CT-2 (2.9 µg)	0.57 + .01	5.6
Bovine trypsin (1.6 µg)	0.14 + .003	2.5

[a] Number in parentheses indicates amount of protein in the 3.0 ml reaction volume.

[b] Means (+ S.D.) from triplicate determinations.

Table 10. Amino acid composition of the anionic trypsin-like
           enzyme and the major cationic trypsin (CT-2) from
           larvae of A. megatoma and comparison with bovine
           trypsin.

	Residues per mole		
Amino acid	PMSF-Anionic trypsin	PMSF-CT-2	Bovine trypsin[a]
Lys	8	12	14
His	4	6	3
Arg	8	6	2
Asp	12	12	22
Thr	16	12	10
Ser	28	18	33
Glu	12	12	14
Pro	12	12	9
Half cys	8	9	12
Gly	24	24	25
Ala	20	12	14
Val	24	27	17
Met	4	6	2
Ile	8	12	15
Leu	4	12	14
Tyr	8	12	10
Phe	4	6	3
Trp	ND[b]	ND[b]	4
Total residues	204	210	223
MW	24,400	25,534	24,000

[a] From Walsh and Neurath (1964).

[b] Not determined.

Finally, the amino-acid compositions of the anionic trypsin and CT-2 were determined and compared with that of bovine trypsin (Table 10). The general compositions of the three enzymes were similar. However, the ratio of dicarboxylic amino acids (Asp + Glu) to basic amino acids (Lys + Arg + His) was about 2:1 for bovine trypsin compared with about 1:1 for the two trypsins from A. megatoma. Although amide N was not determined, the cationic trypsin (CT-2) from A. megatoma did have a higher number of basic amino acid residues than did the anionic larval trypsin. This may partly explain its cationic nature.

Properties of the Chymotrypsins

The pH activity curves for the anionic chymotrypsin and CC-1 against BTpNA were similar (Fig. 8B). Maximum activity for both enzymes occurred in a broad, mildly alkaline range from about pH 7.5 to pH 10.0. Below pH 6.0 and above pH 10.0 the activity was significantly reduced.

The BTpNA activity of the anionic and cationic chymotrypsins of A. megatoma was completely inhibited by PMSF and by TPCK, a site-directed chymotrypsin inhibitor (Table 11). TLCK activated the mixture of CC-2 and CC-3 apparently by preventing tryptic (CT-1) degradation of the chymotrypsin during the preincubation period. SBTI caused partial inhibition of all of the enzymes. OVTI had little effect on the anionic chymotrypsin but partially inhibited CC-1. The chelator EDTA had little effect.

The kinetic constants for the hydrolysis of BTpNA by the anionic chymotrypsin and CC-1 were compared and found to be very similar (Fig. 9A). The K_m values for this substrate were 5.5 x 10^{-4}M and 3.2 x 10^{-4}M, respectively. The V_{max} values were 1.7 x 10^{-6}M -min^{-1} for the anionic enzyme compared to 1.4 x 10^{-6}M -min^{-1} for the cationic enzyme. The kcat values for the 2 enzymes were calculated to be 150 min^{-1} and 93 min^{-1}, respectively.

Properties of the Aminopeptidase

The effect of pH on the hydrolysis of LpNA by the aminopeptidase from A. megatoma is shown in Fig. 8C. This enzyme had a much narrower activity range than the larval trypsins or chymotrypsins. The optimum activity was at pH 7.5.

When examined electrophoretically, the aminopeptidase moved in the anionic gel system with a mobility of 0.07 (Fig. 3F). The protein did not stain very readily with either the aniline blue black or coomassie blue. However, two closely spaced protein bands

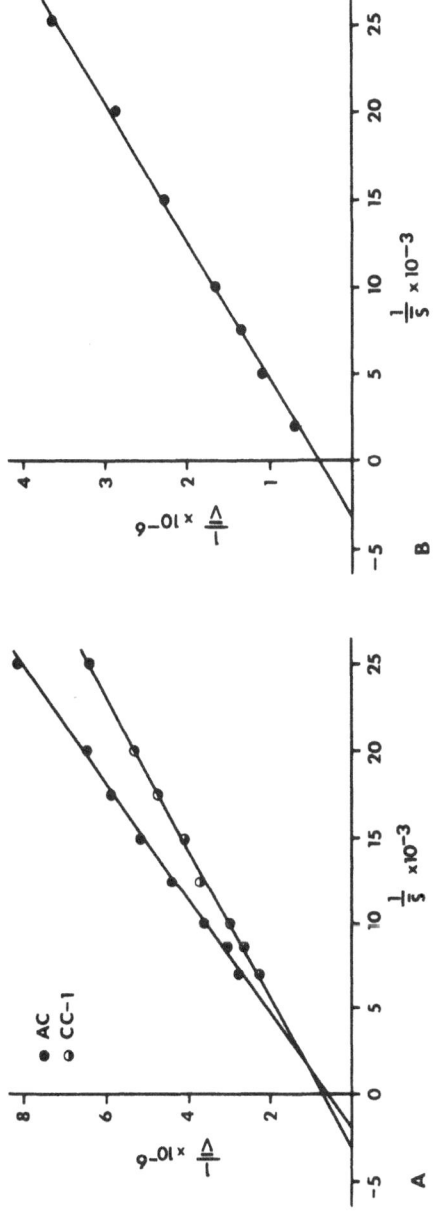

Fig. 9 Lineweaver–Burke plots of average velocity ($\bar{v}$, moles liter^{-1} min^{-1}) against average substrate concentration ($\bar{S}$, moles liter^{-1}) for: A – BTpNA hydrolysis by the anionic chymotrypsin (AC) and the major cationic chymotrypsin (CC-1); and B – LpNA hydrolysis by the aminopeptidase fraction.

Table 11. The effect of selected inhibitors on the hydrolysis of
           BTpNA by the chymotrypsin-like enzymes from larvae of
           A. megatoma.

| Inhibitor | Final concentration[a] | Mean inhibition (%) | | |
		Anionic	CC-1	CC-2+ CC-3
PMSF	1 mM	99	99	94
TLCK	0.1 mM	22	8	50[b]
TPCK	0.1 mM	100	100	100
SBTI	1 µg/ml	40	23	52
OVTI	1 µg/ml	8[b]	34	ND[c]
EDTA	1 mM	5	7	0

[a]In a 2.0 ml reaction volume. Enzyme was first preincubated with
   inhibitor in a total volume of 200 µl for 60 min at 30°C prior to
   addition of substrate.
[b]Activated by this value over control.
[c]Not determined.

were discernible and together made up the band at 0.07. When the
gel was incubated with LβNA and stained with fast garnet, a strong
positive reaction was obtained in the zone at 0.07 (Fig. 3G).
Because of diffusion of enzyme and product in the gel during the
incubation period, it was not possible to determine if both proteins
in the gel had aminopeptidase activity. However, when the stain
had faded for several weeks in some of the gels, it appeared that
both closely spaced proteins had reacted with the substrate. Thus,
the aminopeptidase fraction is not homogeneous but composed of at
least two enzymes with nearly identical mobilities.

Kinetic parameters for the hydrolysis of LpNA by the aminopeptidase fraction were estimated from double reciprocal plots of
v vs $\bar{S}$ (Fig. 9B). For this substrate a K_m of 3.2×10^{-4}M (range
$2.8-3.7 \times 10^{-4}$M), a V_{max} of 2.4×10^{-6}M-min^{-1} (range 2.2-2.7 x
10^{-6}M -min^{-1}), and a kcat of 3700 min^{-1} were calculated.

DISCUSSION

The results of this study indicated the presence of a mixture
of endopeptidases and exopeptidases in the larval midguts of
A. megatoma. The inhibitor studies indicated that serine proteinases
were present and that trypsin-like and chymotrypsin-like enzymes were
responsible for most of the proteinase activity. There was no
significant EDTA-sensitive proteinase activity present. This
indicates a difference between this species and larvae of the
keratinolytic T. bisselliella in which two EDTA-sensitive

metalloproteinases were demonstrated (Ward, 1975b; c). Seven (4 major and 3 minor) anionic trypsin-like proteinases were also demonstrated in larvae of T. bisselliella (Ward, 1975a). However, these enzymes were not inhibited by SBTI, OVTI, or lima bean trypsin inhibitor and thus are different in their reactivity compared with the trypsin-like enzymes of A. megatoma.

Most digestive proteinases isolated from insects behave as anions at neutral or slightly alkaline pH (Gooding and Huang, 1969; Gooding, 1972; Miller et al., 1974; Ward, 1975a; Kunz, 1978). However, cationic proteinases have also been demonstrated (Yang and Davies, 1971; Ward, 1975a; Giebel et al., 1971). The digestive enzymes of A. megatoma were predominantly cationic, including the three cationic chymotrypsins, two trypsins, and the carboxypeptidase fraction. The designation of the enzymes as anionic or cationic is convenient for purposes of categorization; however, the net charge on various trypsins (and presumably chymotrypsins) is not thought to be an important factor in the overall reactivity or specificity of the enzyme (Walsh and Wilcox, 1970).

Trypsins

Trypsins are serine proteinases that cleave intact protein chains on the carboxyl side of basic L-amino acids such as arginine or lysine (Walsh and Wilcox, 1970). Through the use of ester substrates with analogous structures, trypsin-like digestive activity has been found in essentially all examined insect species. The electrophoretically pure anionic trypsin-like enzyme and CT-1 and CT-2 from A. megatoma resembled mammalian trypsins in general amino acid compositions, molecular weight, and several additional characteristics. All three enzymes of A. megatoma hydrolyzed TAME, BAEE, and BApNA but none of the other substrates. The enzymes were completely inhibited by PMSF and TLCK indicating that both serine and histidine participate at active centers. The trypsins were also inhibited by SBTI (especially CT-2 and CT-2) and to a lesser extent by OVTI but not by the metal chelator. In addition, the estimated MW's (between 20,000 - 25,000) and the pH activity curves of the trypsins of A. megatoma resembled those of trypsin-like enzymes isolated from other insects (Gooding and Rolseth, 1976; Kunz, 1978).

Although there were no significant differences between CT-1 and CT-2 in the properties studied, the anionic trypsin differed from CT-1 and CT-2 in having a sharper pH activity curve, a lower K_m for BApNA, a different response to SBTI, and most significantly, a much lower level of proteinase activity against hide powder azure. The cleavage specificity of the three trypsins from A. megatoma towards intact protein chains has not been determined. Some trypsin-like enzymes from insects have many characteristics (i.e.,

esterolytic activity) of mammalian trypsins but give significantly
different digests when incubated with proteins of known sequence
(Zwilling et al., 1972). Thus, the difference in proteinase activity between the anionic and the two cationic trypsins in A. megatoma indicated a probable difference in the substrate or cleavage
specificity of these enzymes.

Chymotrypsins

Chymotrypsins cleave intact protein chains on the carboxyl
side of aromatic amino acids (Walsh and Wilcox, 1970). Although
these enzymes are less well studied in insects than the trypsins,
several preparations with chymotrypsin-like activity have been
described. Gooding and Huang (1969) isolated an anionic BTEE-
hydrolyzing enzyme with a molecular weight of 20,800 from the
predatory beetle, Pterostichus melanarius Illiger. Two fractions
with chymotryptic-like activity were isolated from Apis mellifera
L. (Giebel et al., 1971). One fraction hydrolyzed ATEE and gave
cleavage peptides with insulin A chain identical to that of α-
chymotrypsin. The second fraction hydrolyzed $GP_\beta NA$ but not ATEE
and gave a different peptide cleavage pattern. Knecht et al. (1974)
isolated a chymotrypsin-like (BTEE) enzyme from Locusta migratoria
(L.). The enzyme had a MW of 17,000 and was inhibited by PMSF and
TPCK. A chymotrypsin-like enzyme with a MW of 35,500 from
Glossinia morsitans morsitans Westwood hydrolyzed BTEE and ATEE
but not BTpNA (Gooding and Rolseth, 1976) and a single anionic
enzyme with ATEE activity was found in T. bisselliella (Ward,
1975a).

The activities of the four chymotrypsin-like enzymes found in
A. megatoma resembled those of mammalian chymotrypsin. All four
enzymes hydrolyzed ATEE, BTEE, BTpNA, and GPpNA and all were completely inhibited by PMSF and TPCK. The anionic chymotrypsin and
CC-1 had similar molecular weights and similar kinetic properties
and pH activity curves against BTpNA; however, the proteinase
activities of these enzymes from A. megatoma have not been studied.

Carboxypeptidases

As a result of tryptic and chymotryptic action on intact
proteins, peptides with carboxyl terminal basic and aromatic amino
acids are generated. Carboxypeptidase A (the enzyme that cleaves
carboxyl terminal aromatic amino acids) and carboxypeptidase B
(the enzyme that cleaves carboxyl terminal basic amino acids) can
be demonstrated in the crude midgut homogenate of A. megatoma.
The carboxypeptidase activity (HPLA) in A. megatoma is cationic in
nature and elutes from the CM-Sephadex in the region of CT-1.
Further studies on the isolation of this enzyme(s) are in progress.

Aminopeptidase

Mammalian aminopeptidases have molecular weights of about 300,000 and are most active between pH 7.0-8.0 (Wachsmuth et al., 1966; Himmelhoch, 1969). The aminopeptidases studied in insects generally have much lower molecular weights (ca. 100,000) but are most active in a similar pH range (Ward, 1975b; Gooding and Rolseth, 1976). The aminopeptidase of A. megatoma had an estimated MW of 126,000 and was most active at pH 7.5.

In the 7.5% anionic gel system the aminopeptidase fraction of A. megatoma had a mobility of 0.07. In T. bisselliella two major zones of aminopeptidase activity were detected at Rm 0.19-0.25 and 0.42-0.50 in a similar gel system (Ward, 1975a; 1975d). Each zone contained several aminopeptidase bands with nearly identical mobilities. Similarly, the mobilities of the two aminopeptidase bands of A. megatoma were nearly identical.

The dependence of aminopeptidase from mammalian systems for metal ions has been demonstrated (Marks et al., 1968; Himmelhoch, 1969). Preliminary studies have shown that the aminopeptidase of A. megatoma is partially inhibited by EDTA and completely inhibited by 1,10-phenanthroline (Baker, unpublished data). This indicates the probable requirement of a metal ion for maximum enzyme activity.

Role in Keratin Digestion

In developing an assay for keratinase, keratin azure, a dye-treated wool, was obtained from Sigma. In preliminary studies, the wool fibers were cut into 1-2 mm lengths and incubated in 50 mM tris pH 8.0 with the purified trypsins from A. megatoma. CT-1 and CT-2 released measurable amounts of dye in a 2 hr incubation at 30°C. The anionic trypsin was inactive. Since there was also a minor amount of dye released following incubation with bovine trypsin in this assay, it is not certain if CT-1 and CT-2 actually hydrolyzed the keratin. There may be contaminating proteins in the keratin azure that would preclude its use as a specific substrate for keratinolytic enzymes.

SUMMARY

The complex of endopeptidases and exopeptidases from the mid-guts of larvae of A. megatoma was partially resolved and co-purified using ion exchange, gel filtration, and affinity chromatographic techniques. Three electrophoretically-pure trypsin-like enzymes (one anionic and two cationic) were isolated. The trypsins were similar in their molecular weights, pH optima, and in their esterolytic and amidase activities, but differed in their response to certain inhibitors and especially in their proteinase activities.

Four chymotrypsin-like enzymes (one anionic and three cationic) were
isolated. They were similar in their molecular weights, esterolytic
and amidase properties, and pH optima. An aminopeptidase fraction
with a molecular weight of about 126,000 and a pH optimum of 7.5 was
also isolated. Electrophoretically the aminopeptidase fraction
consisted of two protein bands with nearly identical mobilities.
Both bands had aminopeptidase activity. Carboxypeptidase activity
was present and was cationic in nature but was not further characterized.

ACKNOWLEDGEMENT

        I would like to thank Stanley M. Woo, Biological Technician,
for excellent technical assistance during these studies, for the
statistical analyses of the kinetic data, and for the photography.
I would also like to thank Dr. Karl J. Kramer, U.S. Grain Marketing
Research Center, Manhattan, Kansas 66502, for the amino acid
analyses of the purified trypsins.

REFERENCES

Andrews, P., 1964, Estimation of the molecular weights of proteins
    by Sephadex gel-filtration, Biochem. J. 91:222.
Baker, J. E., 1974, Influence of nutrients on the utilization of
    woolen fabric as a food for larvae of Attagenus megatoma (F.)
    (Coleoptera: Dermestidae), J. Stored Prod. Res. 10:155.
Baker, J. E., 1975, Protein utilization by larvae of the black
    carpet beetle, Attagenus megatoma, J. Insect Physiol. 21:613.
Baker, J. E., 1976, Properties of midgut proteases in larvae of
    Attagenus megatoma, Insect Biochem. 6:143.
Baker, J. E., 1977a, Growth and development of the black carpet
    beetle on the laboratory diet, Ann. Entomol. Soc. Amer. 70:
    296.
Baker, J. E., 1977b, Substrate specificity in the control of
    digestive enzymes in larvae of the black carpet beetle, J.
    Insect Physiol. 23:749.
Baker, J. E., 1978, Midgut clearance and digestive enzyme levels
    in larvae of Attagenus megatoma following removal from food,
    J. Insect Physiol. 24:133.
Barr, A. J., Goodnight, J. H., Sall, J. P., and Helwig, J. T.,
    1976, A user's guide to SAS 76, SAS Institute, Raleigh, North
    Carolina.
Davies, R. C., Riordan, J. F., Auld, D. S., and Valle, B. L., 1968,
    Kinetics of carboxypeptidase A. I. Hydrolysis of carbobenzoxyl-
    glycyl-L-phenylalanine, benzoylglycyl-L-phenylalanine, and hip-
    puryl-DL-β-phenyllactic acid by metal substituted and acetylated
    carboxypeptidases, Biochemistry 7:1090.
Davis, B. J., 1962, Disc electrophoresis. II. Method and applica-
    tion to human serum proteins, Ann. N.Y. Acad. Sci. 121:404.

Erlanger, B. F., Edel, F., and Cooper, A. G., 1966, The action of chymotrypsin on two new chromogenic substrates, Arch. Biochem. Biophys. 115:206.

Erlanger, B. F., Kokowsky, N., and Cohen, W., 1961, The preparation and properties of two new chromogenic substrates of trypsin, Arch. Biochem. Biophys. 95:271.

Folk, J. E., Piez, K. A., Carroll, W. R., and Gladner, J. A., 1960, Carboxypeptidase B. IV. Purification and characterization of the porcine enzyme, J. Biol. Chem. 235:2272.

Fraenkel, G., and Blewett, M., 1946, The dietetics of the clothes moth, Tineola bisselliella Hum., J. Exp. Biol. 22:156.

Giebel, W., Zwilling, R., and Pfleiderer, G., 1971, The evolution of endopeptidases - XII. The proteolytic enzymes of the honeybee (Apis mellifica L.), Comp. Biochem. Physiol. 38B:197.

Gold, A. M., 1965, Sulfonyl fluorides as inhibitors of esterases - III. Identification of serine as the site of sulfonylation on phenylmethane-sulfonyl α-chymotrypsin, Biochemistry 4:897.

Gooding, R. H., 1972, Digestive processes of haematophagous insects - II. Trypsin from the sheep ked Melophagous ovinus (L.) (Hippoboscidae, Diptera) and its inhibition by mammalian sera, Comp. Biochem. Physiol. 43B:815.

Gooding, R. H., and Huang, C. T., 1969, Trypsin and chymotrypsin from the beetle Pterostichus melanarius, J. Insect Physiol. 15:325.

Gooding, R. H., and Rolseth, B. M., 1976, Digestive processes of haemotophagous insects. XI. Partial purification and some properties of six proteolytic enzymes from the tsetse fly Glossina morsitans morsitans Westwood (Diptera: Glossinidae), Can. J. Zool. 54:1950.

Himmelhoch, S. R., 1969, Leucine aminopeptidase: A zinc metallo-enzyme, Arch. Biochem. Biophys. 134:597.

Hummel, B. C. W., 1959, A modified spectrophotometric determination of chymotrypsin, trypsin, and thrombin, Can. J. Biochem. Physiol. 37:1393.

Knecht, M., Hagenmaier, H. E., and Zebe, E., 1974, The proteases in the gut of the locust, Locusta migratoria, J. Insect Physiol. 20:461.

Kunitz, M., 1947, Crystalline soybean trypsin inhibitor. II. General properties, J. General Physiol. 30:291.

Kunz, P. A., 1978, Resolution and properties of the proteinases in adult Aedes aegypti (L.), Insect Biochem. 8:169.

Lee, H., and Wilson, I. B., 1971, Enzymic parameters: Measurement of V and Km, Biochim. Biophys. Acta 242:519.

Lowry, O. H., Rosebrough, N. J., Farr, A. L., and Randall, R. J., 1951, Protein measurement with the Folin phenol reagent, J. Biol. Chem. 193:265.

Mallis, A., Burton, B. T., and Miller, A. C., 1962, The attraction of salts and other nutrients to the larvae of fabric insects, J. Econ. Entomol. 55:351.

Marks, N., Datta, R. K., and Lajtha, A., 1968, Partial resolution of brain arylamidases and aminopeptidases, J. Biol. Chem. 243:2882.

McClure, W. O., Neurath, H., and Walsh, K. A., 1964, The reaction of carboxypeptidase A with hippuryl-DL-β-phenyllactate, Biochemistry 3:1897.

Miller, J. W., Kramer, K. J., and Law, J. H., 1974, Isolation and partial characterization of the larval midgut trypsin from the tobacco hornworm, Manduca sexta, Johannson (Lepidoptera: Sphingidae), Comp. Biochem. Physiol. 48B:117.

Morihara, K., Oka, T., and Tsuzki, H., 1967, Multiple proteolytic enzymes of Streptomyces fradiae. Production, isolation, and preliminary characterization, Biochim. Biophys. Acta. 139:382.

Ornstein, L., 1962, Disc electrophoresis. I. Background and theory, Ann. N.Y. Acad. Sci. 121:321.

Powning, R., and Irzykiewicz, H., 1962, Studies on the digestive proteinase of clothes moth larvae (Tineola bisselliella) II. Digestion of wool and other substrates by Tineola proteinase and comparison with trypsin, J. Insect Physiol. 8:275.

Racusen, D., 1967, Double-disc electrophoresis of proteins, Nature 213:922.

Schwert, G. W., and Takenaka, Y., 1955, A spectrophotometric determination of trypsin and chymotrypsin, Biochim. Biophys. Acta 16:570.

Shaw, E., 1967, Site-specific reagents for chymotrypsin and trypsin, in: "Methods in Enzymology," Vol. 2, Colowick, S. P., and Kaplan, N. O. ed., Academic Press, New York.

Wachsmuth, E. D., Fritze, I., and Pfleiderer, G., 1966, An aminopeptidase occurring in pig kidney. I. An improved method of preparation, physical and enzymic properties, Biochemistry 5:169.

Walsh, K. A., and Neurath, H., 1964, The primary sequence of bovine trypsinogen. Proc. Natl. Acad. Sci. 52:884.

Walsh, K. A., and Wilcox, P. E., 1970, Serine proteases, in: "Methods of Enzymology," Vol. 19, S. P. Colowick and N. O. Kaplan, ed., Academic Press, New York.

Ward, C. W., 1975a, Resolution of proteases in the keratinolytic larvae of the webbing clothes moth, Aust. J. Biol. Sci. 28:1.

Ward, C. W., 1975b, Properties and specificity of the major metal chelator-sensitive proteinase in the keratinolytic larvae of the webbing clothes moth, Biochim. Biophys. Acta 384:215.

Ward, C. W., 1975c, Properties and specificity of a second metal chelator-sensitive proteinase in the keratinolytic larvae of the webbing clothes moth, Aust. J. Biol. Sci. 28:439.

Ward, C. W., 1975d, Aminopeptidases in webbing clothes moth larvae. Properties and specificities of enzymes of highest electrophoretic mobility, Aust. J. Biol. Sci. 28:447.

Waterhouse, D. F., 1952a, Studies on the digestion of wool by insects. VII. Some features of digestion in three species of dermestid larvae and a comparison with Tineola larvae, Aust. J. Sci. Res. 5B:444.

Waterhouse, D. F., 1952b, Studies on the digestion of wool by insects. VI. The pH and oxidation-reduction potential of the alimentary canal of the clothes moth larvae (Tineola bisselliella (Humm.)), Aust. J. Sci. Res. 5B:178.

Waterhouse, D. F., 1958, Wool digestion and mothproofing, in: "Advances in pest control research," Vol. 2, R. L. Metcalf, ed., Interscience Publishers, Inc., New York.

Yang, Y. J., and Davies, D. M., 1971, Trypsin and chymotrypsin during metamorphosis in Aedes aegypti and properties of the chymotrypsin, J. Insect Physiol. 17:117.

Zwilling, R., Medugorac, I., and Mella, K., 1972, The evolution of endopeptidases - XIV. Non-tryptic cleavage specificity of a BAEE-hydrolyzing enzyme (β-protease) from Tenebrio molitor, Comp. Biochem. Physiol. 43B:419.

INSECT DIETETICS: COMPLEXITIES OF PLANT-INSECT INTERACTIONS

John C. Reese

Department of Entomology and Applied Ecology
University of Delaware
Newark, Delaware 19711

A phytophagous insect which can freely move from one place
to another must locate a plant before attempting to feed upon it.
Stimulated by Professor Gottfried Fraenkel's historic paper
(Fraenkel, 1959), much excellent work has been done on how insects
locate hosts and are repelled by non-host species, often through the
effects of various allelochemics. Still another aspect is how
suitable the plant is to the insect once it has settled down to
feed. The chosen plant must be .capable of supporting growth,
development and reproduction, if it is to be a suitable host. The
feeding insect must ingest food "that not only meets its nutritional requirements, but is also capable of being assimilated and
converted into the energy and structural substances required for
normal activity and development" (Beck, 1972; Beck and Reese, 1976).
This concept, termed insect dietetics by Beck (1972), is thus
considerably broader than classical nutrition which deals with
specific nutrient requirements of a species.

The literature dealing with various aspects of insect dietetics
has expanded tremendously in recent years (see recent books by
Rosenthal and Janzen [1979] and Maxwell and Jennings [1980]). In
this paper I will briefly summarize a few key aspects of insect
dietetics literature and cite some examples from my work.

PHEROMONES

The production of pheromones by phytophagous insects is tied
at least indirectly to host plant suitability for supporting insect
reproduction. The more direct relationships have been clouded by
controversy and by the complexity of the situation (the latter may

317

explain, in part, the former). Hardee (1970), for example, thought
that pheromone production by male boll weevils was dependent upon
the weevils ingesting cotton squares or bolls, and yet Mitlin and
Hedin (1974) were able to demonstrate de novo biosynthesis from
acetate, mevalonate, and glucose.

Hendry et al. (1975) also presented evidence for direct interactions between an insect's host plant and pheromone production.
Specifically, they presented data suggesting that oak leaf roller
moth (Archips semiferanus) females emit different pheromones when
reared on different hosts. Subsequently, Miller et al. (1976)
identified the sex pheromone of the oak leaf roller moth as a blend
of trans-11-tetradecenyl acetate and cis-11-tetradecenyl acetate,
and that the blend (67:33), trans: cis) did not vary significantly
regardless of whether the females had been reared on various oak
species or even on semi-synthetic diet. Hendry (1976) rebutted by
stating that the semi-synthetic diet was still a diet containing
plant consituents (pinto beans) and thus dietary sources of pheromones or pheromone precursors. Hindenlang and Wichmann (1977) then
published a retraction of the earlier work they had done with Hendry,
saying they could not detect tetradecenyl acetates in highly purified
oak leaf extracts. Recently, however, Hendry et al. (1980) have once
again stated that they have evidence for insects obtaining pheromones from their host plants. This time, myrcene (a compound in
Ponderosa pines) was labeled with deuterium. Male pine bark beetles
apparently convert myrcene to two of their pheromones, ipsdienol
and ipsenol, as these compounds show an isotopic abundance of
deuterium very similar to that of the original myrcene.

METABOLISM AND CHRONIC EFFECTS

Important as behavioral aspects are to insect dietetics, if
ovipositional "errors" are not "punished" in some way, this chemical
barrier would probably quickly break down. A species would find
that a larger and larger number of species of plants are suitable.
Thus, the role of physiological inhibitors and the chronic effects
of these substances becomes increasingly apparent. However, there
are cases of larval growth rates not closely following host plant
choice (Smiley, 1978), and so the picture is obviously not as simple
as we might like to imagine.

Examples of resistant varieties retaining their resistance for
long periods of time suggest that such resistance is in fact due
to a number of separate mechanisms. Individuals within an insect
population that can detoxify a chemical resistance factor will have
an adaptive advantage and thus probably increase in proportion to the
rest of the population. Thus, resistance will soon break down. If,
on the other hand, a resistant plant contains factors which affect
both behavior and metabolism, then two mutations, one behavioral

and one metabolic, would have to arise simultaneously before a
resistance biotype could arise (Beck and Schoonhoven, 1980;
Erickson and Feeny, 1974).

SEQUESTRATION

There are a number of cases in which an insect's host plant is
more suitable because the insect is able to sequester either actual
compounds or metabolic products of the plant compounds as defensive
substances. When disturbed, larvae of the sawfly Neodiprion sertifer
discharge a defensive material through their mouths. The material
consists of certain pine resins sequestered in two diverticular
pouches in the foregut (Eisner et al., 1974). Another excellent
example of this type of plant-insect interaction is the sequestration
of cardiac glycosides by monarch butterfly larvae as a method of
defense against bird predation on the adults (Brower, 1969; Roeske
et al., 1976; Rothschild, 1972).

APHID FEEDING DETERRENTS IN SORGHUM

Although there is still much to be done in order to understand
the dietetics of the greenbug, Schizaphis graminum, recent experiments have shed some light on the behavioral aspects (Dreyer et al.,
1980). Although the greenbug has only recently been a problem
on sorghum, it is making its presence felt. Some sorghum lines
are relatively resistant to the greenbug. Resistance appears to
include both non-preference (antixenosis [Kogan and Ortman, 1978])
and antibiosis. Further, the resistant lines appear to be more
tolerant than the susceptible ones are.

Work to determine the nature of barley resistance to the greenbug implicated a role for benzyl alcohol (Juneja et al., 1972; 1975).
However, the presence of benzyl alcohol in the plant has not been
confirmed (Starks, personal communication).

Of the four biotypes now recognized (Starks and Burton, 1977),
biotype C is the most destructive one on sorghum. The feeding
habits vary from biotype to biotype. Thus, one cannot be sure
cells encountered during probing do not play an important role in
the suitability of a particular line as a host. Further, it is not
known for biotype C whether feeding takes place only in the phloem,
or may also take place in the cells.

With this background in mind, a bioassay procedure was developed
to evaluate feeding deterrent effects of sorghum constituents
(Dreyer et al., 1980). In the isolation work the bioassay was used
as a guide to follow the feeding deterrent activity of the extracts
during the course of fractionation. The bioassay consisted of
feeding aphids a synthetic diet to which various compounds or plant

extracts were added at known concentration. The percentage of aphids
that would feed on the diet was used as a measure of the feeding
deterrent properties of the test substances. Two disadvantages of
the bioassay are that it can only be used on water-soluble substances and that it only measures feeding deterrent properties.
Neither growth nor reproductive inhibition is measured.

Little is known about the feeding habits of biotype C greenbugs. Thus, it appeared to be justified to use whole plant extracts
of sorghum, rather than trying to get enough phloem sap to work with.
The resistant line IS-809, and a commercially available line G-449-
GBR, were used in most experiments. The biologically active materials that we isolated were soluble in acetone, methanol, and water.

Fresh plant material was crushed and promptly extracted with
methanol or acetone. Extractions done this way avoid the hydrolysis
of the cyanohydrin glucoside, dhurrin, known to be present in certain
sorghum lines. Sorghum is also known to contain β-glucosidase which
will hydrolyze dhurrin to yield free HCN and p-hydroxybenzaldehyde
(Mao and Anderson, 1967; Kojima et al., 1979). These extraction
procedures should have very quickly inactivated such enzymes as β-
glucosidase. A very important step in the extraction procedure was
the adsorption of the extracts on XAD-2. Aromatic materials tend
to be adsorbed on XAD while sugars and aliphatic compounds are not
(Loomis et al., 1979). Compounds adsorbed on XAD can be recovered
by elution with methanol. Aqueous sorghum extracts inhibited greenbug feeding; aqueous extracts that had been through XAD had no
feeding deterrent activity; the active fraction was recovered by
elution with methanol. Thus, it appeared that the feeding deterrent
materials were aromatic and very polar.

Chromatographic separations (Sephadex LH-20) of the XAD-retained
material failed to give a clear separation of activity, suggesting
that at least two or more biologically active components were present in the extracts. Similar results were obtained with various
other techniques. Working up the column fraction enabled us to
isolate p-hydroxybenzoic acid, p-hydroxybenzaldehyde, dhurrin,
luteolin-7-glucoside, and procyanidin.

Each of these materials was bioassayed over a series of concentrations in order to obtain its ED_{50} (that concentration of
material in the diet which caused a 50% reduction in the measured
response (feeding in this case) when compared to the controls).
Of these materials, dhurrin (ED_{50} = 0.16%), p-hydroxybenzaldehyde
(ED_{50} = 0.13%), and procyanidin (ED_{50} = 0.08%) were particularly
effective (Table 1 and Fig. 1). Luteolin-7-glucoside (no
activity at 1.0%) was judged inactive. Compounds such as tannic
acid and benzylalcohol (implicated, as discussed earlier in
greenbug resistance in barley) were also effective in reducing the
percentage of feeding greenbugs (Table 1).

Table 1. Percentage of greenbugs feeding on diets containing plant allelochemics

Concentration	Dhurrin	Tannic Acid	Benzyl Alcohol	P-hydroxy-Benzaldehyde
0.0	85.7%	82.6%	70.5%	85.7%
0.02		25.1		75.0
0.05		5.7		
0.08	57.2			
0.1	50.1	9.0	40.1	63.0
0.15	48.6			
0.2	37.6	15.5	13.5	31.3
0.5	22.4	2.3	7.5	3.3
1.0	10.9	2.2	6.4	1.2

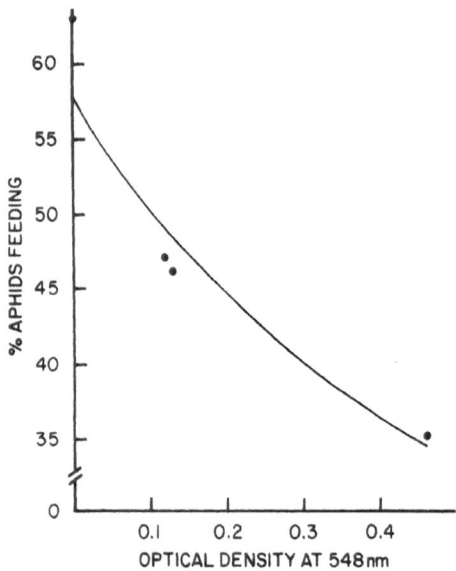

Fig.1   Relationship of greenbug feeding to optical density at 548
         mm (a measure of procyanidin content).

The greenbug shows a relatively high tolerance to the commercially available cyanohydrin glycoside, amygdalin (ED_{50} = 0.50%),
compared to dhurrin, suggesting that the aldehyde hydrolysis product
of the glucoside played the major role in dhurrin activity.
Further, the aglycone of amygdalin, benzaldehyde had an ED_{50} of
0.40% whereas p-hydroxybenzaldehyde the aglycone of dhurrin had
an ED_{50} of 0.13%. Although the extraction procedure was designed
to prevent dhurrin hydrolysis, we have no real evidence to indicate
whether the p-hydroxybenzaldehyde is present in the free form or
not.

Of the compounds isolated, procyanidin may be the major feeding
deterrent in resistant lines of sorghum, at least in more mature
plants. First, dhurrin levels are known to decrease with increasing
maturity of the plant (Hogg and Ablgren, 1943; Woodhead and Bernays,
1977). Second, the procyanidin fractions had relatively high
biological activity in our bioassay. Further, there was a good
correlation between procyanidin level and level of feeding deterrency
in crude extracts of four sorghum lines (Fig. 1).

BOLLWORM FEEDING BEHAVIOR IN COTTON

Insect dietetics not only involves interactions between different plants and an insect, but also between different parts of a
particular plant. In work done on the chemistry of host plant
resistance in cotton (Waiss et al., 1977; Chan et al., 1978a),
two intriguing questions have arisen. The first is that since
early stage lepidopterous larvae tend to be more sensitive to
cotton plant allelochemics than older larvae (Shaver and Parrott,
1970), what part of the plant is fed on during the most sensitive
portion of the larval stage of Heliothis zea? The second question
is, how can we explain the fact that growth inhibition due to cotton
compounds incorporated into artificial diet can be achieved at
concentrations of 1/10 those known to occur in plants (Waiss et
al., 1977; Chan et al., 1978a; Shaver et al., 1977; Shaver and
Lukefahr, 1971)? In an attempt to answer these questions, we investigated the parts of the cotton square eaten by H. zea larvae
during the early stages of larval growth (Reese et al., 1980).

A total of 9,294 cotton square from field- and greenhouse-
grown plants were dissected. The whole plants were also carefully
examined for damage. Both glanded and glandless cottons were
utilized in our experiments, but no real differences could be detected.

More than half of all larvae were found in squares that were
< 6 mm in diameter (measured at the widest point of the calyx)
during the first 8 days of the larval stage. On terminal shoots
from plants grown in the field more than half of all larvae found
were in squares < 4 mm.

A total of 25.6% of all larvae were found on the leaves. Many
of these larvae were in very small leaves that were still partially
rolled up.

Heavily damaged squares (those having > 75% damage to one or
more parts of the square) were almost always < 2 mm in diameter
(Table 2). A large number of the damaged squares were hollowed out.
In the field, they quickly dried out and were shed by the plant.
These very small larvae were not capable of doing much damage to
larger squares. However, large numbers of squares suffered at
least light damage.

For all of the size categories of squares, the anthers and the
filaments were the parts of the square that received ≥ 20% damage
most often (Table 3). These parts of the square were not eaten to
the exclusion of the other parts, but damage to bracts, calyx, and
petals was mainly due to entry holes chewed by the larvae; many
holes did not penetrate all the way to the anthers. A large number

Table 2. Numbers of squares receiving damage by degree of damage
          and size of square.

	S q u a r e S i z e			
	<2mm	2-4mm	5-6mm	7-10mm
Light Damage[a]	228	331	373	366
Medium Damage[b]	74	46	37	20
Heavy Damage[c]	867	74	27	11

[a]Light damage = most heavily damaged part of the square was 1-24%
destroyed.

[b]Medium damage = most heavily damaged part of the square was 25-74%
destroyed.

[c]Heavy damage = most heavily damaged part of the square was 75-100%
destroyed.

Table 3. Numbers of squares receiving $\geq$ 20% damage to various
          parts of the square.

	S q u a r e S i z e			
	<2mm	2-4mm	5-6mm	7-10mm
Bracts	22	3	0	3
Sepals	34	22	8	2
Petals	45	19	11	1
Receptacles	62	20	15	7
Ovules	67	17	19	12
Ovary	71	18	16	11
Stigma	64	20	16	9
Style	65	15	13	8
Anthers	108	52	40	26
Filaments	94	41	26	21
Staminal Column	74	26	13	12
Corolla Tube	35	4	5	5

of holes that penetrated through the lower part of the calyx
terminated in the ovary and lower portions of the anthers; holes
in the upper portion of the sepals tended to lead to damaged anthers.
Shaver et al. (1977) made similar observations with H. virescens;
larvae that fed on portions of the ovary had entered through the
receptacle.

Of various plant parts examined, the anthers were the lowest
in condensed tannin (Chan et al., 1978b) a growth-inhibiting compound isolated from cotton. Also, it took a great deal of anther
tissue to inhibit the growth of H. virescens, compared to various
other tissues (Chan et al., 1978a). Thus, it appears that in
partial answer to the second question, the larvae are probably
not ingesting the same concentration of certain allelochemics as
measured by analysis of whole squares. In answer to the first
question, since most of the larvae are in tiny squares in the terminal of the plant, and are hollowing out tiny squares, concentrating
on the anthers of slightly larger squares, during the most sensitive
part of the larval stage, we should search for an unfavorable
chemical composition in the tiny squares, especially in developing
anther tissue. This could be doubly important due to the apparent
feeding site of the larvae, and to the greater sensitivity of small
larvae to biologically active compounds. Also, the longer exposure (due to slowed growth) to parasitoids, predators and adverse
weather could reduce the larval population significantly.

INTERACTIONS BETWEEN ALLELOCHEMICS AND NUTRIENTS

Many of the deleterious effects of plant allelochemics may be
due, at least in part, to various interactions between these
allelochemics and essential nutrients. It is important not only
to consider the presence of nutrients, but whether or not they are
truly available to the insect. Although quite a number of examples
of such interactions may be cited from the vertebrate literature
(Reese, 1979), little work has been done with insects.

An excellent example of such interactions involves allelochemics that are closely related structural analogs of essential
nutrients. L-canavanine is apparently similar enough to L-arginine
to be incorporated into proteins. However, such proteins do not
seem to function as normal proteins and thus L-canavanine is deleterious to the insect (Rosenthal et al., 1976; Vanderzant and
Chremos, 1971; Isogai et al., 1973a; 1973b; Dahlman and Rosenthal,
1975). A less apparent mechanism must underlie the observation
that increased levels of cholesterol spared the adverse effects of
certain diterpene acids (Elliger et al., 1976).

Protease inhibitors interact with nutrients in that they
inhibit the ability of the insect to digest proteins to amino acids

in order to be passed across the gut wall. The effects various
protease inhibitors have been reviewed (Ryan, 1973; Ryan and Green,
1974). Soybean trypsin inhibitors have various adverse effects on
development and protease activity in Tribolium castaneum (Birk and
Applebaum, 1960), and can cause increased adult mortality in
Sitophilus oryzae (Su et al., 1974).

In a few cases, an interaction between nutrients and non-
nutrients in the sense of allelochemic effects on assimilation or
efficiency of conversion of assimilated material has been demonstrated. For example, gossypol decreased assimilation by Heliothis
zea larvae, but had no effect on utilization by H. virescens larvae
(Shaver et al., 1970). Similarly, sinigrin reduced assimilation
by Papilio polyxenes asterius (Erickson and Feeny, 1974). We
studied the effects of a number of plant allelochemics on growth
and development in the black cutworm (Agrotis ipsilon) (Reese and
Beck, 1976a;b;c). We found that various phenolics can reduce
growth by inhibiting ingestion, assimilation, or efficiency of
conversion of assimilated food, or some combination of these effects.
Thus, a non-nutrient may have striking effects on the nutritional
status of an organism by inhibiting the utilization of essential
nutrients. Clearly, the investigation of the effects of plant
allelochemics on phytophagous insects is a very important part of
insect dietetics.

While working in association with A. C. Waiss, Jr., at the
Western Regional Research Center (USDA) in Berkeley, California,
I examined the effects of several compounds implicated in host
plant resistance to Heliothis zea. Maysin is a flavone glycoside
that has been isolated from corn silks and has been shown to inhibit
H. zea growth (Waiss et al., 1979). The primary mechanism by which
maysin reduces growth in H. zea appears to be through a reduction
in ingestion (Table 4). Although the ED_{50} of maysin is about 0.13%,
it took 0.25% maysin in the diet to reduce H. zea growth to 50% of
the controls in the nutritional index experiment. Apparently
starting the nutritional index experiment with 1 day old larvae
rather than newly hatched ones was sufficient to lower the sensitivity to maysin a good deal. Similar relationships of larval age
to toxicity of other compounds have been observed (Shaver and
Parrott, 1970).

Pinitol is a compound isolated from soybeans that can inhibit
the growth of H. zea (Dreyer et al., 1979). Pinitol was tested at
the ED_{50} level of 0.7% but once again was relatively ineffective
when 1 day old larvae were used. Newly hatched larvae were, however,
quite sensitive (Table 4). Although the efficiency of conversion
was very slightly reduced, the primary factor was certainly a
reduction in ingestion.

Table 4. Effects of maysin and pinitol on nutritional indices and related parameters in _Heliothis zea_ larvae (expressed as percent of controls).

	0.15% Maysin[a]	0.25% Maysin[a]	0.7% Pinitol[a]	0.7% Pinitol[b]
Dry Wt. Eaten	89.7%	42.1%	92.0%	44.5%**
Initial Fresh Wt. of Larvae	100.0	100.0	100.0	100.0
% Dry Matter of Larvae	102.3	94.9*	103.1	84.2**
Dry Wt. Gain	91.3	50.0**	93.7	41.8**
Dry Wt. of Feces	91.1	40.2**	89.1	38.1**
% Dry Matter of Feces	104.4	111.7	105.4	131.2**
AD[c]	97.3	109.7	105.2	131.7**
ECD[d]	106.4	110.8	96.5	72.0**
ECI[e]	103.1	116.6	101.2	90.7

[a] Newly hatched larvae were placed on control diet for 24 hrs, then transferred to fresh experimental or control diet, and allowed to feed for 10 days.

[b] Newly hatched larvae were placed directly on experimental or control diet, and allowed to feed for 11 days.

[c] AD = assimilation.

[d] ECD = efficiency of conversion of assimilated material.

[e] ECI = efficiency of conversion of ingested food.

* Significant at $P<0.05$ level.

** Significant at $P<0.01$ level.

Tannins have been known for many years to be capable of forming complexes with other compounds, particularly proteins. Feeny (1968) found that oak leaf tannin reduced the growth of winter moth larvae (_Operophtera brumata_) and subsequently showed that oak

leaf tannin forms a hydrolysis-resistant complex with casein
(Feeny, 1969). Since these findings it has been widely assumed
that growth-inhibiting effects of tannins in insects are due to the
formation in the gut tract of tannin-dietary protein complexes that
are not readily digested. Additionally, it is supposed that many
digestive enzymes may be complexed, further reducing the rate of
assimilation across the gut wall. Several recent papers have, however, shown that the picture is not this clear for at least some
species of insects. For example, Fox and Macauley (1977) working
with 13 Eucalyptus spp. with a wide range of concentrations of
condensed tannins, found that tannins did not appreciably reduce the
availability of nitrogen in Paropsis atomaria larvae. In a series
of experiments using in some cases rather high levels of tannin,
Bernays (1978) found no evidence for a reduction in digestion due to
dietary tannin in several grasshopper species. In Locusta migratoria
hydrolysable tannin had a deleterious effect, however, by damaging
the midgut epithelium. Chan et al. (1978b) have isolated a condensed
tannin from cotton with a molecular weight of about 4850. This tannin inhibits the growth of Heliothis virescens, but experiments with
condensed tannin-casein or polyamide complexes showed no reduction in
antibiotic activity (Chan et al., 1978b), suggesting that its growth-
inhibiting ability involves more than the ability to complex with gut
tract proteins. Recent experiments by Lawson and Klug (personal
communication) shed some additional light on the situation. They
found that when Anisota senatoria feeds on Quercus bicolor it has a
nitrogen utilization efficiency of 39.4%. However, when Pieris rapae
feeds on Barbarea vulgaris, its nitrogen utilization efficiency is
36.17% (Slansky and Feeny, 1977). Presumably the tannin level of
Quercus bicolor (an oak species) is far higher than Barbarea
vulgaris (yellow racket), and yet the nitrogen utilization is
comparable for the herbivores feeding on these plants. It seems
unlikely therefore, that the oak tannins are blocking assimilation
in A. senatoria.

In H. zea nutritional index experiments conducted at the
Western Regional Research Center, I examined the effects of tannins.
Using the cotton condensed tannin (Chan et al., 1978b), I found that
H. zea growth can be reduced by cotton condensed tannin (Table 5).
Once again, growth does not appear to be reduced by any appreciable
reduction in assimilation; the primary factor is a reduction in the
amount the larvae eat. This does not necessarily correspond to non-
preference, since this was not a choice-test situation. It simply
shows that for some reason the larvae consume less of the tannin-
containing diet. Schuster, on the other hand, found that H.
virescens larvae consumed more high tannin cotton leaf tissue than
normal tannin tissue (personal communication).

In the case of the winter moth, tannins may very well complex
with dietary proteins, thus reducing assimilation. The assumption

Table 5.   Effects of cotton condensed tannin on nutritional indices
           and related parameters in <u>Heliothis</u> <u>zea</u> larvae (expressed
           as percent of controls).   Newly hatched larvae were
           placed on control diet for 24 hrs, then transferred to
           fresh experimental or control diet and allowed to feed for
           10 days.

	Condensed Tannin Concentration			
	0.0%	0.1%	0.15%	0.2%
Dry Wt. Eaten	100.0%	101.5%	52.1%[**]	33.6[**]
Initial Fresh Wt. of Larvae	100.0	111.4	100.0	100.0
% Dry Matter of Larvae	100.0	100.8	83.6[**]	96.9
Dry Wt. Gain	100.0	105.5	46.7[**]	29.2[**]
Dry Wt. of Feces	100.0	94.3	46.2[**]	34.8[**]
% Dry Matter of Feces	100.0	112.0	109.3	152.2[**]
AD	100.0	111.1	136.2[**]	92.6
ECD	100.0	96.0	65.4[**]	96.0
ECI	100.0	104.3	83.3[*]	83.1[*]

that this is how tannins inhibit growth in other species needs to be
carefully re-examined.   In the experiments cited above and in my
work with <u>H. zea</u>, mechanisms other than a reduction in assimilation
appear to be operating.

MOISTURE LEVEL

     Although stored products insects may be capable of strict water
conservation, insects which live on growing plant tissue require
relatively high moisture levels (Waldbauer, 1962; 1964; 1968).   Too
much water, on the other hand, may dilute nutrients to a deleterious
degree.   House (1965) found that <u>Celerio</u> <u>euphorbiae</u> larvae tend to
compensate for dilution by eating more.   Dilution may also cause

an increase in efficiency of conversion, as with <u>Prodenia</u> <u>eridania</u> (Soo Hoo and Fraenkel, 1966a; 1966b). Using various lepidopterous larvae, Feeny (1975) also found that the efficiency of conversion of assimilated food decreased with decreasing moisture levels of various host plant species. In addition, moisture level and efficiency of conversion of both the nitrogenous and the caloric contents of food were directly related in <u>Hyalophora</u> <u>cecropia</u> larvae (Scriber, 1977). Likewise, black cutworm larvae exhibited a decreasing efficiency of conversion with decreasing moisture levels in an artificial diet (Reese and Beck, 1978). However, the optimal moisture level in terms of dry weight gained by the larvae was quite different from that for efficiency of conversion, due to the interaction between efficiency and the actual amount of dry material the larvae ingested.

CONCLUSION

Many aspects of insect dietetics are key factors in the suitability of a given plant species as a host for a particular insect. The insect may obtain from its host compounds which will be crucial to successful reproduction. If the plant contains the essential nutrients for the insect but the utilization of these nutrients is blocked by allelochemics, growth may be slowed, exposing the insect to predators and parasitoids for a longer period of time. Defensive substances may be sequestered by the insects, thus enhancing the chances of the insect being able to carry on "normal activity and development". If the insect will not feed on the plant, plant nutrients cannot be utilized. If deleterious compounds are unevenly distributed in the plant, the insect can perhaps avoid their parts during the early, sensitive period of growth. Some of our assumptions on how allelochemics and nutrients interact need to be looked at closely. The concept of insect dietetics is clearly not a simple one, but may provide a useful framework for furthering our understanding of the chemistry of plant-insect interactions.

ACKNOWLEDGEMENTS

I thank Dr. J. G. Rodriguez and Dr. T. K. Wood for their encouragement and suggestions. Published with the approval of the Director of the Delaware Agricultural Experiment Station as Miscellaneous Paper No. 897, Contribution No. 492 of the Department of Entomology and Applied Ecology, University of Delaware, Newark, Delaware.

REFERENCES CITED

Beck, S. D., 1972, Nutrition, adaptation and environment, <u>in</u>: "Insect and Mite Nutrition", J. G. Rodriguez, ed., North Holland Publishing Co., Amsterdam, pp. 1-6.
Beck, S. D., and Reese, J. C., 1976, Insect-plant interactions: Nutrition and metabolism, <u>Recent</u> <u>Adv</u>. <u>Phytochem</u>. 10:41-92.

Beck, S. D., and Schoonhoven, L. M., 1980, Insect behavior and
    plant resistance, in: "Breeding Plants Resistant to Insects",
    F. G. Maxwell and P. R. Jennings, ed., John Wiley and Sons,
    New York, pp. 115-135.
Bernays, E. A., 1978, Tannins: An alternative viewpoint, Entomol.
    Exp. Appl. 24:244-253.
Birk, Y., and Applebaum, W., 1960, Effect of soybean trypsin inhi-
    bitors on the development and midgut proteolytic activity of
    Tribolium castaneum larvae, Enzymologia 22:318-326.
Brower, L., 1969, Ecological chemistry, Sci. Amer. 220:22-29.
Chan, B. G., Waiss, A. C., Jr., Binder, R. G., and Elliger, C. A.,
    1978a, Inhibition of lepidopterous larval growth by cotton
    constituents, Entomol. Exp. Appl. 24:294-300.
Chan, B. G., Waiss, A. C., Jr., and Lukefahr, M., 1978b, Condensed
    tannin, an antibiotic chemical from Gossypium hirsutum, J.
    Insect Physiol. 24:113-118.
Dahlman, D. L., and Rosenthal, G. A., 1975, Non-protein amino acid-
    insect interactions. 1. Growth effects and symptomology of
    L-canavanine consumption by tobacco hornworm, Manduca sexta
    (L.), Comp. Biochem. Physiol. 51A:33-36.
Dreyer, D. L., Chan, R. G., Waiss, A. C., Jr., Hartwig, E. E., and
    Beland, G. L., 1979, Pinitol, a larval growth inhibitor for
    Heliothis zea in soybeans, Experientia 35:1182-1183.
Dreyer, D. L., Reese, J. C., and Jones, K. C., 1980, Chemical basis
    of insect-plant interactions. II. Aphid feeding deterrents
    in sorghum; Bioassay, isolation and characterization, Submitted
    for publication.
Eisner, T., Johnesse, J. S., Carrel, J., Hendry, L. B., and Meinwald,
    J., 1974, Defensive use by an insect of a plant resin, Science
    184:996-999.
Elliger, C. A., Zinkel, D. F., Chan, B. G., and Waiss, A. C., Jr.,
    1976, Diterpene acids as larval growth inhibitors, Experientia
    32:1365-1365.
Erickson, J. M., and Feeny, P. P., 1974, Sinigrin: A chemical
    barrier to the black swallowtail butterfly, Ecology 55:103-111.
Feeny, P. P., 1968, Effect of oak leaf tannins on larval growth of
    the winter moth, Operophtera brumamta, J. Insect Physiol. 14:
    805-817.
Feeny, P. P., 1969, Inhibitory effect of oak leaf tannins of the
    hydrolysis of proteins by trypsin, Phytochem. 8:2119-2126.
Feeny, P. P., 1975, Biochemical coevolution between plants and
    their insect herbivores, in: "Coevolution of Animals and
    Plants", L. E. Gilbert and P. H. Raven, ed., University
    of Texas Press, Austin, pp. 3-19.
Fox, L. R., and Macauley, B. J., 1977, Insect grazing on Eucalyptus
    in response to variation in leaf tannins and nitrogen, Oecologia
    29:146-162.
Fraenkel, G. S., 1959, The raison d 'etre of secondary plant sub-
    stances, Science 129:1466-1470.

Hardee, D. D., 1970, Pheromone production by male boll weevils as
    affected by food and host factors, Contr. Boyce Thompson
    Inst. 24:315-322.
Hendry, L. B., 1976, Insect pheromones: Diet related? Science 192:
    142-145.
Hendry, L. B., Piatek, B., Browne, L. E., Wood, D. L., Byers, J. A.,
    Fish, R. H., and Hicks, R. A., 1980, In vivo conversion of a
    labelled host plant chemical to pheromones of the black beetle
    Ips paraconfusus, Nature284:485.
Hendry, L. B., Wichmann, J. K., Hindenlang, D. M., Mumma, R. O., and
    Anderson, M. E., 1975, Evidence of the origin of insect sex
    pheromones: Presence in food plants, Science 188:59-67.
Hindenlang, D. M., and Wichmann, J. K., 1977, Re-examination of
    tetradecenyl acetates in oak leaf roller sex pheromone and
    in plants, Science 195:86-89.
Hogg, P. G., and Ablgren, H. L., 1943, Environmental breeding and
    inheritance studies of hydrocyanic acid in Sorghum vulgare
    var. sudanense, J. Agric. Res. 67:195-210.
House, H. L., 1965, Effects of low levels of the nutrient content
    of a food and of a nutrient imbalance on the feeding and the
    nutrition of a phytophagous larva, Celerio euphoribae (Linnaeus)
    (Lepidoptera:Sphingidae), Can. Entomol. 97:62-68.
Isogai, A., Chang, C., Murakoshi, S., and Suzuki, A., 1973a, Screen-
    ing search for biologically active substances to insects in
    crude drug plants, J. Agr. Chem. Soc. Japan 47:443-447.
Isogai, A., Murakoshi, S., Suzuki, A., and Tamura, S., 1973b,
    Isolation from "Astragali Radix" of L-canavanine as an inhibi-
    tory substance to metamorphosis of silkworm, Bombyx mori L.,
    J. Agr. Chem. Soc. Japan 47:449-453.
Juneja, P. S., Gholson, R. K., Burton, R. L., and Starks, K. J.,
    1972, The chemical basis for greenbug resistance in small
    grains. I. Benzyl alcohol as a possible resistance factor,
    Ann. Entomol. Soc. Amer. 65:961-964.
Juneja, P. S., Pearcy, S. C., Gholson, R. K., Burton, R. L., and
    Starks, K. J., 1975, Chemical basis for greenbug resistance
    in small grains. II. Identification of the major neutral
    metabolite of benzyl alcohol in barley, Plant Physiol. 56:385-
    389.
Kogan, M., and Ortman, E. F., 1978, Antixenosis - A new term proposed
    to define Painter's "nonpreference" modality of resistance,
    Bull. Entomol. Soc. Amer. 24:175-176.
Kojima, M., Poulton, M. E., Thayer, S. S., and Conn, E. E., 1979,
    Tissue distributions of dhurrin and of enzymes involved in its
    metabolism in leaves of Sorghum bicolor, Plant Physiol. 63:
    1022-1028.
Loomis, W. D., Lile, J. D., Sandstom, R. P., Burbott, A. J., 1979,
    Absorbent polystyrene as an aid in plant enzyme isolation,
    Phytochem. 18-1049-1054.

Mao, C. H., and Anderson, L., 1967, Cyanogensis in Sorghum vulgare -
    II. Partial purification and characterization of two β-gluco-
    sidases from Sorghum tissues, Phytochem. 6:473-483.
Maxwell, F. G., and Jennings, P. R., ed., 1980, "Breeding Plants
    Resistant to Insects", John Wiley and Sons, New York, 683 pp.
Miller, J. R., Baker, T. C., Carde, R. T., and Roelofs, W. L., 1976,
    Reinvestigation of oak leaf roller sex pheromone components and
    the hypothesis that they vary with diet, Science 192:140-143.
Mitlin, N., and Hedin, P. A., 1974, Biosynthesis of grandlure, the
    pheromone of the bollweevil, Anthonomus grandis, from acetate,
    mevalonate, and glucose, J. Insect Physiol. 20:1825-1831.
Reese, J. C., 1979, Interactions of allelochemicals with nutrients
    in herbivore food, in: "Herbivores: Their Interaction with
    Secondary Plant Metabolites", G. A. Rosenthal, and D. H.
    Janzen, ed., Academic Press, New York, pp. 309-330.
Reese, J. C., and Beck, S. D., 1976a, Effects of allelochemics on
    the black cutworm, Agrotis ipsilon: Effects of p-benzoquinone
    hydroquinone, and duroquinone on larval growth, development,
    and utilization of food, Ann. Entomol. Soc. Amer. 69:59-67.
Reese, J. C., Beck, S. D., 1976b, Effects of allelochemics on the
    black cutworm, Agrotis ipsilon: Effects of catechol, L-dopa,
    dopamine, and chlorogenic acid on larval growth, development,
    and utilization of food, Ann. Entomol. Soc. Amer. 69:68-72.
Reese, J. C., Beck, S. D., 1976c, Effects of allelochemics on the
    black cutworm, Agrotis ipsilon: Effects of resorcinol, phloro-
    glucinol, and gallic acid on larval growth, development, and
    utilization of food, Ann. Entomol. Soc. Amer. 69:999-1003.
Reese, J. C., Beck, S. D., 1978, Interrelationships of nutritional
    indices and dietary moisture in the black cutworm (Agrotis
    ipsilon) digestive efficiency, J. Insect Physiol. 24:473-479.
Reese, J. C., Chan, B. G., Malm, N. R., and Waiss, A. C., Jr., 1980,
    Bollworm feeding behavior on cotton, Submitted for publication.
Roeske, C. M., Seiber, J. N., Brower, L. P., and Moffitt, C. M.,
    1976, Milkweed cardenolides and their comparative processing
    by monarch butterflies, Recent Adv. Phytochem. 10:93-167.
Rosenthal, G. A., and Janzen, D. H., eds., 1979, "Herbivores:
    Their Interaction with Secondary Plant Metabolites", Academic
    Press, New York, 718 pp.
Rosenthal, G. A., Janzen, D. H., and Dahlman, D. L., 1976, A novel
    means of dealing with L-canavanine, a toxic metabolite, Science
    192:256-258.
Rothschild, M., 1972, Some observations on the relationship between
    plants, toxic insects and birds, in: "Phytochemical Ecology",
    J. B. Harborne, ed., Academic Press, New York, pp. 1-12.
Ryan, C. A., 1973, Proteolytic enzymes and their inhibitors in
    plants, Ann. Rev. Plant Physiol. 24:173-196.
Ryan, C. A., Green, T. R., 1974, Proteinase inhibitors in natural
    plant protection, Rec. Adv. Phytochem. 8:123-140.

Scriber, J. M., 1977, Limiting effects of low leaf-water content
on the nitrogen utilization, energy budget, and larval growth
of Hyalophora cecropia (Lepidoptera:Saturniidae), Oecologia
(Berl.) 28:269-287.

Shaver, T. N., Garcia, J. A., and Dilday, R. H., 1977, Tobacco
budworm: Feeding and larval growth on component parts of
cotton flower buds, Environ. Entomol. 6:82-84.

Shaver, T. N., and Lukefahr, M. J., 1971, A bioassay technique
for detecting resistance of cotton strains to tobacco bud-
worms, J. Econ. Entomol. 63:1274-1277.

Shaver, T. N., Lukefahr, M. J., and Garcia, J. A., 1970. Food
utilization, ingestion, and growth of larvae of the bollworm
and tobacco budworm on diets containing gossypol, J. Econ.
Entomol. 62:1544-1546.

Shaver, T. N., and Parrott, W. L., 1970, Relationship of larval
age to toxicity of gossypol to bollworms, J. Econ. Entomol.
63:1802-1804.

Slansky, F., Jr., and Feeny, P., 1977, Stabilization of the rate of
nitrogen accumulation by larvae of the cabbage butterfly on
wild and cultivated food plants, Ecol. Monogr. 47:209-228.

Smiley, J., 1978, Plant chemistry and the evolution of host
specificity: New evidence from Heliconius and Passiflora,
Science 201:745-747.

Soo Hoo, C. F., Fraenkel, G., 1966a, The selection of food plants
in a polyphagous insect, Prodenia eridania, J. Insect Physiol.
12:693-709.

Soo Hoo, C. F., and Fraenkel, G., 1966b, The consumption, digestion,
and utilization of food plants by a polyphagous insect,
Prodenia eridania (Cramer), J. Insect Physiol. 12:711-730.

Starks, K. J., and Burton, R. L., 1977, Determining biotypes, cul-
turing, and screening for plant resistance, USDA Tech. Bull.
No. 1556, 12 pp.

Su, H. C., Speirs, R. D., and Mahany, P. G., 1974, Trypsin inhibitors: Effects on the rice weevil (Coleoptera: Curculionidae)
in the laboratory, J. Ga. Entomol. Soc. 9:86-87.

Vanderzant, E. S., and Chremos, J. H., 1971, Dietary requirements
of the boll weevil for arginine and the effect of arginine
analogues on growth and on the composition of body amino acids,
Ann. Entomol. Soc. Amer. 64:480-485.

Waiss, A. C., Jr., Chan, B. G., and Elliger, C. A., 1977, Host plant
resistance to insects, in: "Host Plant Resistance to Pest",
P. A. Hedin, ed., ACS Symposium Series 62, American Chemical
Society, Washington, D.C., 286 pp.

Waiss, A. C., Jr., Chan, B. G., Elliger, C. A., Wiseman, B. R.,
McMillian, W. W., Widstrom, N. W., Zuber, M. S., and Keaster,
A. J., 1979, Maysin, a flavone glycoside from corn silks with
antibiotic activity toward corn earworm, J. Econ. Entomol. 72:
256-258.

Waldbauer, G. P., 1962, The growth and reproduction of maxillecto-
    mized tobacco hornworms feeding on normally rejected non-
    solanaceous plants, Entomol. Exp. Appl. 5:147-158.

Waldbauer, G. P., 1964, The consumption, digestion, and utilization
    of solanaceous and non-solanaceous plants by larvae of the
    tobacco hornworm, Protoparce sexta (Johan.) (Lepidoptera:
    Sphingidae), Entomol. Exp. Appl. 7:253-269.

Waldbauer, G. P., 1968, The consumption and utilization of food by
    insects, Adv. Insect Physiol. 5:229-288.

Woodhead, S., and Bernays, E., 1977, Changes in release rates of
    cyanide in relation to palatability of sorghum to insects,
    Nature 270:235-236.

TECHNIQUES AND APPLICATIONS OF MEASUREMENTS OF CONSUMPTION AND

UTILIZATION OF FOOD BY PHYTOPHAGOUS INSECTS

Marcos Kogan and Jose R. P. Parra

Section of Economic Entomology
Illinois Natural History Survey and Illinois
    Agricultural Experimenta Station
University of Illinois
Urbana, Illinois 61801

Departamento de Entomologia
Escola Superior de Agricultura Luiz de Queiroz
Universidade de São Paulo
Piracicaba, Brazil

Applications of measurements of consumption and utilization of food by insects are at the interface between alimentary physiology and host selection behavior. It is, therefore, in the basic fields of nutrition, community ecology, and behavior, and in the applied fields of host plant resistance and biological control that one finds studies using consumption indices as analytical criteria.

Most studies conducted to date used utilization indices as a measure of the adequacy of diets to insects (e.g., Waldbauer, 1964; Kogan, 1972; Latheef and Harcourt, 1972; Baker and Schwalde, 1975; Bailey, 1976; Bhat and Bhattacharya, 1978; Ram and Bhattacharya, 1978). The effect of specific diet components such as water, inorganic and organic nutrients and secondary plant metabolites may have subtle effects on insect behavior and physiology. These effects are better revealed by measuring the relationships between the a-mount of food ingested and the conversion of that food into body matter. In community ecology it is useful to consider the relative fitness of a complex of phytophagous species with regard to host plants. As an extension of these concepts to agricultural entomology, intake and utilization analyses have been useful in investigations of mechanisms of plant resistance to insects and in the classification of levels of resistance.

METHODS IN FOOD INTAKE AND UTILIZATION

Much progress has been made in quantitative nutritional analysis in insects after publication of Waldbauer's (1968) review on consumption and utilization of food by insects. That review helped systematize the field and established a uniform terminology. Since then several extensive studies have been conducted on phytophagous insects reared either on natural or artificial diets, or on changes in consumption rates in insects affected by various types of stress such as temperature, parasitism, moisture or crowding.

A summary of the most widely used parameters in utilization analyses was presented by Klein and Kogan (1974) using a formal notation for computerized computations. The formulas and symbols used in that paper are presented in Table 1 and 2, restricted only for computations based on dry weight values. These formulas are based on gravimetric measurements of animals, food, and feces. In addition to the formulas in Table 1 others may be necessary for computations using non-gravimetric methods. Several of these methods have been proposed using nutritionally neutral dyes, radioisotopes, and calorimetric procedures. A comparative analysis of the five most commonly used methods was performed by Parra and Kogan (submitted). The final section of this paper presents a summary of this comparative study and a discussion of its principal conclusions.

APPLICATIONS OF MEASUREMENTS OF CONSUMPTION AND UTILIZATION PARAMETERS

Consumption and utilization indices have been used in alimentary physiology to determine the role of specific diet components. Reese (1978) and Reese and Beck (1978) made extensive use of these indices to analyze the effect of allelochemics, mainly phenolics, on growth and development of black cutworm, Agrotis ipsilon (Hufnagel), larvae. It is apparent from these studies that the number of hydroxyls and their relative position (o or p) differentially affected ingestion, utilization (ECI) and digestibility (AD).

The nitrogen content of diets for phytophagous insects or factors that may result in inadequate utilization of the dietary nitrogen greatly affect the fitness of phytophagous insects in relation to their host plants (McNeil and Southwood, 1978). Nitrogen utilization analyses have been used to interpret adaptation of insects feeding on low and on high nitrogen plants (Slansky and Feeny, 1977; Scriber, 1979), and have suggested a certain degree of pattern development in the association of oligophagous or polyphagous species with herbaceous or aboreous host species (Scriber and Feeny, 1979).

Lipid utilization by Pieris brassicae L. fed cabbage leaves and a meridic diet was analyzed by Kastari and Turunen (1977). It was

Table 1. Basic parameters used in quantitative nutritional analyses—symbols, meanings, and computations. For explanation of computational parameters refer to Table 2.

Symbol[a]	Meaning	Computations
$AWD_{r,p}$	Animal dry-weight	$AWF_{r,p} \times ADWR_p$ [a]
$FCD_{r,p}$	Food consumed	$\left(FO_{r,p} \times \dfrac{CD,_{r,p}}{CO_{r,p}} \right) - FD_{r,p}$
$WGD_{r,p}$	Weight gain	$AWD_{r,p+1} - AWD_{r,p}$
$ECID_{r,p}$	Efficiency of conversion of ingested food	$WGD_{r,p}/FCD_{r,p}$
$ECD_{r,p}$	Efficiency of conversion of digested food	$WGD_{r,p}/(FCD_{r,p} - FES_{r,p})$
$AD_{r,p}$	Approximate digestibility	$(FCD_{r,p} - FES_{r,p})/FCD_{r,p}$
$TWGD_r$	Total animal weight gain	$MAXAWD_r - AWD_{r,1}$
$MAWD_r$	Mean animal weight	$\displaystyle\int_{p=1}^{NP_r} AWD_{r,p}\, dp/ND_r$
$TFCD_r$	Total food consumed	$\displaystyle\sum_{p=1}^{NP_r} FCD_{r,p}$
CID_r	Consumption index	$TFCD_r/(MAWD_r \times ND_r)$
GRD_r	Growth rate	$TWGD_r/(MAWD_r \times ND_r)$
$RGRD_r$	Mean relative growth rate	$(\ln MAXAWD_r - \ln AWD_{r,1})/ND_r$

[a]The D appended to the standard symbols ECI, CI, and GR indicates that computations are based on dry weights. Subscript: r = serial number of animal or group of animals in replication r; p = serial number of period in replication; a period is a full larval instar, a fraction of an instar or a day.

Table 2. Parameters used as input for computations of indices and
          values in quantitative nutritional analyses described in
          Table 1.

Symbol	Meaning
$AWF_{r,p}$	Animal fresh-weight at start of period
$FO_{r,p}$	Fresh-weight of food given at start of period
$FN_{r,p}$	Fresh-weight of food removed at end of period
$FD_{r,p}$	Dry-weight of food removed at end of period
$CO_{r,p}$	Fresh-weight of control diet at start of period[a]
$CN_{r,p}$	Fresh-weight of control diet at end of period
$CD_{r,p}$	Dry-weight of control diet
$FES_{r,p}$	Dry-weight of feces at end of period
$ADWR_{p}$	Animal dry-weight ratio[b]
ND_{r}	Number of days in all periods except last one
NP_{r}	Number of periods
$MAXAWD_{r}$	Maximum animal weight

[a] Control diets are usually 1/2 leaves, the other 1/2 being the
food given to the insects, or artificial medium handled exactly
as the artificial medium used in the experiments.

[b] $ADWR_p = AWD_c/AWF_c$ where AWD_c and AWF_c are the dry and fresh-
weights of control animals which are kept under identical
environmental conditions and receive the same food as the test
animals.

apparent that natural diets that contained phospholipids provided a more balanced diet in terms of lipid composition.

Utilization indices and growth parameters have been used to rank leguminous hosts of the Mexican bean beetle, Epilachna varivestis Mulsant. An array of 22 plants, varieties and lines of soybean and several Phaseolus species, was classified by cluster analysis as preferred, intermediate and marginal hosts. Various effects of the host plants on insect growth and development could be interpreted on the basis of the nutritional qualities of the plants (Kogan, 1972).

These examples represent but a small sample of the numerous applications of consumption, utilization, and growth analyses. Indeed Bhattacharya and Pant (1976) updated a table originally published by Waldbauer (1968) which compiled much of the published information on intake and utilization in insects. Waldbauer's table contained data on 44 species, resulting from studies in about 35 research papers published between 1920 and 1966. Bhattacharya and Pant added some 33 species, based on 17 papers published between 1967 and 1974. There is still a great deal more to be learned about the alimentary physiology of insects and consumption and utilization analyses are likely to be increasingly used in research in this field. Data acquisition, however, is a limiting factor for wider application of these techniques. Results are still too variable and interpretation of data is often difficult. For these reasons various alternative techniques for measuring the fundamental parameters have been proposed and successfully used in some cases. There have not been, however, comparative studies of these various techniques, and criteria for their selection have not been established. In 1978 we undertook to test four basic techniques for measuring intake and utilization and compare these techniques with the standard gravimetric method. The details of this study are presented elsewhere (Parra and Kogan, submitted) but a summary of our results and conclusions is presented in the next section.

TECHNIQUES FOR MEASURING CONSUMPTION AND UTILIZATION

All the indices used in intake and utilization analyses basically require measurement of five sets of parameters: a) animal weight at time t_1, b) animal weight at time t_2, c) weight of food provided to animal at t_1, d) weight of food remaining at t_2, and e) weight of feces accumulated during interval t_2-t_1. Several methods have been used to acquire these data. These methods generally fall into five categories: a) gravimetric, b) colorimetric, c) radioisotopic, d) calorimetric, and e) enzymatic. Gravimetry is the direct measurement of all parameters using a balance. All other methods rely upon indirect measurements of indicators of consumption.

Some of these indicators are expected to enter the metabolism of insects; others pass through the digestive tract unchanged and consumption is measured by the concentration of the indicator in the feces.

The colorimetric methods that have been tested with insects used as indicators chromic oxide (Cr_2O_3) (McGinnis and Kasting 1964a;b) or the red pigment Calco Oil Red (COR) (Daum et al., 1969). Both methods require accurate weighing of animals at t_1 and t_2 and the quantitative incorporation of the indicator (0.1% w/v COR, 4% w/v Cr_2O_3) into the medium. The COR method proceeds with successive acetone extractions of the dye from the animal, the medium, and the feces and the spectrophotometric analysis of the extracts at 510 mu. The Cr_2O_3 method requires oxidation of the insoluble Cr_2O_3 to the more soluble state. This is accomplished with a digestion mixture containing sodium molybdate, concentrated sulfuric acid and perchloric acid (McGinnis and Kasting, 1964a). The mixture needs to be handled with extreme care as perchloric acid fumes may accumulate in exhaust ducts of hoods and become explosive. An all glass rotary micro Kjeldahl system (Kontes Scientific Glassware, New Jersey) was used with success in our studies. Instead of the colorimetric determination using diphenyl carbazide we used the direct measurement of chromium by atomic absorption spectroscopy.

Radioisotopes have been used to some extent to measure food consumption and utilization. We used ^{14}C-glucose at an activity level in the medium of 2.6×10^6 cpm. Experiments were conducted in a closed system to permit collection of the expired $^{14}CO_2$ in an appropriate scintillation cocktail (Fig. 1 A., B.). After acetone extraction of labeled tissues from the animal and extraction of labeled fractions from the feces, activity was measured in three aliquots from each sample in a liquid scintillation counter. Readings were converted to dpm's and results expressed as percentages of labeled materials. Computations of ECI, ECD, and AD followed Waldbauer (1968).

Quite often researchers are interested in the caloric equivalents of the food ingested and in an analysis of the efficiency of energy conversion. We used an oxygen bomb calorimeter. Careful calibration of the apparatus was done by burning a pellet of benzoic acid. Corrections were made for the unburned fuse wire and the nitric acid produced. Use of the bomb calorimeter requires accurate weighings of all processed samples, thus there are no real savings in weighing times. Calorimetry is not an alternative to gravimetry; these are in fact complementary methods. Formulas for computations of energy conversion factors can be found in Waldbauer (1968).

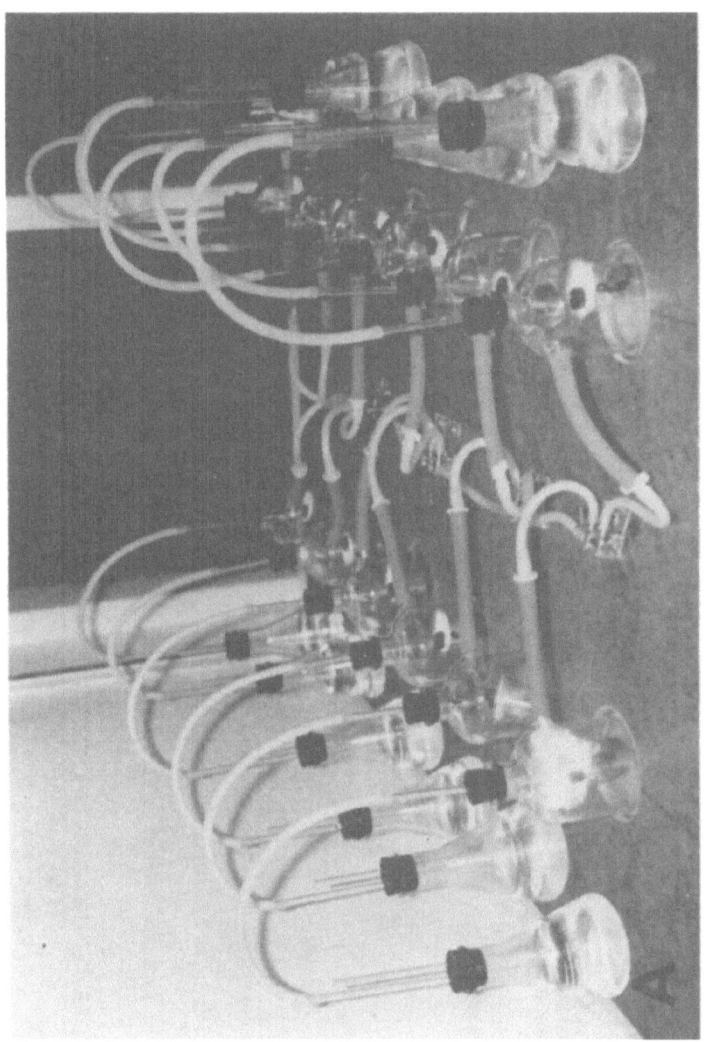

Fig.1A. Assembly of bell jars and air-inlet bottles used in enclosed system for the radioisotopic method.

Fig.1B. Detail of the bell jar with a rearing cup containing the ^{14}C-glucose medium and one soybean looper larva. The expired CO_2 was quantitatively collected in an appropriate scintillation cocktail.

An additional method, that had not been included in our comparative analysis, is the use of products of catabolism that can be detected in excreta. These products behave as natural markers that may be extracted and analyzed. Bhattacharya and Waldbauer (1969) used uric acid in investigations of utilization values in insects feeding on grain. This method was particularly convenient in investigation of insects whose excreta are mixed with the food. Battacharya and Pant (1976) presented an excellent review and discussion of these and several other methods used in consumption and utilization in insects.

EXPERIMENTAL

We used in our comparative study larvae of the soybean looper, Pseudoplusia includens (Lepidoptera: Noctuidae). Insects were obtained from a culture maintained in the laboratory at Urbana, since 1972. Each year field individuals are introduced into the colony to maintain its genetic diversity. Experiments were conducted using the same basic artificial medium used in the maintenance of the colony. This medium is a slightly modified version of the cabbage looper medium of Henneberry and Kishaba (1966), and it is presented in Table 3. Experiments were conducted with recently emerged larvae and carried through pupation. Details of the experimental techniques, preparation of media, and analytical procedures and detailed results are presented in Parra and Kogan (submitted).

COMPARATIVE ANALYSIS OF METHODS

The five methods used in these analyses were: 1) standard gravimetric, 2) Calco Oil Red, 3) Cr_2O_3, 4) ^{14}C-glucose, and 5) calorimetric. For each indirect method (2, 3, and 4), and for the calorimetric method, measurements were also taken by standard gravimetry. Thus, all the analyses have been conducted by comparing the results obtained by gravimetry, using the regular artificial medium, with the results obtained with the indirect methods and with calorimetry. In addition, each method had an internal control: the gravimetric measurements obtained with larvae reared on the media containing the various additives required for performance of the specific indirect method.

The potential effect of the additives or that of the particular experimental condition was measured by comparing all gravimetric measurements using the experimental media with gravimetric measurements using the standard medium. These comparisons for weight gain, food consumed, CI, ECD, and AD are shown in Figures 2-6 respectively. The shaded bar diagrams represent the same parameters measured by the given indirect method or by calorimetry. Food consumed by calorimetry was measured gravimetrically; thus, only one bar is shown in Figure 3 for the calorimetric method.

Table 3. Composition of the alfalfa-meal/wheat-germ diet modified
          from that of Henneberry and Kishaba (1966), used to rear
          the soybean looper.

Agar	50.0 g
Water	3080.0 ml
Vitamin free casein	126.0 g
Alfalfa leaf meal	54.0 g
Sucrose	96.0 g
Wheat germ	108.0 g
Wesson's salt mixture	36.0 g
Alphacel	18.0 g
4 M KOH	18.0 ml
Wheat germ oil	10.0 ml
Vitamin mixture (Vanderzant modification)	36.0 g
Sorbic acid	6.8 g
Methyl-p-hydroxybenzoate	6.8 g
Ascorbic acid	13.0 g
Aureomycin	0.5 g
Formaldehyde (10%)	15.0 ml

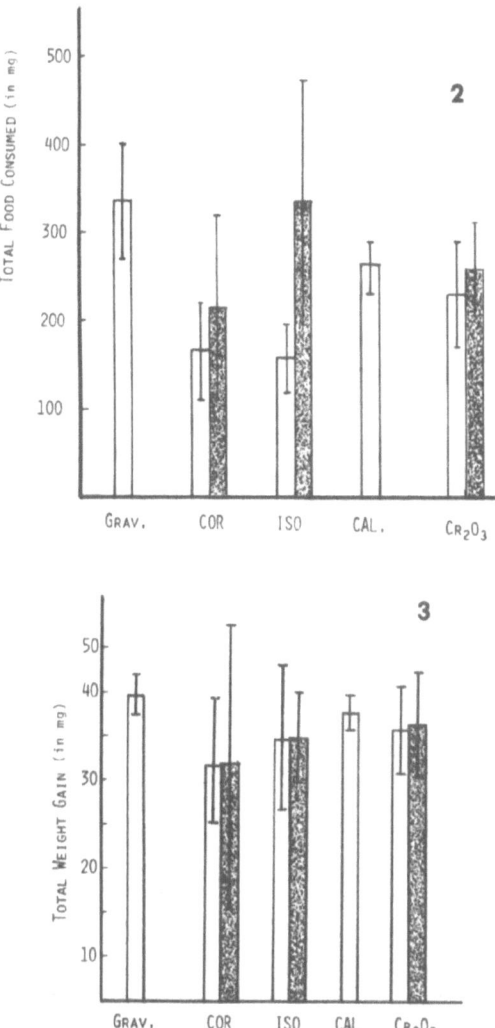

Fig. 2-6 Graphs of weight gain, food consumed, ECI, ECD and AD
respectively for soybean looper larvae feeding on artificial medium. Measurements made by gravimetric (open
bars) and by the Calco Oil Red (COR), isotopic (ISO),
calorimetric (CAL) and the chromic oxide (Cr_2O_3) methods.
The non-gravimetric methods are represented by the stippled
bars. Vertical lines show standard deviations. (Based
on data from Parra and Kogan, submitted).

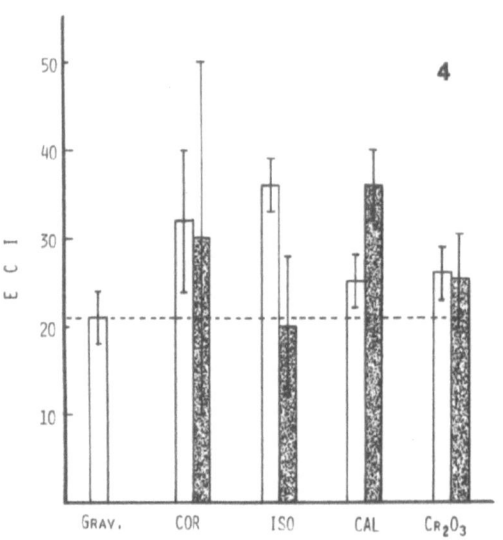

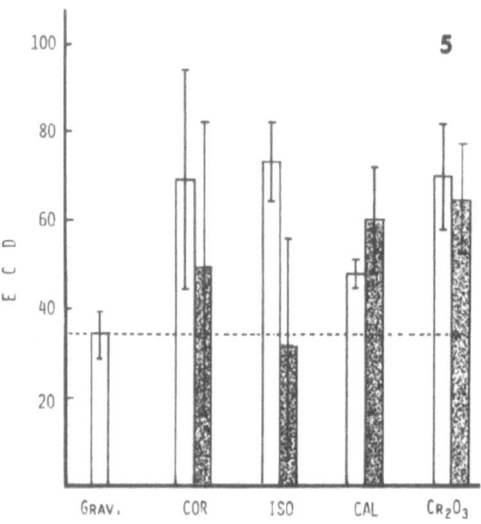

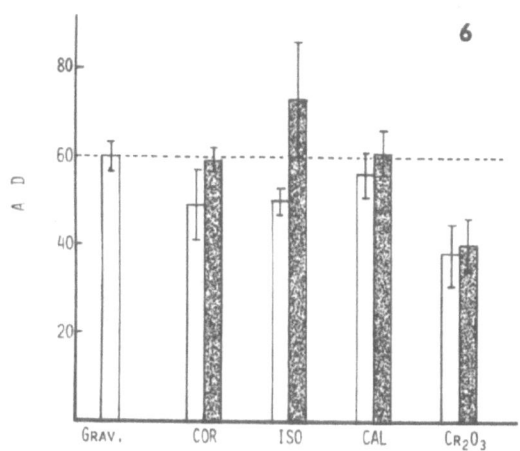

Main sources of variation in experiments of this nature include: a) the inherent individual variability among insects in a
population, b) fluctuations of moisture content in media; c) differences in feeding behavior resulting from the effect of additives
in the diets, d) differences in the utilization of the diet after
ingestion, e) differences in instrumentation, and finally f) differences in manipulation of the samples.

Effect of diet quality could be measured by comparisons with
results obtained with the standard artificial medium.   It was
apparent that larvae ate more of the medium without additives
of any other medium.   The Calco Oil Red (COR) was apparently detrimental to the larvae as larvae reared on a medium containing 0.1%
COR were about 24% lighter than those reared on the standard
artificial medium.   The other media had little effect on the
larvae. Food consumption, however, was substantially reduced in
all indirect methods.   It was apparent that larvae ate less of
these media but utilized them more efficiently as weight gain was
not greatly affected (except by COR).

Assuming that gravimetry is the most direct method, gravimetric results should closely approximate the true values of the parameters; therefore, these values were used as a measure of accuracy. Diagrams in Figures 2-6 provide a visual indication of deviation from accuracy when one compares the height of the clear and the shaded bars within each pair. ECI, ECD, and AD were consistently accurate when measured by the Cr_2O_3 method. All other methods departed substantially from their gravimetric estimates. Of particular interest here are the very low ECD values measured by the radio-isotopic method, which can be explained by the fact that the expired CO_2 was explicitly used in these computations. Variations in utilization measured by the calorimetric method may suggest that efficiencies in energy conversion were not directly comparable to efficiencies of nutrient utilization.

The effect of instrumental errors and error introduced through manipulation of samples measured as $(1 - CV)$ x 100 where CV is the coefficient of variation or the ratio of the standard deviation to the mean. This criterion is an estimate of the precision of measurements. Precision of most gravimetric measurements was of the order of 80 percent. Thus, departures much greater than 80 percent would indicate an undesirable lack of precision in the measurements. Both the Cr_2O_3 and the calorimetric methods were very precise, the Calco Oil Red method was less than 35 percent precise and the radioisotopic method was adequate (ca. 60 percent) for ECI but dropped to ca. 20 percent for ECD. This result was probably due to difficulties in quantitatively collecting the expired CO_2.

CONCLUSIONS

The time required to process samples by the indirect methods varied from 6 (radioisotopic) to 18 times (Cr_2O_3) more than the time required by standard gravimetry. All indirect methods required the use of an analytical balance in addition to more expensive equipment used in specific determinations (spectrophotometers, scintillation counter, etc.). We gained little if any, precision in our indirect measurements, and accuracy was also affected by most indirect procedures. It seems, therefore, improbable that one may justify the use of non-gravimetric procedures whenever gravimetric procedures can be used. Under certain experimental conditions where food consumed cannot be directly measured, alternative methods may be justified. Our results indicate that use of Calco Oil Red should be made judiciously as, at least for the soybean looper, there was a detrimental effect. Whenever caloric equivalents are needed one must resort to a bomb calorimeter. In this case instead of making caloric measurements for all specimens it may be sufficient to obtain dry weights of all parameters and compute caloric equivalents of aliquots.

The value of data on intake and utilization in understanding insect-food interaction is unquestionable. There are, however, no shortcuts. Experiments involving gravimetric measurements of insect growth and food consumption are painstakingly tedious, but all the alternatives that we tested were less precise, some rather inaccurate and most were exceedingly more expensive to run.

REFERENCES CITED

Bailey, C. G., 1976, A quantitative study of consumption and utilization of various diets in the Bertha armyworm, Mamestra configurata (Lepidoptera: Noctuidae), Can. Entomol. 108:1319-1326.

Baker, J. E., and Schwalde, C. P., 1975, Food utilization by larvae of the furniture carpet beetle, Anthrenus flavipes, Entomol. Exp. Appl. 18:213-219.

Bhat, N. S., and Bhattacharya, A. K., 1978, Consumption and utilization of soybean by Spodoptera litura (Fabricius) at different temperatures, Indian J. Entomol. 40:16-25.

Bhattacharya, A. K., and Waldbauer, G. P., 1969, Quantitative determination of uric acid in insect feces by lithium carbonate extraction and the enzymatic - spectrophotometric method, Ann. Entomol. Soc. Amer. 62:925-927.

Bhattacharya, G. B., and Pant, N. C., 1976, Studies on the host plant relationships: Consumption and utilization profile in insects, Proceedings of the National Academy of Science, India 46:274-301.

Daum, R. J., Mckibben, G. M., Davich, T. B., and McLaughlin, R., 1969, Development of the bait principle for boll weevile control: Calco Oil Red N-1700 dye for measuring ingestion, J. Econ. Entomol. 62:370-375.

Henneberry, T. J., and Kishaba, A. N., 1966, Cabbage loopers, in: "Insect Colonization and Mass Production", C. N. Smith, ed., Academic Press, New York, pp. 461-478.

Kastari, T., and Turunen, S., 1977, Lipid utilization in Pieris brassicae reared on meridic and natural diets: Implications for dietary improvement, Entomol. Exp. Appl. 22:71-80.

Klein, I., and Kogan, M., 1974, Analysis of food intake, utilization, and growth in phytophagous insects - A computer program, Ann. Entomol. Soc. Amer. 67:295-297.

Kogan, M., 1972, Intake and utilization of natural diets by the Mexican bean beetle, Epilachna varivestis - A multivariate analysis, in: "Insect and Mite Nutrition. Significance and Implications in Ecology and Pest Management", J. G. Rodriguez, ed., North Holland, Amsterdam, pp. 107-126.

Latheef, M. A., and Harcourt, D. G., 1972, A quantitative study of food consumption, assimilation, and growth in Leptinotarsa decemlineata (Coleoptera: Chrysomelidae) on two host plants, Can. Entomol. 104:1271-1276.

McGinnis, A. J., and Kasting, R., 1964a, Colorimetric analysis of chromic oxide used to study food utilization by phytophagous insects, J. Agr. Food Chem. 12:259-262.

McGinnis, A. J., and Kasting, R., 1964b, Comparison of gravimetric and chromic oxide methods for measuring percentage utilization and consumption of food by phytophagous insects, J. Insect Physiol. 10:989-995.

McNeil, S., and Southwood, T. R. E., 1978, The role of nitrogen in the development of insect/plant relationships, in: "Biochemical Aspects of Plant and Animal Coevolution", J. B. Harborne, ed., Academic Press, London, 435 pp.

Parra, J. R. P., and Kogan, M., 1980, Comparative analysis of methods for measurements of food intake and utilization using the soybean looper, Pseudoplusia includens and artificial medium, Entomol. Exp. et Appl., in press.

Ram, S., and Bhattacharya, A. K., 1978, Consumption of soybean by Diacrisia obliqua Walker, Indian J. Entomol. 40:335-336.

Reese, J. C., 1978, Chronic effects of plant allelochemics on insect nutritional physiology, Entomol. Exp. et Appl. 24:625-631.

Reese, J. C., and Beck, S. C., 1978, Interrelationships of nutritional indices and dietary moisture in the black cutworm (Agrotis ipsilon) digestive efficiency, J. Insect Physiol. 24:437-479.

Scriber, J. M., 1979, Post-ingestive utilization of plant biomass and nitrogen by Lepidoptera: Legume feeding by the southern armyworm, N. Y. Entomol. Soc. 87:141-153.

Scriber, J. M., and Feeny, P., 1979, Growth of herbivorous caterpillars in relation to feeding specialization and to the growth form of their food plants, Ecol. 60:829-850.

Slansky, F., Jr., and Feeny, P., 1977, Stabilization of the rate of nitrogen accumulation by larvae of the cabbage butterfly on wild and cultivated food plants, Ecol. Monogr. 47:209-228.

Waldbauer, G. P., 1964, The consumption, digestion and utilization of solanaceous and non-solanaceous plants by larvae of the tobacco hornworm, Protoparce sexta (Johan.) (Lepidoptera: Sphingidae), Entomol. Exp. et Appl. 7:253-269.

Waldbauer, G. P., 1968, The consumption and utilization of food by insects, Adv. Ins. Physiol. 5:229-288.

(The index is limited to organisms, structures, chemical compounds and actions upon which the symposium papers were focused.)

V